Flat Roof Construction Manual

MATERIALS
DESIGN
APPLICATIONS

SEDLBAUER
SCHUNCK
BARTHEL
KÜNZEL

Birkhäuser
Basel

Edition Detail
Munich

Authors

Klaus Sedlbauer
Prof. Dr.-Ing. Dipl.-Phys.
Fraunhofer Institute for Building Physics, Stuttgart/Holzkirchen/Kassel
University of Stuttgart, Chair of Building Physics

Eberhard Schunck
Prof. (retd.) Dipl.-Ing. Architect
Munich University of Technology, Chair of Construction Engineering

Rainer Barthel
Prof. Dr.-Ing.
Munich University of Technology, Chair of Structural Design

Hartwig M. Künzel
Dr.-Ing.
Fraunhofer Institute for Building Physics, Hygrothermal Department, Holzkirchen/Stuttgart

Assistants:
Matthias Beckh, Dipl.-Ing.; Christian Bludau, Dipl.-Ing.; Mark Böttges, Dipl.-Ing.; Philip Leistner, Dr.-Ing.; Eberhard Möller, Dipl.-Ing.; Zoran Novacki, Dipl.-Ing.; Lutz Weber, Dr. rer. nat.; Wolfgang Zillig, Dr.

Specialist articles:
Christian Schittich, Dipl.-Ing. Architect (introduction)
Editor-in-chief, DETAIL, Munich

Ulrich Max, Dr.-Ing. (fire)
Ingenieurbüro für Brandsicherheit AGB, Bruchsal; lecturer in fire protection, University of Stuttgart, Chair of Building Physics

Consultants:
Theodor Hugues, Prof. (retd.) Dr.-Ing. (construction details)
Munich University of Technology, Chair of Design, Construction Engineering & Building Materials

Hartwig J. Richter (construction details)
Roofing master and assessor, Traunreut

Michael Wichmann (flat roof construction)
Assessor, CAD-point, Oranienburg

Editorial services

Editors:
Cornelia Hellstern, Dipl.-Ing.; Sandra Leitte, Dipl.-Ing.; Johanna Billhardt, Dipl.-Ing.

Editorial assistants:
Carola Jacob-Ritz, MA; Peter Popp, Dipl.-Ing.; Irene Stecher; Cosima Strobl, Dipl.-Ing. Architect

Drawings:
Ralph Donhauser, Dipl.-Ing.; Marion Griese, Dipl.-Ing.; Martin Hämmel, Dipl.-Ing.; Daniel Hajduk, Dipl.-Ing.; Elisabeth Krammer, Dipl.-Ing.; Dejanira Ornelas Bitterer, Dipl.-Ing.

Translation into English:
Gerd H. Söffker, Philip Thrift, Hannover

Proofreading:
Roderick O'Donovan, Vienna (A)

Production & layout:
Roswitha Siegler, Simone Soesters

Reproduction:
ludwig:media, Zell am See (A)

Printing & binding:
Aumüller Druck, Regensburg

Bibliographic information published by the German National Library:
The German National Library lists this publication in the Deutsche Nationalbibliografie; detailed bibliographic data is available on the Internet at http://dnb.d-nb.de.

This book is also available in a German language edition (ISBN 978-3-0346-0580-9)

Publisher:
Institut für internationale Architektur-Dokumentation GmbH & Co. KG, Munich
www.detail.de

Distribution:
Birkhäuser GmbH
P.O. Box 133, CH-4010 Basel, Switzerland
Tel. 0041-61-568 98 00; Fax. 0041-61-568 98 99
e-mail: sales@birkhauser.ch
www.birkhauser.com <http://www.birkhauser.com>

Printed on acid-free paper produced from chlorine-free pulp.
TCF∞

ISBN: 978-3-0346-0658-5 (hardcover)

9 8 7 6 5 4 3 2 1

Contents

Preface

The world's natural resources are dwindling. This fact concerns not only oil and gas, but increasingly also more specialised raw materials such as indium, geranium and antimony, the prices for which have risen at an unprecedented rate in recent years. And climate change continues; it is too late to stop it, at best we can only slow it down. The consequences are already affecting all walks of life and everyday routines – in the way we design, construct, use and recycle our buildings, for instance.

Topics such as lower energy consumption and environmental protection are being discussed more and more in public. And they are now being joined by other aspects such as the closing of materials cycles and the ecological assessment of the entire phase of building utilisation. The concept of sustainability is increasingly becoming intrinsic to modern construction.

All this adds up to a property market and a building industry that are under growing pressure to minimise the environmental influences of construction. Despite this, 30–40 % (depending on which report you read) of our energy resources are consumed in the operation of buildings.

If we project all this on to roof design, then the dominating issue is energy consumption. Thermal insulation measures, as the most effective means of saving resources, are being employed in many variations, for both new-build and refurbishment projects. For example, the transmission heat losses through a roof provided with 20 cm insulation can be reduced by about 80 % in comparison to an uninsulated roof. And the payback times for such undoubtedly sensible measures are only a few years, although they should always be shaped by a certain pragmatism. For instance, where flat roofs are converted into pitched roofs during refurbishment projects because the latter allegedly exhibit better insulation characteristics, or where flat roofs are not planned in the first place because they are allegedly vulnerable to serious damage, especially with thick insulation, then we run the risk of losing part of our architectural culture. If we were to allow flat roofs to disappear gradually from our built environment, then we would lose an ecological potential that has never been fully exploited, i.e. the use of roof surfaces as a climatic micro-cosmos or simply as a water retention basin. As a result, we would find ourselves in a curious situation: ecological opportunities would go unused for ecological reasons – a completely unjustified state of affairs because technology provides us with a wealth of tried-and-tested solutions for building functioning flat roofs with an excellent level of thermal insulation. It is for that reason that the *Flat Roof Construction Manual* focuses on the subject of designing energy-saving flat roofs free from damage.

To begin our study of flat roofs, Part A outlines their evolution. Flat roofs were already well established in many cultures and many climatic conditions before they started to spread across Central Europe and North America as well. Additional usable floor space on the roof plus advantages for fire protection were the main reasons for their new-found popularity. However, it was not until waterproofing materials reached a certain stage of development and codes of practice for the design and construction of flat roofs were drawn up in the 1960s did we achieve the basis for good-quality flat roofs in cold and wet regions, too. From time to time, the flat roof was merely a fashion, one that led to the "roofs dispute" of the early modern movement. But the dispute has been resolved; today, flat roofs and pitched roofs exist side by side as equal partners.

Part B "Structure" investigates the structural aspects of flat roofs. Besides the various loads acting on the flat roof, the chapters of this section discuss potential primary structures and their optimisation. The whole range of structural materials – concrete, metals, timber and glass – is also presented here for the reader.

Part C "Building physics" is a comprehensive and well-founded presentation of the building physics principles that affect flat roofs. The chapter on thermal insulation explains the facets of energy-saving construction on the basis of steady-state and non-steady-state thermal conditions and shows a number of practical solutions.

In buildings, moisture is the most significant cause of damage, and this is especially true for flat roofs. One of the main sources of this are errors in the design and construction of components. This means that sustainable building without damage cannot be even contemplated without a hygrothermal evaluation of the flat roof design. Non-steady-state moisture processes within the construction are therefore illustrated and explained.

The section on fire focuses on selecting the most suitable materials, classifying the fire resistance of various forms of construction and the legislative requirements.

But sustainable building includes much more besides just energy efficiency issues, e.g. socio-cultural factors such as the comfort of users. Therefore, sound insulation is also considered in detail in this book. The health, well-being and, in workplaces, productivity of building occupants are the primary criteria of acoustics. And the aim is not just to reduce noise or prevent noise pollution, but rather to create suitable conditions for users proactively. Aspects such as airborne sound insulation to protect against external noise, impact and structure-borne sound insulation to control internal noise, and sound absorption within rooms all play critical roles here. The individual criteria are discussed in relation to flat roof design and practical solutions are proposed.

Part D "Design principles" describes the most important materials used these days for the individual layers in a flat roof construction. The catalogue of materials extends from waterproofing (bituminous, plastic, elastomeric and liquid) to insulating materials, glass and impermeable concrete, which tends to play a special role in practice. Additional layers, e.g. those necessary in green roofs, are also discussed.

Having explained the individual layers of the construction, these are then combined to create the customary forms of construction for flat roofs. The main variants are: positioning the waterproofing above the insulation, below the insulation or between layers of insulation.

Other important subjects in this section are green roofs and roofs designed for foot or vehicular traffic. Such roofs provide supplementary uses, but they do place greater demands on the construction. Forms of flat roof construction encountered less frequently in practice - impermeable concrete, glass and metal - are also examined here. These roofs may satisfy certain architectural or engineering needs but do represent a challenge in terms of building physics. Practical solutions for joints and junctions, rooflights, safety features and drainage are among the further topics covered in Part D. Care and maintenance as well as the refurbishment of flat roofs – as the logical extension of design and construction – are also introduced to the reader.

Part E "Construction details" complements the chapter on forms of construction. The clear drawings illustrate the essentials of the constructional and building physics requirements. The details so important to the various forms of flat roof construction, e.g. edges, junctions with walls, penetrations and drainage, are dealt with here in depth.

The examples of construction details shown in this part of the book should, however, be understood as illustrating principles and not universally applicable solutions. They are intended to explain in a practical way how the comprehensive requirements placed on different flat roof designs can be solved, and therefore should be regarded as a starting point for everyday design and detailing tasks.

The projects shown in Part F "Case studies" demonstrate the diverse design options for flat roofs. The projects were selected in the first instance according to design and architectural aspects, but the diversity of potential forms of construction and materials was another priority. The examples have been taken from different locations with correspondingly different conditions with respect to climate, technical regulations and standards of building. Once again, the details in this section therefore do not represent universally applicable solutions, but rather details that must be adapted to suit each respective situation.

To conclude this preface, I would like to thank all my colleagues in this field and all the institutions and persons whose competence and dedication have contributed to the production of this book.

Klaus Sedlbauer
June 2010

Part A   Introduction

Fig. A   Roof edge detail, temple in Lhasa, Tibet (CN)

The evolution of the flat roof

Christian Schittich

A 1

A 2

Roofs of earth and loam – archetypal forms

At the start of the 20th century, as the flat roof gradually started to spread across Central Europe too, its most devoted advocates, great names such as Adolf Loos or Le Corbusier, turned it into almost a myth. Not just because it rendered possible the cubist building form so sought after in those days for aesthetic reasons, but primarily because of its usefulness. Because, used sensibly, a horizontal termination to a building returns to its occupants that area consumed by the building itself in the first place. Flat roofs are more than just the indispensable upper termination to a building, more than just protection against the weather or other outside dangers. Whenever technically feasible or wherever climatic conditions allow, people have always used the flat roof as a welcome extension to their living space, as an ancillary and circulation area, or as a terrace, but also for lighting and ventilating the rooms below. Traditionally, flat roofs are found mainly in hot regions with low rainfall, and are less common in climate zones with moist, hot conditions or heavy snowfalls. The building materials available locally is another important reason for this. For wherever rainfall is scarce, supplies of timber are often in short supply as well. And the construction of a flat roof consumes much less of that valuable resource than is the case with complex pitched roof structures, and if necessary much smaller cross-sections will often suffice.

For example, just one layer of naturally crooked, short branches is adequate as a loadbearing layer for a short span, or just a few joists if the span is longer. On top of this, depending on local availability, sticks or brushwood, reeds, bamboo or dried palm leaves, but in some places even stone flags. Usually also a layer of sand, leaves or pine cones to regulate the interior climate and moisture levels. Waterproofing is assured with the materials available in the immediate vicinity: earth and loam, which is compacted, smoothed flat and often also impregnated. Nevertheless, traditional flat roofs require permanent maintenance. In many regions that work is usually carried out once a year, after the rainy season. Over the centuries, people have devised ingenious waterproofing methods. For example, in Sana'a, the capital of Yemen, where traditionally the *qadath* is used, a type of waterproof screed made from a mixture of water, lime and basaltic lava plus a sophisticated impregnation of cooking fat (Figs. 1 and 2).
For the Pueblo people of south-western USA it is the need for constant expansion that is crucial as well as the consumption of materials and the usability of the space. This is because their houses, the pueblos, are based on an additive system: individual rooms, roughly equal in size, are added horizontally or vertically, like a modular system, to meet family or other needs (Fig. 7). The specific uses of the various rooms can be varied at any time, room functions swapped around.

A 1 Traditional flat roof in Yemen: laying the short crooked branches over tree trunks and …
A 2 … laying the loose fill.
A 3 House in Lhasa, Tibet (CN)
A 4 A house of the Hunza people, Kashmir (PK)
A 5 Roofscape of the monastery complex at Laprang, Qinghai (CN)
A 6 House, Sada (YE)
A 7 Taos Pueblo, New Mexico (USA)

A 3

A 4

"One worker uses his metal scraper to push materials from each pile, in a mix ratio of approx. 60 % ash to 40 % lime, to a first pair of workers. These two ... start to crush their portions and at the same time mix them to create a more or less homogeneous mass. Once this mixture has achieved a certain granular consistency it is ... pushed across to the next pair of workers until it has ... become a creamy, relatively dry paste. It takes about an hour for one portion ... to pass through this system. The monotonous rhythm of the hammering is accompanied by storytelling or, more frequently, by the antiphonal songs of the workers

... Five or six workers squat ... on the ground tamping the *qadath* [a traditional screed] brought in to them with sharp-edged stones in a semicircular pattern. At the same time this unfinished surface is constantly splashed with water from a ... straw brush, the *meknesse*. It takes one man about two hours to tamp one square metre ... The surface of this first layer is rough and marked by the pattern of the tamping action. After being allowed to dry for 2–3 days, the process is repeated with a second layer ... During the 7–10 days it takes for the *qadath* to cure, the surface is frequently wetted with limewash (two

handfuls of lime dissolved in a bucket of water), again using the *meknesse* ... Once the screed has set, the surface is polished with palm-sized stones; these stones are family heirlooms handed down from generation to generation ... The now almost white surface of the *qadath* is treated with hot fat to give it its final, waterproof, satin-like finish ... The floor finish produced upon completion of these numerous manual processes has now acquired its key properties; it is waterproof, abrasion-resistant, mildly elastic and ... will last for well over a century."

Jan Martin Klessing [1]

The houses of the Hunza people in the part of Kashmir belonging to Pakistan are typical of the barren regions between the Caucasus and the Himalayas (Fig. 4). Their small, densely packed, one-room houses are entered via a square opening in the roof, which also serves to admit daylight and air and allows smoke to escape from the inside. The opening can be closed if necessary.

On steep mountainsides flat roofs are often the only level areas available outside. They are used for all types of domestic work, occasionally for threshing grain and even as bedrooms on hot summer nights. But they are especially important for drying fruit because the inhabitants of barren mountainous regions rely on an elaborate stock-piling lifestyle owing to the long cold winters and the relatively small numbers of livestock that can be kept.

Flat roofs are also widespread in Tibet, for the aforementioned reasons. The flat roofs of the small farmhouses in rural districts – waterproofed with a flattened layer of loam – serve as an extension to living and working areas, and there are often rooftop terraces at different levels. Flat roofs can also be found on the splendid town-houses of Lhasa, Xigazê or Gyanzê (Fig. 3), where they also function as pathways. The great monastery complexes are particularly impressive; here, several thousand monks live in town-like structures made up of closely packed cubic buildings. In the centres of these complexes it is only the shrines and temples that are crowned with golden roofs in the Chinese style. But these roofs are not just designed to provide protection against the weather; the pitched roof is artificially elevated to create a symbol.

A 5

A 6

A 7

"We come now to treat of Pavements, which also partake somewhat of the Nature of Coverings … Those which are open to the Air ought to be raised in such a Manner, that every ten Foot may have a Declivity of, at least, two Inches, to throw off the Water, … if the Pavement is to be upon Rafters, cover them over with Boards, and upon them lay your Rubbish or Fragments of Stone a Foot high, and beaten together, and consolidated with the Rammer. Some are of Opinion, that under these we ought to lay Fern, or Spart, to keep the Mortar from rotting the Timber."

Leon Battista Alberti [2]

A 8

A 9

A 10

A 11

A 12

Renaissance and Baroque: the first flat roofs in Central Europe

Whereas the flat roof has been in use in many countries around the Mediterranean, in Asia and in the Americas since time immemorial, it remained an insignificant building form in Central and Northern Europe for thousands of years. It was not until the Renaissance, as unambiguous, geometrically straightforward building forms and facades started to replace the Gothic architecture of pointed arches and flying buttresses, that people started to express a wish to see a horizontal upper termination to a building. But as true flat roofs were still very complex in terms of their building technology, clay tile and stone flag roofs with a shallow pitch were concealed behind tall parapets und balustrades, which often appeared to be autonomous components. At the same time, leading architects such as Leonardo da Vinci or Leon Battista Alberti started experimenting with the feasibility of flat roofs. The legendary Hanging Gardens of Babylon too, which the Greeks had included in their Seven Wonders of the World, had again inspired the imaginations of scholars and architects since the early days of the Renaissance, motivating them to creative interpretations and attempts at reconstruction (Fig. 8). For example, Pope Pius II had hanging gardens built for his palace in his "ideal city", Pienza, in 1462. This was followed by numerous copies throughout Italy, culminating in the development on the Borromean Islands in Lake Maggiore, where around 1630 Count Carlo III Borromeo started to transform the Isola Bella into a system of 10 garden terraces (Fig. 9). Not long afterwards, in the heyday of the Baroque, we start to see the first rooftop gardens north of the Alps. But they remained sporadic, confined to expensive prestigious structures because of the enormous amount of work required to provide the thick insulating and drainage layers made from expensive waterproofing materials such as copper, lead and tar. One impressive example that still survives is the garden that the Prince-Archbishop of Passau had built for his palace around 1700 – a spacious south-facing terrace on the north bank of the River Inn, with trees and bushes in tubs, flower beds and fountains (Fig. 10).

In a treatise written around 1722, Dresden's Councillor of Building and Commerce, Paul Jacob Marperger, urged the use of flat roofs – which he called terraces – not purely as luxury commodities for edification, but primarily for purely practical reasons, for the general public as well. Even though his ideas at that time remained essentially utopian, he dedicated himself passionately to the universal adoption of usable flat roofs and therefore anticipated many of the arguments that would be voiced 200 years later by the flat roof advocates in the "roofs dispute" of the early modern movement. Marperger listed aesthetics among his reasons as well as the saving of timber or the reduced fire risk, and mentioned the diverse usage options or the gain in space for building owners "when instead of a tall roof his house would have another storey". [3]

Wood-cement roofs and early reinforced concrete

However, the crucial breakthrough in the feasibility of such ideas came about 100 years later in the form of the wood-cement roof, developed by master cooper Samuel Häusler from Silesia. His inexpensive design involved bonding together several layers of oil paper with pitch or tar in situ and subsequently covering these with sand and gravel (Figs. 13 and 14). This type of roof became quickly established in the second half of the 19th century, primarily for ancillary buildings in the cities, also because of the much lower fire risk when compared with pitched roofs with their timber roof structures, and the relatively good thermal insulation properties. At the same time, in the age of Romanticism, the rooftop garden was gaining in importance. In 1867 the Royal Master Mason of Berlin, Carl Rabitz, recommended the adoption of flat roofs in his brochure entitled "Natural Roofs of Volcanic Cement" – also because of the possibility of creating rooftop gardens. Using wonderful illustrations, he describes balmy summer evenings on his own terrace with the wine flowing under a sky of stars (Fig. 11). The next key impulse in the evolution of the flat roof was provided by reinforced concrete in its

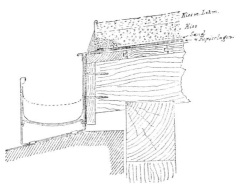

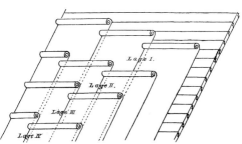

A 13   A 14

early forms, which enabled not only the simple construction of flat suspended floors and roofs, but in the eyes of the avant-garde also demanded an architecture to match the material. One of the pioneers in the use of the new material, the French engineer François Hennebique, demonstrated all the constructional and structural possibilities of the new material with technical virtuosity on his own house at Bourg-la-Reine near Paris (1900–1904). However, despite the cantilevering storeys and garden terraces at various levels, the design language was still essentially that of the 19th century; other architects of this period were employing a much more radical architecture, e.g. the young Tony Garnier in his designs for his "ideal city", the "Cité industrielle" (Fig. 15).
In 1907 the Swiss architects Otto Pfleghard and Max Haefli together with the engineer Robert

Maillart designed the Schaffhausen-Thurgau Sanatorium in Davos with flat roofs and sun terraces on the topmost floor (Fig. 12). With its uncompromisingly modern architectural language, the building was well ahead of its time. As early as 1899, these architects had designed a hospital in reinforced concrete which Sigfried Giedion later described as follows: "It is certainly also the first time that flat roofs (asphalt) with internal drainage have been used for residential buildings." [5]

A 8   The Hanging Gardens of Babylon, Athanasius
       Kircher, engraving, 17th century
A 9   Isola Bella in Lake Maggiore, J. B. Fischer von
       Erlach, copperplate engraving, 1721
A 10   Garden terrace of the Prince-Archbishop's palace
       in Passau (D), c. 1700
A 11   Rooftop garden of the Royal Master Mason of Berlin,
       Carl Rabitz, Berlin (D), c. 1867
A 12   Schaffhausen-Thurgau Sanatorium in Davos (CH),
       1907, Otto Pfleghard and Max Haefli
A 13   Roof edge detail for a wood-cement roof
A 14   Layers of oil paper for a wood-cement roof
A 15   Residential district, Cité industrielle project, 1917,
       Tony Garnier

A 15

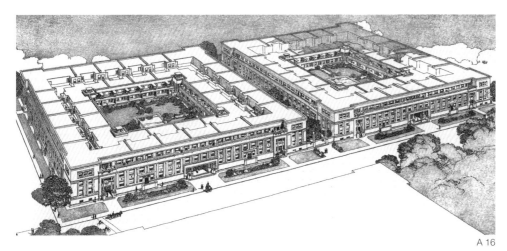

A 16

Skyscrapers and bituminous felt: the flat roof asserts itself in Chicago

But the flat roofs mentioned above remained individual, exclusive examples. In the second half of the 19th century it was the USA, and principally Chicago, that took the lead in the use of flat roofs on a large scale.
After the devastating fire of 1871, in which about 18 000 buildings were destroyed, an unprecedented economic boom brought about a massive expansion of the city. Prices and a shortage of land resulted in a very dense urban layout characterised by building right up to the boundaries of the plots and taller and taller buildings. Most of the new buildings were purely utility structures which for financial reasons, but also because of the reduced fire load compared with pitched roofs with steel or timber supporting structures, were finished off with a flat "lid".
On the early skyscrapers with their facades still employing the language of classicism, this horizontal building termination was not only aesthetically desirable, but also offered space

for the building services installations that were now starting to appear. The construction of flat roofs was made possible by the development of bituminous felt which employed bitumen as the waterproofing material. Bitumen is a waste product obtained during the distillation of crude oil and had been available in the USA since the mid-19th century.

The horizontal plate becomes a design element

A few decades later, Frank Lloyd Wright transferred the ideas of the Chicago School from the offices and department stores of the cities to the small buildings of the American suburbs. Right from his early designs, we see a diverse range of reasons for the use of flat roofs in his unparalleled output – from the accessible rooftop terrace to the reinforced concrete slab cantilevering for architectural reasons (Fig. 17) and the simple cubes of his "Usonian" houses (Wright himself remarked that pitched roofs had been avoided on those houses for financial

reasons). The publication of his output by the Berlin-based Wasmuth publishing house in 1910 ensured his decisive influence on a whole generation of architects and artistic groups in Europe, from De Stijl to Walter Gropius and Mies van der Rohe. Wright designed the clearly structured blocks for the Lexington Terrace Apartments project in Chicago in 1901. With their stepped terraces around a central courtyard and access from each apartment, this design certainly anticipates the "stepped house", i.e. a house with terraces, even though the internal layouts paid little attention to orientation (Fig. 16).
Just over 10 years later, it was Adolf Loos who boasted that his Scheu House in Vienna (1912) represented the first stepped house built in Central Europe (Fig. 19). He had certainly collected ideas for this during his earlier travels around the Aegean and North Africa, even if he does deny this: "The Orient did not even enter my thoughts when I designed this house. All I meant was that it would be a great convenience to be able to stride out from the bedrooms,

A 17

"One must ask oneself why terraces have been common for thousands of years in the Orient and why they have not been used in our climes. The answer is simple: The forms of building construction known hitherto could only realise the flat roof and the terraces in frost-free regions. Since the invention of the wood-cement roof (gravel roof) and since the use of asphalt, the flat roof and hence the terrace is also possible. The flat roof has been the dream of architects for four centuries. This dream became a reality in the middle of the 19th century. But most architects didn't know what to do with the flat roof."

Adolf Loos [6]

A 18

which are on the first floor, onto a large, common terrace. Anywhere, whether in Algiers or Vienna." [6]

Furthermore, in this and other designs by Loos, the great purist, who throughout his life opposed the use of senseless ornamentation, aesthetic considerations of course also play a decisive role in the shaping of his strictly cubic building form. But this did not make him popular with his fellow citizens. The irritated public of Vienna, still ruled by the Kaiser at this time, missed the accustomed roof and complained in no uncertain terms about this architectural affront.

During that period the flat roof became more quickly established on industrial buildings, which were seen as utility structures where appearance counted less than economics. Just how wide the scope for interpretation was at the start of the 20th century, even among progressive architects, is impressively demonstrated by, on the one hand, Hans Poelzig's expressionistic, monumental Werder Mill (1906), with its massive walls, and, on the other, Walter Gropius' Fagus Factory (1911–14) in Alfeld an der Leine, which with its set-back loadbearing structure and softened corners already heralded the start of the modern movement (Fig. 20).

A 19

A 16   Lexington Terrace Apartments project, 1901, Frank Lloyd Wright
A 17   Kaufmann Residence ("Fallingwater"), Mill Run (USA), 1937, Frank Lloyd Wright
A 18   Yahara Boat Club project, 1902, Frank Lloyd Wright
A 19   Scheu House, Vienna (A), 1912, Adolf Loos
A 20   Fagus Factory, Alfeld an der Leine (D), 1911, Walter Gropius, Adolf Meyer

A 20

"… reinforced concrete has a hostile enemy: expansion, the risk of cracking. In order to overcome the risk of cracking, it is advisable to plan hanging gardens on the roofs. Why? Because they retain a certain level of moisture and protect against expansion. Furthermore, it is incredibly pleasant for the human soul to rest among living greenery on the roof."

Le Corbusier [8]

A 21

A 22

The early days of the modern movement

In the years before World War I it was only a few avant-garde architects who experimented with the new rational cubist language. And we certainly cannot speak of a uniform style – partly due to the continued presence of strong movements such as historicism, traditionalism and Art Nouveau. But in the immediate post-war years the collapse of the old order resulted in a fertile breeding ground for new ideas. And more than just a few architects and clients were now of the opinion that a new architecture was needed to reflect the new political and social structures. Painting, too, provided decisive momentum and contributed crucially to the establishment of the International Style.

For example, the members of the Dutch group De Stijl, a circle of artists and architects formed in 1917, transformed the abstract geometry of the painters Piet Mondrian and Theo van Doesburg into three-dimensional, neo-plastic concepts. Gerrit Rietveld's Schröder House in Utrecht (1924), with its space-forming plates intersecting at right-angles, can be regarded as the most rigorous example of this interpretation (Fig. 21). But eight years prior to that, Robert van't Hoff, another member of the group, had designed Henny Villa in Huis ter Heide (Fig. 22) – a profoundly modern piece of architecture for the Europe of that period. In terms of both its interior layout and its reinforced concrete outer shell, which is dominated by the overhanging flat roof, it takes its themes from the designs of Frank Lloyd Wright, whom Van't Hoff had met during a trip to the USA.

In the Soviet Union it was the Constructivists Malevich, Tatlin and Chernikhov who employed this new architectural language in their bold Utopian designs. In France, besides Le Corbusier it was primarily Robert Mallet-Stevens and André Lurçat, pupils of Josef Hoffmann, who helped the modern movement to achieve a breakthrough. And in Italy Giuseppe Terragni, with his Novocomum apartment block and the Casa del Fascio in Como. In the case of all these architects, whose common stylistic feature is a cubist architecture devoid of ornamentation, architectural considerations were among the main reasons for choosing the flat roof. But

there were other, equally important, reasons that led to the spread of the flat roof in the early days of the modern movement, e.g. new construction techniques that required the new forms, illustrated in exemplary fashion by Mies van der Rohe's design for a reinforced concrete office block dating from 1922. New interior layouts also played a role. Wright's idea of the unconstrained interior layout quickly found favour in Europe and was developed further by Mies van der Rohe to create the flowing space, the most rigorous realisation of which was his Barcelona Pavilion (Fig. 23). It is the cantilevering roof plate that emphasizes the continual transition from interior to exterior in the layout below.

Prefabrication und the "roofs dispute"

Other architects, in addition to Adolf Loos and Le Corbusier, focused on the utilisation of the roofs. Richard Döcker published his book *Terrassentyp* (terrace type) in 1929, the prime aim of which was to illustrate the necessity for sun terraces in hospitals and also the pleasantness of rooftop gardens on private buildings, using his own buildings and projects as examples. For many architects the firm belief in technical progress and rapid developments in industrialisation were important reasons for choosing the flat roof. It was precisely the urgent need for housing for the masses that convinced the avant-garde that roofs – in keeping with the

A 23

"… one must come to a totally different solution. It is necessary for the roof to slope inwards, for it to carry the snow over the entire winter, and for the meltwater that arises as a result of the central heating to drain away via a downpipe which is no longer external to the building, but rather internal, is possibly located in the middle, i.e. where it is warmest. And that this downpipe extends from the inward-sloping roof surface to a drain at the base of the building, where there is no risk of freezing, and into which, incidentally, pipes from bathrooms and elsewhere discharge."

Le Corbusier [10]

A 24

state of the art – in all developed regions of the world had to be flat and could only be water-proofed with industrially manufactured materials such as sheet metal, asphalt or bituminous felt. The traditional craft-like methods of working were rejected as obsolete and old-fashioned. On this theme, Franz Schuster wrote the following in 1927 in the magazine *Das Neue Frankfurt*: "It would contradict all labour economics and the technical spirit of our age, which again and again entice us to build our houses from large-format elements, if we were to place the old hand-crafted pitched roof, with its posts, ties, rafters, battens, tiles and many, many nails, on top of the house walls set up with the few turns of a crane." [9] The euphoria for the prefabri-

cated building was enormous. Many experiments were carried out, but the great breakthrough eluded the experimenters.
The "roofs dispute", which had been smouldering for some time, finally burst into flame with the Weißenhof Estate in Stuttgart, which was built in 1927 to demonstrate the new way of building. Architects were divided into two camps. The question "Flat roof or pitched roof?" became a question of attitude. Whereas one group regarded the flat roof as the symbol of the new age, the new technology, and regarded it as indispensable, the more traditions-oriented group regarded the pitched roof as the expression of the indigenous roots of building and polemicised vehemently against the cubist constructions in Stuttgart.

Le Corbusier's pair of semi-detached houses for the Deutscher Werkbund exhibition in Stuttgart proved to be a convincing declaration of his architectural philosophy, which he summarised in his famous book *Five Points of a New Architecture* (Fig. 25). Hardly any architect of the modern movement propagated the design and use of flat roofs as decisively as he. Le Corbusier was convinced that "it is human instinct to climb up to the roof of a house" [11], and asked: "Does it not truly offend all logic when a whole urban surface … remains unused?" [12] Rooftop gardens, one essential demand in his *Five Points*, were ascribed not only functional, economic and architectural attributes, but purely technical and constructional ones as well.

A 21   Schröder House, Utrecht (NL), 1924, Gerrit Rietvelt
A 22   Henny Villa, Huis ter Heide (NL), 1919,
       Robert van't Hoff
A 23   German Pavilion, Barcelona (E), 1929,
       Ludwig Mies van der Rohe (reconstructed
       1983–86)
A 24   Sketches for comparing new and conventional
       forms of construction, 1929, Le Corbusier
A 25   Rooftop terrace, semi-detached houses,
       Weißenhof Estate, Stuttgart (D), 1927, Le Corbusier

A 25

Forms of construction in the early modern movement

In the early years of the modern movement, when the flat roof was already a permanent feature of the language of progressive architects, there was still a deep rift between the architectural desires of the planners on the one hand and the constructional and technical possibilities on the other. Many new waterproofing materials and patents appeared on the market, but reliable experience was lacking at that time. So at the start of the 1920s the wood-cement roof continued to prevail, although reinforced concrete was gradually taking over from timber sheathing on timber joists as the loadbearing structure. From the building physics viewpoint in particular, there were still major problems to overcome. Thermal insulation was generally minimal and often attached inside, and the problem of thermal bridges was only scantily addressed. In order to avoid condensation, an additional air space beneath the loadbearing structure was frequently provided in the form of a suspended "Rabitz" ceiling (iron wire mesh embedded in gypsum). As external waterproofing, the wood-cement finishes were joined by asphalt and, ever more frequently, roofing felts too.

Flat roof drainage up until the early 1920s was still essentially to the outside, the perimeter, which required a minimal fall in one direction. But larger roof surfaces with internal drainage were already being built, e.g. the Schatzalp Sanatorium in Davos (c. 1900). Extremely enlightening with respect to the state of the art during the 1920s are the results of a poll that Walter Gropius carried out in 1926 among leading international architects for the journal *Bauwelt*. With only a few exceptions, all believed that they had the constructional problems under control. However, the many different views regarding the sequence of layers and the design of various junction details clearly reveal the great uncertainty still prevailing at that time. Otto Haesler and Peter Behrens were still firmly committed to the wood-cement roof, which had been rejected by others because the poor ventilation frequently resulted in rotting of the wood. Josef Hoffmann criticised the "inadequate durability of the layers of felt and shortcomings in detecting flaws". [13] Heavyweight roof structures with 30–40 mm cork or "Torfoleum" (compressed, impregnated peat boards) as thermal insulation plus two or three layers of roofing felt as waterproofing was the recommendation. The brothers Bruno and Max Taut swore by a special asphalt-saturated roof canvas together with asphalt board or mastic asphalt. There were also many different opinions regarding edge details and junctions with rising masonry.

The purist details of Mies van der Rohe, designed for their aesthetic effect only, with only a minimal upstand along the edge of the roof, certainly occupy a special position in this dispute (Fig. 29).

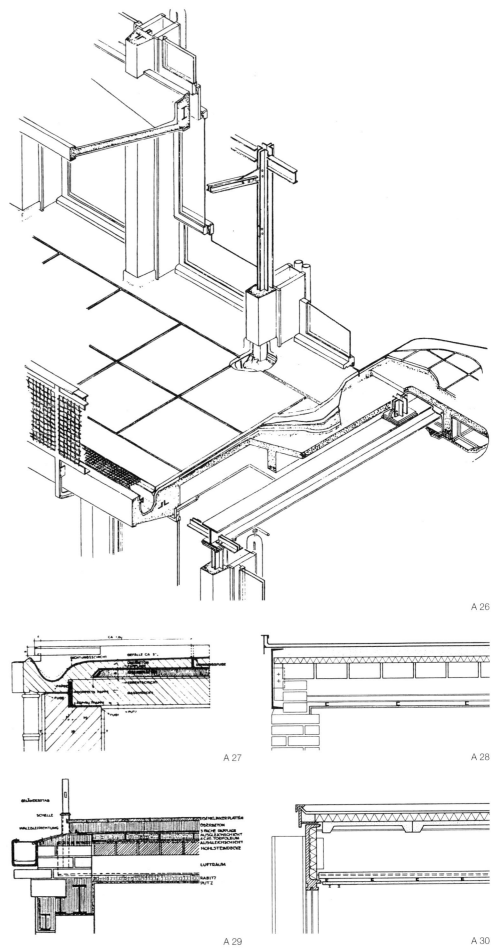

A 26

A 27

A 28

A 29

A 30

A 31

A 32

A 33

The Frankfurt standard for small dwellings was published in 1927. It dealt in detail with flat roof constructions and can be regarded as one of the first guidelines for such roofs (Fig. 27).

The global spread of the flat roof through the International Style

In the years following World War II, as the International Style was enjoying its triumphal procession around the world, the flat roof associated with this architecture suddenly became the norm. But the technical problems had by no means been fully eradicated. In Germany and many other countries the years of reconstruction and economic prosperity resulted in huge numbers of quickly erected buildings for the masses whose aesthetic and technical shortcomings would do permanent damage to the reputation of the flat roof.

On the other hand, many dedicated architects exploited the architectural and functional options of the flat roof to the full. For example, Ludwig Mies van der Rohe used the idea of the external loadbearing structure for his Crown Hall (1956) on the IIT Campus in Chicago (Fig. 34). The flat

roof of the building seems to float, suspended from four welded solid-web steel girders, above the space below, which has no intervening columns and therefore remains fully flexible and universal in its usage. He managed to create the apparently completely detached floating roof in his design for the National Gallery in Berlin (1968), which is in the form of a steel grillage. During the same period, the architects of the modular systems, large structures and stepped buildings so typical of the 1960s and 1970s made systematic use of the opportunities presented by the flat roof because its horizontal form was essential to achieving flexible, additive systems or the private open area in front of every dwelling. Good examples of this are the Metastadt System of Richard J. Dietrich (1965 onwards; Fig. 31) and the Olympic Village in Munich (1972), where owing to the separation of road traffic and pedestrians additional public thoroughfares and circulation zones are provided on vast flat roof structures (Fig. 32).

A few years later, architects with a technological bent, such as Lord Norman Foster with his Sainsbury Centre for Visual Arts in Norwich (1978), or Michael Hopkins with his Patera System (1984), attempted to resolve the constructional differ-

ence between roof and facade and construct the entire building envelope, including the roof covering, exclusively from industrially manufactured components. One purely concrete roof without any waterproofing at all was ventured by Heinz Isler on his own house (1964) in Burgdorf, Switzerland, which is protected by a dense, natural covering of plants and is still working well today (Fig. 33).

A 26 Isometric section through sun terrace, hospital in Waiblingen (D), 1928, Richard Döcker
A 27 Roof edge detail according to the Frankfurt standard for small dwellings, 1927
A 28 Proposal for roof edge detail for accessible flat roof, Erich Mendelsohn
A 29 Roof edge detail, Crown Hall, IIT, Chicago (USA), 1953, Ludwig Mies van der Rohe
A 30 Dachrand, Farnsworth House, Illinois (USA), 1950, Ludwig Mies van der Rohe
A 31 Model of Metastadt System, Wulfen (D), 1975, Richard J. Dietrich
A 32 Olympic Village, Munich (D), 1972, Heinle & Wischer
A 33 Isler's own house, Burgdorf (CH), 1964, Heinz Isler
A 34 Crown Hall, IIT, Chicago (USA),1956, Ludwig Mies van der Rohe

A 34

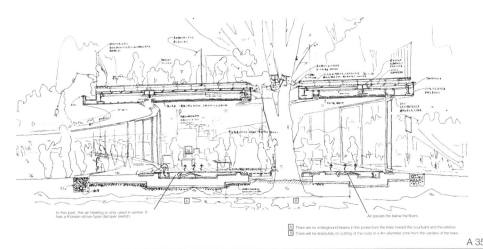

In this part, the air heating is only used in winter. It has a Korean-stove-type damper switch.

Air passes the below the floors.

A There are no underground beams in the zones from the trees toward the courtyard and the exterior.
B There will be absolutely no cutting of the roots in a 4m-diameter zone from the centers of the trees.

A 35

A 36

A 37

The flat roof in contemporary architecture

Flat or pitched? Until well into the 1980s picking the right roof form represented virtually a confession of faith for architects. Only those who advocated the horizontal building termination were regarded as modern and hence up to date, whereas the proponents of the inclined variation saw themselves quickly forced into the corner of the "old school". Today, however, the "roofs dispute" is long since a thing of the past. Flat roofs and pitched roofs exist side by side on equal terms. At the same time, new materials and forms of construction are increasingly diluting the difference between the two forms. More and more, roof and walls merge into a uniform building envelope. However, wherever true flat roofs are used, their designers, especially in the case of prestigious structures, increasingly turn them into a fifth facade. Dominique Perrault, for example, covered the entire roof surface of Berlin's velodrome plus the areas of the facade defined by the loadbearing structure with what was at that time (1997) a new type of metal fabric for architectural applications. Besides the surprising visual unity, he also achieved interesting effects caused by the strong reflections of the incident light. For the MAXXI Museum in Rome (2010), the British-Iraqi architect Zaha Hadid used the elevation of the building in order to underscore the dynamic of the sculptural building form. She divided up the roof surfaces with glass ribbons that give the impression of movement and at the same time allow daylight to illuminate the interior of the museum below. In Lille the French architects Jean-Marc Ibos and Myrto Vitart created a flat roof of glass that lies like a reflective pond on the urban square in front of the Musée des Beaux-Arts (1997; Fig. 36). Massimiliano Fuksas, on the other hand, upgraded the roofs to his engineering centre (2004) in Maranello, northern Italy, with real water and therefore at the same time achieved a natural climatic effect for the adjacent offices.
Other designers have surprised us with unconventional uses of their roof surfaces. For example, Takaharu and Yui Tezuka used the entire oval roof to their kindergarten in Tokyo (2007) as a huge playground (see pp. 196–197). A similar

A 38

concept was pursued by the Berlin architects Armand Grüntuch and Almut Ernst with their grammar school in Dallgow-Döberitz (2005; see pp. 191–195). Although a less extrovert design, it offers a more diversified platform for experiences with numerous spatial references. Riken Yamamoto converted the entire horizontal termination to his university buildings in Saitama, Japan, into a huge maze-like garden (1999; Fig. 37). The nearby terminal in Yokohama, completed three years later, was also turned into an urban space for the public (Fig. 35). So the flat roof returns the land consumed by the building to the urban space and roofs coalesce into a morphological unity.

References
[1] Klessing, Jan Martin: Traditional flat roofs in the Yemen: the rehabilitation of the "Samsarat al-Mansurah" in Sana'a. Detail 5/1997, pp. 698–702
[2] Alberti, Leon Battista: Ten Books on Architecture. Book III, Chap. XVI. Florence, 1485, (engl. trans: Leoni, James, 1755)
[3] Marperger, Paul Jacob: frontispiece of the treatise on the benefits of flat roofs. Dresden c. 1722, Friedrich Bock and Georg Gustav Wieszner (ed.), Nuremberg, 1930
[4] ibid.
[5] Giedion, Sigfried. In: Das Neue Frankfurt. No. 2, 1928
[6] Loos, Adolf: Das Grand-Hotel Babylon. In: Die neue Wirtschaft. Vienna, 1923
[7] ibid.
[8] Le Corbusier. In: Neue Zürcher Zeitung, 24 Jun 1934. Cited by Bosmann, Jos (ed.): Le Corbusier und die Schweiz. Zurich, 1987
[9] Schuster, Franz. In: Das Neue Frankfurt. No. 7, 1926/27
[10] see ref. [8]
[11] see ref. [8]
[12] Cited by Hoffmann, Ot: Handbuch für begrünte und genutzte Dächer. Leinfelden-Echterdingen, 1987
[13] Hoffmann, Josef. In: Bauwelt, 1926

A 39

Part B   Structure

Fig. B   Company restaurant, Ditzingen (D), 2008, Barkow
         Leibinger Architekten

Loadbearing structure

Matthias Beckh, Mark Böttges,
Eberhard Möller, Zoran Novacki

B 1.1

Buildings and structures are exposed to diverse so-called actions. These are, primarily, dead, imposed, wind, snow and ice loads, thermal effects and the effects of fires or different loading conditions during construction.

The purpose of a loadbearing structure is to resist these actions and guarantee the stability, durability and serviceability of the construction in accordance with economic requirements. The adequate design of all loadbearing elements prevents larger deformations or vibrations, which in a worst-case scenario could lead to the collapse of the structure. Deformation limits are important for flat roofs in order to rule out, for example, ponding on the roof or damage to facade elements and non-loadbearing internal walls.

The magnitudes of the various actions are mostly subjected to fluctuations in terms of time and position. And, as the properties of the building materials are subjected to a natural scatter and all components are produced with certain tolerances, the resistances of the loadbearing elements cannot be predicted with 100 % accuracy.

Standards

As a result of the aforementioned uncertainties, it is clear that an economically reasonable safety concept for the loadbearing structure cannot achieve an absolute level of safety but only a reasonably high one. In order to establish a uniform level, the generally recognised codes of practice are grouped together in standards. In 1975 the European Community adopted a programme for harmonising the technical codes of practice for structural design. The Eurocode programme includes the following standards, most of which consist of several parts:
- EN 1990 (Eurocode 0): Basis of structural design
- EN 1991 (Eurocode 1): Actions on structures
- EN 1992 (Eurocode 2): Design of concrete structures
- EN 1993 (Eurocode 3): Design of steel structures
- EN 1994 (Eurocode 4): Design of composite steel and concrete structures

- EN 1995 (Eurocode 5): Design of timber structures

The European standards (Euronorms, EN) allow Nationally Determined Parameters (NDP) to be defined for a number of points. The corresponding points plus additional stipulations and explanations are defined in the National Annex (NA) to each Euronorm. Currently, the Euronorms mostly represent alternatives to the regulations applicable in the Member States, but are intended to replace these gradually.

The Euronorms and the majority of national standards are based on a safety concept derived from probabilities, which take into account the different statistical distributions of the respective variables plus the scatter of actual measurements. In doing so, the safety concept is related to both the effects to which a loadbearing structure is subjected (actions) and also to the loaded components themselves (resistances).

Characteristic values

The required level of safety is defined for every individual basic variable in terms of both actions and resistances by way of characteristic values and additional partial safety factors.

Characteristic values for actions are specified based on the probability of their occurrence (Fig. B 1.2). To do this, they are classified according to their duration as follows:
- Permanent actions, e.g. dead loads and fixed items that are normally stationary. Permanent actions, which exhibit only limited scatter and can be determined with some accuracy, are generally taken into account by means of statistical mean values.
- Variable actions, e.g. imposed, wind and snow loads, can generally vary depending on position and location. The characteristic values for these duration-dependent, changing effects are usually specified in such a way that the probability of their not being exceeded over a period of one year is 98 %. Related to an assumed period of use of 50 years, this means that, on average, the value will be reached or exceeded once in that time. The characteristic values of variable actions

B 1.1 Cité du Design, Saint-Étienne (F), 2010,
LIN Finn Geipel and Giulia Andi
B 1.2 Safety concept according to the Eurocodes:
schematic representation of the probability density
of action F and resistance R
F_k Characteristic value of a (variable) action
F_d Design value of an action
γ_F Partial safety factor for actions
R_k Characteristic value of a resistance
R_d Design value of a resistance
γ_M Partial safety factor for a component property
B 1.3 Characteristic values for self-weights

therefore refer to loads that can realistically occur during the lifetime of a structure. From this we can conclude that with a much shorter period of use, e.g. temporary structures or situations during construction, the wind and snow loads can be reduced.
- Accidental actions, e.g. impact, fire, explosion, and seismic actions

Partial safety factors
During the design of a structure, the characteristic values are multiplied by partial safety factors in order to guarantee an adequate level of safety. This factor is generally 1.35 for dead loads, and 1.5 for imposed, wind and snow loads. When verifying the serviceability of a structure, e.g. to limit deflection, the partial safety factor for dead loads is 1.0. Furthermore, the simultaneous action of various loads, e.g. snow and wind, must be considered. But as it is assumed that several loads do not act simultaneously with their full characteristic values, the standards stipulate rules and specify reduced combined safety factors for such situations.

Strength and stiffness values
On the resistance side, the characteristic values are established by way of loading tests; the strength and stiffness values of all common building materials have been determined in extensive test series. For example, the minimum value of the yield stress of steel forms the characteristic value for the strength of this material. Owing to the different material properties of timber in the axial, radial and tangential directions, different values have been defined depending on loading type and direction, e.g. the 5 % fractile of the ultimate stress of timber in tension. The level of safety is defined by way of partial safety factors for the resistance variables, too. These coefficients depend on the scatter of the results from the loading tests. The characteristic values are divided by these coefficients to bring about a reduction. For calculating the design value of a steel component the partial safety factor is 1.1, for a structure made of timber (i.e. a natural material) it is 1.3.

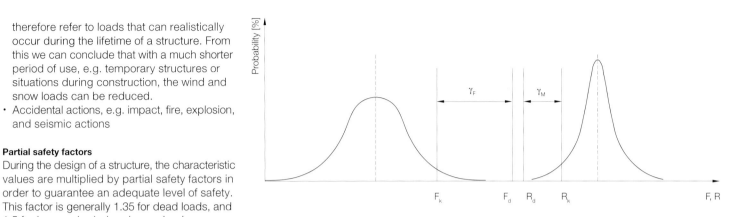

B 1.2

Metals	[kN/m³]
Aluminium	27
Lead	114
Copper	89
Brass	85
Steel & wrought steel	78.5
Zinc	72

Concrete	[kN/m³]
Reinforced concrete (DIN 1045)	25
Aerated concrete, density class 0.40–0.80 (DIN 4223)	5.2–9.5
Lightweight concrete	
• reinforced	5–20
• plain	9–21

Timber	[kN/m³]
Softwood, general	5
Hardwood	7–11
Particleboard	6
Plywood	6

Roof waterproofing, insulation, surface finishes	[kN/m²]
Bitumen & polymer-modified bitumen roofing felts (DIN 52130 & 52132)	0.04
Bituminous roofing or waterproofing felts, also synthetic roofing felts (incl. adhesive), per layer	0.05
Cold-applied, self-adhesive, tear-resistant bitumen sheeting	0.017
Bitumen & polymer-modified bitumen roofing felts for torching (DIN 52131 & 52133)	0.07
Bitumen roofing felt with metal foil inlay (DIN 18190-4)	0.03
Bitumen roofing felt on glass fleece base (DIN 52143)	0.03
Synthetic roofing felt	0.02
Uncoated bitumen-saturated sheeting (DIN 52129)	0.01
Uncoated bituminous roofing felt	0.02
Thermal insulation (glass or mineral wool), per 10 mm thickness	0.02
Hardboard	0.05
Timber decking, 24 mm	0.14
Chippings (incl. sealing coat)	0.05
Gravel topping, 50 mm deep (classed as imposed load)	1

Metal roof coverings	[kN/m²]
Aluminium (0.7 mm thick sheet aluminium, incl. 24 mm decking)	0.25
Profiled aluminium sheets (0.8 mm thick, 80 mm deep, no decking)	0.04
Profiled aluminium sheets on battens	0.08
Corrugated, trapezoidal & ribbed aluminium sheets	0.05
Locked double welt standing seam roofing of titanium-zinc or copper (0.7 mm thick, incl. sheathing & 24 mm decking)	0.35
Copper roof with double welts (0.6 mm thick sheet copper, incl. 22 mm decking)	0.3
Steel pantiles roof (galvanised steel to DIN 59231)	
• incl. battens	0.15
• incl. sheathing & 24 mm decking	0.3
Profiled steel sheets (with trapezoidal, ribbed or double-rib profiles)	
• 26 mm profile depth, 0.75 mm nom. sheet thickness	0.075
• 121 mm profile depth, 1.50 mm nom. sheet thickness	0.24
Corrugated sheet steel roof	
• galvanised sheet steel (incl. fixing materials to DIN 59231)	0.25
• zinc with batten rolls (incl. 22 mm decking)	0.30

Other roof coverings	[kN/m²]
Plastic corrugated sheets (profile to DIN EN 494), w/o purlins, incl. fixing materials, made from fibre-reinforced polyester resins (density: 1.4 g/cm³)	
• sheet thickness 1 mm	0.03
• with caps to fixings	0.06
Plastic corrugated sheets on battens	0.2
PVC-coated polyester fabric w/o loadbearing structure	
• type 1 (tear strength 3.00 kN/5 cm width)	0.0075
• type 2 (tear strength 4.70 kN/5 cm width)	0.0085
• type 3 (tear strength 6.00 kN/5 cm width)	0.01
Profiled glass, 1 layer	0.27
Profiled glass, 2 layers	0.54

Glazing, incl. framing	[kN/m²]
Normal glass, 5 mm	0.25
Wired glass, toughened safety glass, 6 mm	0.35
Laminated safety glass, 8 mm	0.4
Surcharge for every additional 1 mm thickness of glass	0.03

B 1.3

25

Actions

Actions on structures are dealt with in DIN 1055 and DIN EN 1991 or ÖNORM EN 1991 (Eurocode 1), also SIA 261. In Germany both DIN 1055 and DIN EN 1991 are valid.

Dead loads

Dead loads are permanent actions due to the self-weights of all loadbearing and non-loadbearing parts of the construction, in this case the roof. The self-weights of loadbearing elements, insulation and roof covering are added together to form one action (Fig. B 1.4). An exception to this rule is the self-weight of loose gravel finishes or the substrate on a green roof, which must be considered as variable, i.e. imposed, loads. The loads of building services must also be considered as non-permanent, imposed loads because these could change over the lifetime of the structure.

When verifying the stability of the roof covering against wind uplift, a partial safety factor of 0.9 should be used because in this case the self-weight has a positive effect. Fig. B 1.3 (p. 25) lists characteristic values for the self-weights of building materials and roof coverings. Most of these densities and unit loads have been taken from DIN 1055-1. More accurate values for specific cases can be found in the respective test certificates of the products.

Imposed loads

Imposed loads are actions that occur as a result of the use of a building, e.g. persons, movable items, etc. In Germany these can be determined according to DIN 1055-3 or DIN EN 1991-1-1 (Eurocode 1). Generally, imposed loads do not occur simultaneously with the same intensity on all parts of the loadbearing structure. When designing a structural component, the load distribution should therefore be varied in such a way that it has the most unfavourable effect on that component.

Apart from helicopters, the loads given below are essentially stationary loads without any dynamic effect that could induce vibrations in the component or the building. For roofs where dynamic loads are to be expected, e.g. a helipad, the potential resonance effects must be considered.

Flat roofs can be designed for foot traffic (e.g. rooftop terrace) or for no foot traffic as such, merely the loads of persons carrying out repairs or normal maintenance work. For this latter case, DIN 1055-3 specifies designing for a 1.0 kN point load applied at the most unfavourable point. The application area of this point load should be a square measuring 50 × 50 mm. As this load does not need to be combined with a snow load, it normally has an effect only on components with a very small loaded area, e.g. timber decking.

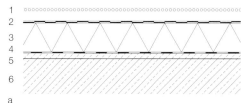

1	Gravel topping, 50 mm	1.00 kN/m²
2	Waterproofing, 2 layers	0.20 kN/m²
3	Thermal insulation	0.10 kN/m²
4	Levelling layer	0.70 kN/m²
5	Concrete laid to falls, 30 mm	
6	Reinforced concrete slab, 200 mm	5.00 kN/m²
	Total	**7.00 kN/m²**

a

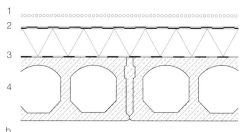

1	Gravel topping, 50 mm	1.00 kN/m²
2	Waterproofing, 2 layers	0.20 kN/m²
3	Thermal insulation with integral falls	0.20 kN/m²
4	Precast concrete elements, 320 mm	4.40 kN/m²
	Total	**5.80 KN/m²**

b

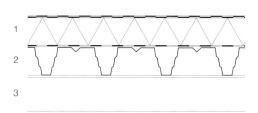

1	Waterproofing, thermal insulation, etc.	0.35 kN/m²
2	Steel trapezoidal profile sheeting	0.20 kN/m²
3	Beams, e.g. HEA 200 @ 3.00 m c/c	0.15 kN/m²
	Total	**0.70 kN/m²**

c

1	Waterproofing, vapour barrier	0.10 kN/m²
2	Thermal insulation, 100 mm	0.10 kN/m²
3	Edge-glued timber elements, 280 mm	1.80 kN/m²
	Total	**2.00 kN/m²**

d

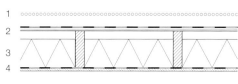

1	Waterproofing, roof decking	0.20 kN/m²
2	Thermal insulation, 100 mm	0.10 kN/m²
3	Timber joists, e.g. 280 × 140 mm @ 600 mm c/c	0.40 kN/m²
4	Vapour check, plasterboard	0.30 kN/m²
	Total	**1.00 kN/m²**

e

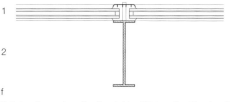

1	Lam. safety glass made from heat-strength. glass, incl. frames	1.50 kN/m²
2	Beams, e.g. IPE 270 @ 2.00 m c/c	0.20 kN/m²
	Total	**1.7 kN/m²**

f

B 1.4

B 1.4   Examples of self-weights of flat roofs with a load-bearing layer of…
     a   reinforced concrete
     b   precast prestressed elements
     c   steel beams with steel trapezoidal profile sheeting
     d   edge-glued timber elements
     e   timber joists
     f   glass

B 1.5   Imposed loads due to rooftop planting
B 1.6   Perpendicular imposed loads for car parks and surfaces with vehicular traffic to DIN 1055-3
B 1.7   Standard loads of helicopters to DIN 1055-3
B 1.8   Average density of snow to DIN EN 1991-1-3
B 1.9   Snow load map for Germany to DIN 1055-5
B 1.10   Characteristic values of snow loads on the ground to DIN 1055-5

An imposed load of 3.0 kN/m² should be assumed for walkways that form part of an escape route. For battens used for fixing roof coverings two 0.5 kN point loads should be applied to the outer quarter-points of the span. For a rooftop terrace, DIN 1055-3 specifies a uniformly distributed imposed load of 4.0 kN/m², which again does not have to be added to the snow load. In addition, the effect of a 2.0 kN point load applied at the most unfavourable position must be investigated.

Neither the Eurocode nor DIN 1055-3 specify loads due to rooftop planting. Fig. B 1.5 lists the loads of normal soil depths that should be considered for green roofs. As with loose gravel finishes, they must be considered as separate imposed loads with the most unfavourable distribution. This takes into account the fact the positions of loose materials can alter over time.

A universally distributed load of $q_k = 2.0 - 3.5$ kN/m², depending on the loaded area, should be assumed for parking surfaces for light vehicles. As an alternative, the designer can examine whether individual axle loads, each $2 \times Q_k = 2 \times 10$ kN, are critical for the design (Fig. B 1.6).

Roofs used as helicopter landing pads must be designed for the loads given in Fig. B 1.7. As helicopter loads are regarded as not essentially stationary, these loads must be multiplied by a vibration coefficient φ; as a rule, φ = 1.4 should be used. As a loose fill on a roof has a damping effect, φ can be reduced depending on the depth of the fill:

$$\varphi = 1.4 - 0.1 \cdot h_{ü} \geq 1.0$$

where:
φ   vibration coefficient [-]
$h_{ü}$ depth of loose fill [m]

Furthermore, roofs used as helicopter landing pads must be designed for a uniformly distributed imposed load of 5.0 kN/m² as well.

Snow loads

The location of the structure, the local climate, the height above sea level and the topography of the site are critical for determining the snow loads. They are also dependent on the form of the structure, the roughness of its surfaces and the thermal insulation properties of the roof. Generally, the density of snow increases with the length of time it is allowed to lie on a surface. Rain falling on top of snow can result in a considerable load on the structure (Fig. B 1.8).

Determining the snow load on the ground to DIN 1055-5
The European concept for determining the basic data for snow loads is based on a 50-year return period. The basis for this is a snow map calibrated for the snow on the ground, which has been drawn up as a result of meteorological observations. The snow map for Germany is a version of the European snow map simplified for practical application. It divides the Federal Republic of Germany into three snow load zones and two sub-zones (Fig. B 1.9).

The characteristic value of the snow load s_k on the ground can be determined depending on the snow load zone in which the site is located and the height of the site above sea level. In Germany the snow load increases with the height above sea level in an approximately parabolic curve (Fig. B 1.10).

The characteristic values in zones 1a and 2a are the result of multiplying the values in zones 1 and 2 by a factor of 1.25.

Higher values may apply in some regions. And for a site 1500 m above sea level, the recommended snow load values should be specified by the authorities responsible for every individual project.

Determining the snow load on the roof to DIN 1055-5
The snow load on the roof must be calculated in accordance with the roof form and the characteristic snow load at ground level. The following equation applies:

$$S = \mu_i \cdot s_k$$

where:
μ_i roof shape coefficient [-]
S characteristic snow load on roof [kN/m²]
s_k characteristic snow load on ground [kN/m²]

On flat roofs the roof shape coefficient is generally μ = 0.8.

Drifting effects can become relevant where structures, e.g. parapets, rise above roof level. In such situations, depending on the height of the parapet or wall, the snow load can increase locally to twice the characteristic snow load at ground level.

Where snow could fall from a higher to a lower roof, the latter must be designed for the extra load caused by such additional snow. According to DIN 1055-5, depending on the difference in height and the pitch of the roof, in such situations the snow load locally can reach four times the characteristic snow load at ground level. It may even be necessary to allow for the impact loads of falling snow masses. However, where the pitch of the higher roof is ≤ 15° or where snowguards can reliably prevent snow from sliding off the higher roof, then these additional loads do not have to be taken into account in the design.

Higher loads that occur when the drainage system is blocked by snow or ice must be taken into account separately.

Type of planting	Typical plants	Weight
Extensive	Mosses, grasses, herbs	0.9–2.0 kN/m²
Intensive	All conventional garden plants except large trees	up to 10 kN/m²

B 1.5

Usage	Area A [m²]	Load per unit area q_k [kN/m²]		Axle load $2 \times Q_k$ [kN]
Traffic and parking areas for light vehicles (total load ≤ 25 kN)	≤ 20	3.5		20
	≤ 50	2.5	or	20
	> 50	2		20

B 1.6

Permissible take-off weight [t]	Standard helicopter load Q_k [kN]
3	30
6	60
12	120

B 1.7

Type of snow	Density [kN/m³]
Freshly fallen snow	1
Snow that has lain for several hours or days	2
Snow that has lain for several weeks or months	2.5–3.5
Wet snow	4

B 1.8

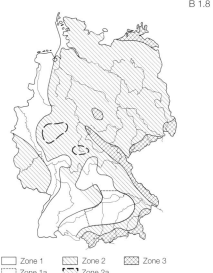

Zone 1   Zone 2   Zone 3
Zone 1a   Zone 2a

B 1.9

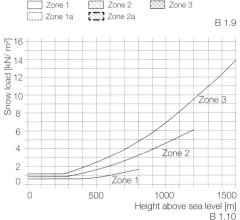

B 1.10

Wind loads

The wind load that acts on a flat roof depends, on the one hand, on natural factors such as wind speed and wind direction, and, on the other, the form, dimensions, surface properties or permeability of the structure exposed to the wind (Fig. B 1.11). The wind can cause pressure, suction and friction forces. The pressure and suction components act perpendicular to the surface of a component, and the tangential friction forces can become critical on long buildings with a certain type of surface property, or free-standing canopies with a low structural depth. In the case of partially open buildings and those with a permeable envelope, an additional internal pressure must be taken into account.

Basic velocity pressure

The density of the air ρ, which depends on height above sea level, air pressure and air temperature, can be used to define a physical relationship between the velocity of the wind and the pressure due to that velocity. The basic velocity pressure q_b for a certain wind velocity v_b can be calculated from the following equation:

$$q_b = 0.5 \cdot \rho \cdot v_b{}^2$$

Taking the air density as $\rho = 1.25$ kg/m^3, this results in:

$$q_b = v_b{}^2/1600$$

where:
q_b basic velocity pressure [kN/m^2]
ρ air density [kg/m^3]
v_b wind velocity [m/s]

Wind maps specify the mean wind velocities v_{ref} over a certain period and the associated velocity pressures q_{ref} (Figs. B 1.12 and 1.13). The characteristic values are mean values measured over a period of 10 min with a 2% probability of their being exceeded within one year. We therefore assume that these values are reached or exceeded once every 50 years. The velocity v_{ref} refers to a height of 10 m above the ground on flat, open terrain.

Peak velocity pressure

Much greater velocity pressures can develop for a few seconds within gusts of wind, and these must also be considered at the design stage. The wind pressures to be assumed must therefore take into account the surrounding terrain, which influences the nature of the gusts, as well as the height of the structure. The stipulations for the velocity pressure to be used are based on the mean peak velocity over a gust duration of 2–4 s. For structures with a maximum height of 25 m, this velocity pressure can be assumed to be constant over the entire height. Fig. B 1.13 lists the corresponding velocity pressures. For structures > 25 m high, the peak velocity pressure should be determined more accurately in accordance with the standards.

Wind pressure and wind suction

The wind pressure w_e acting on the external surface of a structure is:

$$w_e = c_{pe} \cdot q_p(z_e)$$

where:
w_e wind pressure [kN/m^2]
c_{pe} pressure coefficient [-]
$q_p(z_e)$ peak velocity pressure [kN/m^2]
z_e reference height for external pressure [m]

The wind pressure always acts perpendicular to the surface considered. If the calculations result in a wind pressure with a negative sign, then this indicates a wind suction; i.e. positive c_{pe} values represent pressure and negative values suction (uplift on a roof). On flat roofs the suction loads are generally much higher than the pressure loads. All the components must therefore be secured against uplift by attaching them to the supporting structure in some suitable way or by using heavy materials or adding ballast.
The wind pressure can vary considerably across a surface. The standards therefore specify two c_{pe} values: $c_{pe,1}$ is the mean value over a 1 m^2 loaded area and is used for the anchor forces of components affected directly by wind loads; $c_{pe,10}$ is the mean value over an area of 10 m^2 and is used for the parts of the construction that transfer the loads to other parts of the structure. Fig. B 1.17 is a table of c_{pe} values for flat roofs with parapets and so-called sharp, curved or mansard eaves details according to the Eurocode. The magnitude of any ballast necessary can be derived directly from the magnitude of the wind suction for a loaded area of 1 m^2 taking into account the associated safety factors. The standards also contain values for internal pressure, for structures open at the sides and for rooftop structures such as advertising hoardings or external trussed girders.

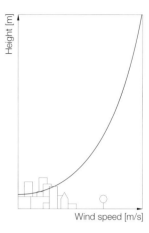

B 1.11

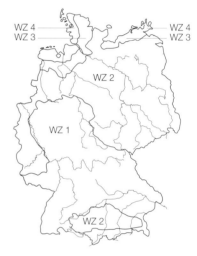

B 1.12

Wind zone	Reference wind velocity vref [m/s]	Associated velocity pressure qref [kN/m²]
1	22.5	0.32
2	25.0	0.39
3	27.5	0.47
4	30.0	0.56

B 1.13

B 1.11   Wind profile: the relationship between wind speed v and height h above ground level
B 1.12   Wind map of Germany to DIN 1055-4
B 1.13   Reference wind velocity and associated velocity pressure to DIN 1055-4
B 1.14   Simplified assumptions for peak (or gust) wind velocity pressures for structures up to 25 m high to DIN 1055-4
B 1.15   Subdivision of flat roof surfaces for determining wind pressure and suction to DIN 1055-4
B 1.16   Beaufort scale for classifying wind speeds and their effects
B 1.17   External wind pressure coefficients for different flat roof areas to DIN 1055-4

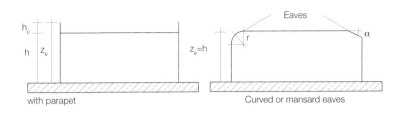

with parapet | Curved or mansard eaves

Wind zone		Velocity pressure q_p [kN/m²] for a building height h of...		
		$h \leq 10$ m	10 m$< h \leq 18$ m	18 m$< h \leq 25$ m
1	Inland	0.5	0.65	75
2	Inland	0.65	0.8	0.9
	Baltic Sea coast & islands	0.85	1	1.1
3	Inland	0.8	0.95	1.1
	Baltic Sea coast & islands	1.05	1.2	1.3
4	Inland	0.95	1.15	1.3
	Baltic & North Sea coasts & Baltic Sea islands	1.25	1.4	1.55
	North Sea islands	1.4	–	–

B 1.14

F   Windward corner
G   Windward edge
H   Windward inner area
I   Leeward inner area

e = b oder 2h
(the smaller value governs
in the calculations)
b   Dimension transverse to wind
direction

B 1.15

Beau-fort	[km/h]	[m/s]	Designation	Description
0	0–1	0–0.3	Calm	Smoke rises vertically
1	1–5	0.3–1.5	Light air	Smoke drifts in the wind
2	6–11	1.6–3.3	Light breeze	Leaves rustle, wind felt on face
3	12–19	3.4–5.4	Gentle breeze	Leaves and small twigs in constant motion, light flags extended
4	20–28	5.5–7.9	Moderate wind	Dust, leaves and loose paper raised, small branches move
5	29–38	8.0–10.7	Fresh wind	Small trees sway, crested wavelets form on inland waters
6	39–49	10.8–13.8	Strong wind	Large branches move, whistling in phone wires, difficult to use umbrellas
7	50–61	13.9–17.1	Very strong wind	Whole trees in motion, inconvenience felt when walking against the wind
8	62–74	17.2–20.7	Gale	Twigs break off trees, walking is difficult
9	75–88	20.8–24.4	Severe gale	Some minor structural damage (e.g. chimney pots and slates removed)
10	89–102	24.5–28.4	Storm	Trees uprooted, structural damage
11	103–117	28.5–32.6	Severe storm	Widespread damage
12	>117	> 32.6	Orkan	Widespread severe damage, very rarely experienced on land

B 1.16

Roof type		Area							
		F		G		H		I [3]	
		$c_{pe,10}$	$c_{pe,1}$	$c_{pe,10}$	$c_{pe,1}$	$c_{pe,10}$	$c_{pe,1}$	$c_{pe,10}$	$c_{pe,1}$
sharp eaves		-1.8	-2.5	-1.2	-2.0	-0.7	-1.2		0.2
									-0.2
with parapet [1]	$h_p/h=0.025$	-1.6	-2.2	-1.1	-1.8	-0.7	-1.2		0.2
									-0.2
	$h_p/h=0.05$	-1.4	-2.0	-0.9	-1.6	-0.7	-1.2		0.2
									-0.2
	$h_p/h=0.1$	-1.2	-1.8	-0.8	-1.4	-0.7	-1.2		0.2
									-0.2
curved eaves [1,5]	$r/h=0.05$	-1.0	-1.5	-1.2	-1.8	-0.4			0.2
									-0.2
	$r/h=0.10$	-0.7	-1.2	-0.8	-1.4	-0.3			0.2
									-0.2
	$r/h=0.20$	-0.5	-0.8	-0.5	-0.8	-0.3			0.2
									-0.2
mansard eaves [2,4]	$\alpha=30°$	-1.0	-1.5	-1.0	-1.5	-0.3			0.2
									-0.2
	$\alpha=45°$	-1.2	-1.8	-1.3	-1.9	-0.4			0.2
									-0.2
	$\alpha=60°$	-1.3	-1.9	-1.3	-1.9	-0.5			0.2
									-0.2

[1] In the case of roofs with a parapet or curved eaves, intermediate h_p/h and r/h values may be obtained by linear interpolation.

[2] In the case of roofs with mansard eaves, values of α between $\alpha=30°$, 45° and 60° may be obtained by linear interpolation. When $\alpha >60°$, cpe values are obtained through linear interpolation between those for $\alpha=60°$ and those for flat roofs with sharp eaves.

[3] In area I, for which positive and negative values are specified, both values must be considered.

[4] The external pressure coefficients given in DIN 1055-4, Tab. 7.4a "External pressure coefficients for pitched and valley roofs", wind direction $\theta = 0°$, areas F and G, should be used for the sloping part of the mansard eaves itself, depending on the angle of the mansard section.

[5] For the curved eaves itself, the external pressure coefficients along the curvature may be obtained by linear interpolation along the curve between the values for the vertical wall and the values for the roof.

B 1.17

Primary structure systems

In the hierarchy of loadbearing systems, the primary structure forms the supporting construction for a flat roof.

An overview of common structural systems is given below, together with information intended to help with the design of the components. The details represent rough guidelines for basic systems for single-storey sheds and are primarily based on structural and constructional aspects. However, the influences of economics, transport and erection have also been considered. Depending on how these factors are weighted, the dimensions in individual cases can deviate considerably from those given here.

Solid-web systems		Depth h	
Simply supported beam		Timber (glulam)	l/17
IPE sections 80–600 mm deep are preferred when using steel. The material is exploited to the full mainly at mid-span. This system is not sensitive to restraints and settlement. Long simply supported beams should always be installed with a camber.		Steel	l/20
		Reinf. concrete	l/12
		Prestr. concrete	l/18
Continuous beam with three spans		Timber (glulam)	l/20
The deflection is smaller than that of a simply supported beam with the same span. The exploitation of the cross-section is more balanced because redistribution results in span and support moments; the span moments are therefore smaller than those of a simply supported beam. Rigid on-site connections should be located at the points of zero bending moment.		Steel	l/25
Portal frame with fixed bases		Timber (glulam)	l/27
This system has multiple degrees of static indeterminacy. The bending moments are distributed over all four corners of the system. The exploitation of the cross-section is more balanced. Owing to the moment at the base of each leg, the foundations must be larger than would be the case with pinned bases.		Steel	l/40–l/30
Two-pin portal frame		Timber (glulam)	l/30–l/20
This system has one degree of static indeterminacy. The bending moments are distributed over the legs, corners and beam. There are no moments at the bases. Rigid on-site connections should be located at the points of zero bending moment.		Steel	l/25
Three-pin portal frame		Timber (glulam)	l/20
This system is statically determinate. It is therefore not sensitive to restraints. The moments in this system are concentrated at the corners of the frame and the sizes of the cross-sections are therefore larger. The three-pin portal frame has advantages during erection and is therefore particularly popular for timber buildings. Transport restrictions usually limit the spans to approx. 40 m.		Steel	l/25
Grillage		Timber (glulam)	l/25–l/18
A grillage is mainly loaded in bending and torsion (with the torsion prevented). The spans of the beams should be more or less equal in both directions. Grillages should always be constructed with a camber.		Steel	l/35–l/25

Open-web systems	Depth h		

Simply supported lattice girder
The individual linear members are loaded in tension or compression only. Lattice girders can therefore be much lighter than solid-web elements. In the case of timber lattice girders it is usually the fasteners that govern the node details. For this and other reasons, the spans possible with timber are shorter than those possible with steel.

Timber (square)	l/9	
Steel	l/15–l/10	

Trussed beam
The trussing beneath the beam reduces the span of the member forming the top chord. The depth of the cross-section can therefore be reduced. The top chord is loaded in compression and therefore counteracts the tension forces in the trussing members. The top chord and the central strut must be restrained or designed to resist lateral buckling.

Timber (glulam)	$h_1 = l/12–l/10$	
	$h_2 = l/40$	
Steel	$h_1 = l/12$	
	$h_2 = l/50–l/35$	

Lattice portal frame with fixed bases
This system has a high degree of static indeterminacy externally. The loads in the individual linear members are distributed over all four corners of the frame. When designed in steel, a three-dimensional form for the lattice frame (e.g. triangular) enables large spans to be realised (e.g. stadiums).

Timber (square)	l/13–l/9	
Steel	l/20–l/10	

Two-pin lattice portal frame
This system has one degree of static indeterminacy externally. There are high compressive forces in the inner chord members at the frame corners. These must be restrained against lateral buckling.

Timber (square)	l/10	
Steel	l/13–l/8	

Three-pin lattice portal frame
This system is statically determinate externally. There are high compressive forces in the lower chord members, which must therefore be restrained against lateral buckling. The closer the axis of the frame is to the line of support, the smaller the individual linear members can be.

Timber (square)	l/9–l/6	
Steel	l/10–l/6	

Grillage of lattice girders
The spans of the girders should be more or less equal in both directions.
Lattice girder grillages should always be constructed with a camber.

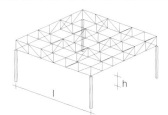

Timber (glulam)	l/16–l/8	
Steel	l/20–l/15	

Space frame
The spans of the girders should be more or less equal in both directions.
Lattice girder grillages should always be constructed with a camber.

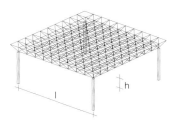

Steel	l/30·l/15	

System optimisation

Depending on the structural system, there are options for minimising the depth of the beam or girder for a given span. Figs. B 1.18–1.22 show various options using the example of a simply supported beam or girder subjected to a uniformly distributed load q; Figs. B 1.23–1.28 show a number of actual applications.

Simply supported beam

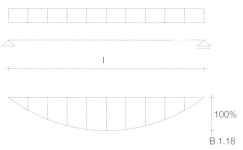

100%

B 1.18

Fixed-end beam

Fixing the beam at both ends enables the bending moments to be reduced to max. 67%.

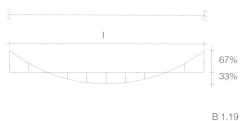

67%
33%

B 1.19

Simply supported beam with cantilever at one end

Moving one support inwards by an amount equal to 30% of the span enables the bending moments to be reduced to max. 36%.

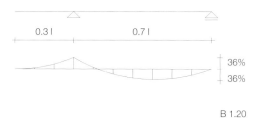

0.3 l   0.7 l

36%
36%

B 1.20

Simply supported beam with cantilever at both ends

Moving both supports inwards by an amount equal to 20% of the span enables the bending moments to be reduced to max. 21%.

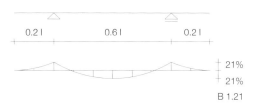

0.2 l   0.6 l   0.2 l

21%
21%

B 1.21

Balanced cantilever

The position of the hinges enables a balanced distribution of moments to be achieved. The aim here is to balance the span and support moments and hence minimise the depth of the cross-section as a whole.

B 1.22

B 1.18–1.22   Various structural systems for a uniformly distributed load q
B 1.23–1.28   A selection of examples showing system optimization in practice

Passenger terminal complex

Bangkok (T), 2006, Murphy/Jahn Architects

· Simply supported lattice girder with cantilevers to both sides
· Depth follows bending moment diagram

B 1.23

Petrol station

Munich (D), 2004, Haack + Höpfner Architekten

· Suspended simply supported beam with cantilevers to both sides
· Span reduced by suspension cables

B 1.24

Shopping centre
London (GB), 2008, Benoy Architects

- Free-form gridshell
- Span reduced by tree-type columns

B 1.25

Wiper-blade factory
Bietigheim-Bissingen (D), 2002, Ackermann & Partner

- Continuous lattice girder
- Structure resolved into individual members to reduce spans and self-weight of structure

B 1.26

Sports hall
Uster (CH), 2000, Camenzind Gräfensteiner

- Balanced cantilever
- Balanced bending moment diagram achieved by controlling the hinge positions in the frame system

B 1.27

Private house
Shizuoka (J), 2001, Shigeru Ban Architects

- Lattice portal frame with fixed bases
- System fixed at base and self-weight reduced by resolving structure into individual members

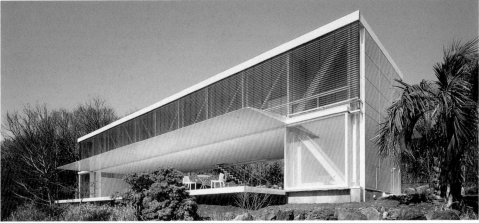

B 1.28

Loadbearing decks

Matthias Beckh, Mark Böttges, Eberhard Möller

B 2.1

The loadbearing deck is the planar roof structure that is supported on the primary structure or, in the case of smaller areas requiring a roof, directly on the walls. The loadbearing deck is usually of reinforced concrete, timber, metal or glass. Besides performing the necessary loadbearing task, it can also provide waterproofing, thermal insulation, sound insulation and fire protection, depending on the form of construction. It must be able to accommodate the loads applied safely and permanently while complying with the deflection criteria. Roofs frequently also fulfil a bracing function, and so the stiffness and robustness of the loadbearing deck, in the form of a plate, must be able to accommodate the horizontal loads applied; otherwise, additional diagonal bracing will be required.

Reinforced concrete

Domes and vaults of concrete (*opus caementitium*) were already being constructed in ancient times by Roman master-builders. Owing to their curved form, a membrane stress state, in which the main forces are compressive, prevails in these structures. In flat roofs and suspended floors, however, bending moments occur which generate a couple consisting of tensile and compressive forces within the concrete cross-section. As the concrete itself cannot accommodate any tensile forces, non-curved loadbearing elements of concrete can only be built in conjunction with reinforcing bars, which accommodate the tensile forces. The first tests on concrete components reinforced with wires were carried out in the mid-19th century. Today, the construction of roofs and suspended floors in reinforced concrete is a standard method covered by the provisions of DIN 1045 "Concrete, reinforced and prestressed concrete structures" and DIN EN 1992-1 (Eurocode 2). Owing to their relatively high self-weight, concrete roofs exhibit excellent sound insulation properties and have a high thermal mass, which has a favourable effect on the interior climate. And fire resistance ratings of F 90 and better are easy to achieve with concrete roofs.
We distinguish between in situ concrete and precast concrete. The formwork for in situ con-

crete components is constructed on the building site, the steel reinforcement is fixed in place and the concrete is poured. There are practically no limits to the shapes that can be formed. Owing to the monolithic form of construction, loads can be carried and transferred in several directions, and a plate effect can generally be assumed without the need for any further verification.
Precast concrete components are produced in a factory and delivered to the building site as prefabricated elements. Eliminating the need to construct formwork, place steel reinforcement and pour concrete on the building site generally leads to cost- and time-savings. The spectrum of precast concrete construction ranges from building systems based on a catalogue of prefabricated components not tied to a particular project to custom components which, although they are produced in the factory, are intended for a specific project. As with in situ concrete, one-off items are possible, and it is usually transport that sets the limits for such elements. A plate effect for stiffening a building can be achieved with precast concrete components by grouting the joints afterwards and forming a ring beam.
Timber-concrete composite construction (see p. 44) and composite slabs making use of steel trapezoidal profile sheeting (see p. 36) represent other ways of using concrete for constructing roofs.

In situ concrete slabs

The preliminary depth of an in situ concrete slab can be obtained by using the simplified formula given in DIN 1045-1, which is based on limiting the slenderness ratio l/d:

$$l_i/d \leq 35$$

where:
l_i equivalent span [m]
d   effective depth (= distance between tension reinforcement and edge of concrete in compression) [m]

The equivalent span depends on the structural system of the loadbearing structure, which in turn takes into account the coefficient α (Fig. B 2.3):

B 2.1 Wiper-blade factory, Bietigheim-Bissingen (D), 2000, Ackermann & Partner
B 2.2 Lifting of the edges of a slab supported on all four sides (dishing)
B 2.3 Coefficient α for determining the equivalent span depending on the structural system according to DIN 1045-1

$l_i = \alpha \cdot l_{eff}$

where:
l_{eff} effective span [m]

In the case of a slab bearing on linear supports, the smaller of the two spans (l_1 or l_2) governs, in the case of a slab bearing on point supports (flat slab), the larger of these two (fig. b 2.3). where suspended floor slabs have to satisfy higher requirements regarding deflection limits, the slenderness ratio chosen should be $l_c/d \leq$ 150. This criterion governs for equivalent spans exceeding about 4.3 m. it should also be maintained on flat roofs because larger deformations resulting from higher slenderness ratios can influence the fall and hence reduce the effectiveness of the roof drainage.

The component depth h is made up of the effective structural depth plus a concrete cover of 20–30 mm.

As solid concrete slabs with a thickness > 300 mm are uneconomic due to the great quantities of materials required, sensible equivalent spans for solid concrete roofs are limited to approx. 6.5 m. The resulting effective span can be greater depending on the structural system (Fig. B 2.3).

According to DIN 1045-1, the minimum depth h for a loadbearing reinforced concrete slab is 70 mm. Further requirements for the component depth result from the fire protection stipulations of DIN 4102-4:

- F 30 A: h = 60 mm (80 mm for a continuous slab)
- F 60 A: h = 80 mm
- F 90 A: h = 100 mm
- F 120 A: h = 120 mm

According to DIN 4102-4 the minimum depth h of a slab supported at individual points is as follows:

- 150 mm for columns with column heads
- 200 mm for columns without column heads

In order to be able to reduce the load due to the self-weight, and hence the quantities of materials, by up to 30%, voids can be formed in slabs, provided they are at least approx. 230 mm deep. Spherical voids have an advantage over tubular ones because they leave concrete webs in both directions so that a biaxial load-carrying effect is possible; spans of up to 15 m are then possible with slab depths of 230 to approx. 600 mm. Voids are not permitted in regions with high shear stresses, e.g. the punching shear zones around columns supporting flat slabs.

When slabs span in two directions and no loads are applied to the perimeter, the deflection kinematics can result in the slab corners lifting visibly from the masonry below (dishing), which can even lead to cracking at the joints (Fig. B 2.2). With continuous slabs it is only the outer corners that are at risk. Stiffening upstand beams or ties back to the floor below represent possible remedies.

Impermeable concrete slabs

It is possible to construct an in situ concrete roof without any further waterproofing if the concrete slab is designed to be impervious to water. The concrete here performs the load-bearing and the waterproofing functions (see "Concretes with high water impermeability", p. 96, and "Impermeable concrete slab as waterproofing layer", p. 104).

Cracks, which can have a negative effect on the impermeability of concrete, occur in concrete when the tensile stresses present exceed the tensile strength of the concrete. However, the loadbearing behaviour of reinforced concrete is characterised by the fact that the reinforcement only experiences significant tensile stresses and contributes to the load-carrying effect after the concrete has cracked. So the aim of design is not to avoid cracks altogether, but rather to limit their size.

Flat roofs belong to loading class 1 (non-pressurised water on horizontal surfaces) of the German "Impermeable Concrete Structures Directive" [1]. For usage class A (e.g. residential and office buildings), this directive stipulates that cracks passing through the complete concrete section are not permissible. Bending cracks are permissible, provided the concrete compression zone has a depth of at least 30 mm or 1.5 times the largest aggregate size. The requirement regarding the minimum depth of the compression zone can be complied with by adequate design for bending. The requirement that cracks passing through the complete concrete section must be avoided is harder to comply with. Such cracks occur when the entire depth of the concrete cross-section is subjected to tensile stresses exceeding the tensile strength of the concrete. This is either the case when external tensile loads are applied or when changes to the volume of the concrete (e.g. due to shrinkage or temperature fluctuations) are prevented by restraints. Such situations can be avoided by prestressing the roof, but the designer should nevertheless aim to minimise any restraint effects in a flat roof made of impermeable concrete.

A simple rectangular plan shape should therefore be preferred to a complicated shape with corner returns. Temperature fluctuations should be minimised by attaching thermal insulation to the outside. On roofs with internal insulation, sliding bearings should be provided to support the roof and compensate for the different expansion between the insulated walls and the roof exposed to external temperature fluctuations. Even on roofs with external insulation, allowing for some form of movement at the supports is advisable, e.g. to ensure that no restraints develop when new concrete shrinks as it cures. If some form of sliding bearing is provided, then it is important to ensure adequate fixed points for the statically determinate horizontal support to the roof. In addition, it should be remembered that in such cases the roof cannot contribute to stiffening the building.

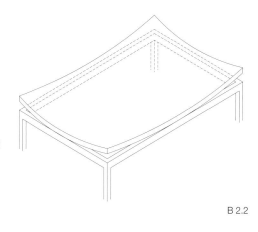

B 2.2

Structural system		Coefficient α
Beam	Slab	
l_i / l_{eff}	l_1, $l_{eff} = l_1$, $l_1 < l_2$	1.0
l_i / l_{eff}	l_1, l_2, $l_{eff} = l_1$, $l_1 < l_2$	0.8
l_i / l_{eff}	l_1, l_2, $l_{eff} = l_1$, $l_1 < l_2$	0.6
l_i / l_{eff}	l_1, l_2, $l_{eff} = l_1$, $l_1 < l_2$	2.4
	l_1, l_2, $l_{eff} = l_2$, $l_1 < l_2$	0.7 (inner bay) / 0.9 (edge bay)

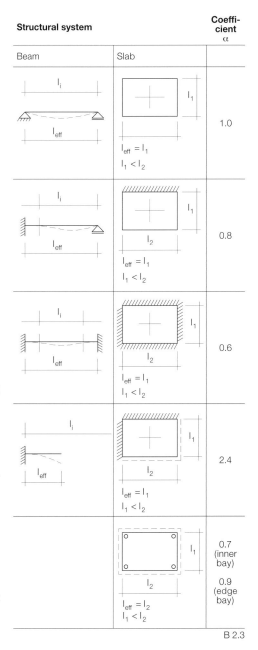

B 2.3

An adequate number of expansion joints must be included in larger roof surfaces. Roofs without external insulation should be regarded more critically in this respect than those with external insulation. The following dimensions should not be exceeded on roofs with no joints and no external insulation [2]:
- 400 m² for roofs on linear supports (slip membrane)
- 900 m² for roofs on point supports (sliding bearings)

In addition, the distance between any corner of the roof and the next fixed point should not be too large [3]:
- up to 15 m for roofs on linear supports
- up to 22 m for roofs on point supports

The depth of a roof with an impermeable concrete slab should not be less than 180 mm [4]. Besides the aforementioned constructional measures, there are concrete technology aspects (e.g. the use of concrete that resists water penetration, the use of cements with a low heat of hydration) and workmanship aspects (e.g. curing) that must be considered. Concretes with a high resistance to water penetration generally contain more water than is necessary for the setting of the cement. This excess water is released into the interior air until the concrete component has reached an equilibrium moisture content, which has a negative effect on the interior climate conditions during that period [5]. Summing up, we can say that the construction of roofs with impermeable concrete can be economically viable but involves much more work during the design stage and that there is a greater risk of flaws than is the case with roofs using conventional waterproofing materials. Checking the imperviousness of the roof after completing the structural carcass and the possibility of repairing potential leaks (e.g. by injection) should be considered during the planning phase.

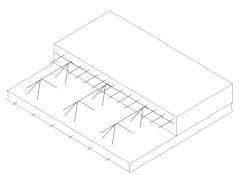

B 2.4

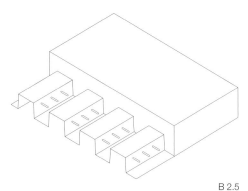

B 2.5

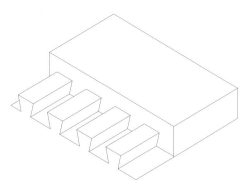

B 2.6

B 2.7

Precast concrete elements without in situ concrete topping
Room-size solid elements without an in situ concrete topping are available for roofs and floors up to a maximum width of 4.5 m. The seamless soffit, which requires no further finishes, is a particular advantage, but the relatively high transport weight is a disadvantage [6]. Such elements are designed in a similar way to in situ concrete slabs.

Precast concrete elements with in situ concrete topping
Lattice girder slab units are produced in widths of up to 3 m. The reinforcement for one-way-spanning roofs is fixed in the prefabricated element at the works. To achieve a two-way-span effect, additional transverse reinforcement can be provided, which is laid in the joints on the building site. The lattice girder cast in at the works acts as reinforcement bonding together the precast and in situ concrete (Fig. B 2.4). Temporary supports every 2–5 m are necessary during erection and concreting. As the top layer of reinforcement is fixed and cast in on site, continuity and plate effects are easy to ensure, even without additional ring beams. The finished slab soffit, interrupted merely by the joints, and the low weight of the precast units are advantages. Owing to the in situ concrete content, progress on site is slower than with precast concrete elements without an in situ concrete topping. The design is similar to that of an in situ concrete slab.

Prestressed elements with in situ concrete topping
Concrete slabs with conventional reinforcement are uneconomic for longer spans. Prefabricated, factory-prestressed lattice girder slab units are therefore recommended for such applications. Such elements result in roofs and floors with a greater slenderness. The prestressing also means that the spacing between temporary supports during erection can be increased.

Steel-concrete composite slabs
In a similar way to the precast concrete element with an in situ concrete topping, steel trapezoidal profile sheeting can be used as permanent formwork with a shear-resistant connection between the steel and the in situ concrete topping. It can therefore perform the function of the bottom layer of reinforcement. The composite effect between steel sheeting and concrete can be achieved by a mechanical interlock (e.g. holes punched in the steel or pins, Fig. 2.5), a continuous friction bond (Fig. 2.6) or by mechanical anchors at the ends. Such slabs are designed according to DIN 18800-5 or DIN EN 1994-1-1. As the composite effect of the individual systems varies considerably, the details are given in National Technical Approvals.
The slab depth is designed in a similar way to an in situ concrete slab, with the effective structural depth d roughly equal to the distance between the top edge of the slab and the centre of gravity of the trapezoidal profile sheeting. The minimum depth of the overall construction

should be 90 mm if the slab is used as a flange for a composite beam or as a stiffening plate. Otherwise, a minimum depth of 80 mm is adequate. In addition to the trapezoidal profile sheeting, a minimum amount of reinforcement (bars or meshes) must be laid in the in situ concrete. In the event of a fire the trapezoidal profile sheeting can accommodate only very limited tensile forces and so the design for the fire situation is frequently critical for determining the additional reinforcement. It is possible to achieve a fire resistance of up to 120 minutes with appropriate additional reinforcement. Temporary supports during erection may be necessary depending on span, depth of in situ concrete topping and manufacturer.

Prestressed concrete hollow-core slabs
These are available as factory-prestressed prefabricated units (Fig. B 2.7). Voids are formed in the concrete during casting to achieve a saving in weight of up to 40% compared to a solid slab. The customary element widths lie between 500 and 1200 mm, the depths depend on the type of unit and are generally between 150 and 400 mm. Spans of up to 18 m are normally possible, but 22 m can be achieved in special cases with slab up to 500 mm deep [7]. Roofs made from hollow-core elements can be used as a stiffening roof plate provided a ring beam is included. The joints between the individual elements are subsequently filled with grout. No temporary supports are required during erection. When several bays of elements are being used, a continuity effect cannot be achieved because the prestressing tendons are normally in the bottom of the units. Cantilevers with a length of up to eight times the slab depth are possible in the direction of the span.
Owing to the saving in weight and the prestressing, greater slenderness ratios are possible than is the case with conventional reinforced concrete slabs; up to l/h = 45 is feasible. However, if a plain soffit without a suspended ceiling is required, then the slenderness should be limited to l/h = 35 [8]. It is not always possible to guarantee the absence of cracks in plaster or jointing compound along the longitudinal joints between the elements [9].
Prestressed concrete hollow-core slabs do not have any conventional reinforcement, instead are reinforced exclusively with prestressing tendons. Design according to DIN 1045-1 or Eurocode 2 is therefore not possible and the respective National Technical Approval must be consulted. There is no reinforcement in the transverse direction, which is why the transverse distribution of large point loads such as heavy vehicular traffic will require the addition of an in situ concrete topping. Fixing heavier loads to the soffit with anchors at a later date can lead to severe damage owing to the delicate nature of these slab elements. Instead, fixings should be cast into the longitudinal joints. Lighter loads, e.g. soffit linings, can be fixed with anchors. However, it should be ensured that the holes for such fixings are drilled along

the axes of the voids so that the prestressing tendons in the webs are not damaged.

Precast double-T elements
Double-T elements are produced with and without prestress. They carry loads in one direction, like the hollow-core slabs, and the transverse distribution of loads beyond an individual element is possible to a limited extent only. Owing to the combination of slab and downstand beam in one component, the use of double-T elements enables long spans to be bridged more economically than with planar elements. However, the depth, 260–900 mm, is generally much greater than, for example, hollow-core slabs with a comparable load-carrying capacity. Spans of up to 25 m are possible with standard prestressed double-T elements [10], but only 17.5 m with the non-prestressed units. As it is essentially the depth of the web and not the depth of the slab that determines the load-carrying capacity of the component, delicate-looking designs that save materials are possible. The component thicknesses are generally dependent on the fire protection requirements. For example, a slab depth of at least 60 mm is necessary to achieve an F 30 A rating, and 100 mm for F 90 A [11]. As with the hollow-core slabs, a plate effect can be achieved by using a ring beam and grouted joints (Fig. B 2.7). Owing to the low depth of the slab, and hence the joints as well, the transfer of shear forces across the joints is poor and there is a risk of cracks forming along the joints, especially on long spans. The addition of an in situ concrete topping can improve the plate effect and transverse distribution of loads compared to the use of the precast units alone. If an in situ concrete topping is planned, then lattice girders are factory-cast into the elements in order to create a bond between the in situ and precast concrete (Fig. B 2.9).

Inverted channel section units
Inverted channel section units are not included in the catalogue of types, but are produced by many companies to suit their own precast concrete systems (Fig. B 2.10). Inverted channel section units do not require any transverse downstand beams when the spacing of the columns matches the system width of the elements. This generally results in a much wider elements than is the case with, for example, double-T elements or prestressed concrete hollow-core slabs. Owing to the larger spans in the transverse direction associated with this, the slab must be deeper and more heavily reinforced than is the case with double-T elements. Transverse distribution of loads beyond the joints is better than with double-T elements because of the greater depth of the joint [12]. Plate effects and erection requirements are similar to those of double-T elements. An additional in situ concrete topping is possible here as well.

B 2.4 Lattice girder slab unit with in situ concrete topping
B 2.5 Steel-concrete composite slab with mechanical bond
B 2.6 Steel-concrete composite slab with friction bond
B 2.7 Prestressed concrete hollow-core unit with grouted joints
B 2.8 Double-T unit with grouted joints
B 2.9 Double-T unit with in situ concrete topping
B 2.10 Inverted channel section unit with grouted joints

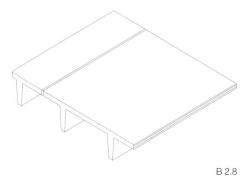

B 2.8

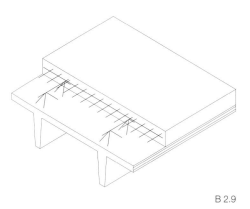

B 2.9

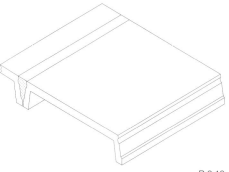

B 2.10

Precast elements of aerated concrete

Owing to the entrapped air, solid slabs of aerated concrete are approx. 65–80% lighter than solid concrete slabs with an identical cross-section. In addition, they can fulfil thermal insulation as well as loadbearing functions. The structural design of reinforced roof and floor units of aerated concrete is covered in DIN 4223-2 and 4223-4 or DIN EN 12602. Owing to the high porosity of the concrete, the reinforcement must be protected against corrosion. The use of reinforced aerated concrete slabs is currently limited to stationary imposed loads ≤ 3.5 kN/m². The transverse distribution of larger loads beyond the element joints is limited and so larger imposed loads can only be carried in conjunction with an additional in situ concrete topping and an appropriate design. However, in the course of revising DIN 1055 it is intended to increase the permissible imposed load to 4.0 kN/m² for rooftop terraces without the need for any additional measures. A corresponding new edition of DIN 4223-4 had not yet been published at the time this book went to press. Slab depths between 100 and 300 mm are available. An l/h ratio of 20–35 can be assumed for the preliminary design of the slab [13]. The maximum permissible span is 7.5 m for the maximum slab depth of 300 mm. Cantilevers may not exceed a length of 1.5 m. With appropriate joints and a ring beam, an aerated concrete roof can be used to provide a plate effect. The maximum spacing of stiffening crosswalls is limited to 35 m in this case. When using the minimum slab depth of 100 mm, a fire resistance rating of F 120 A can be achieved without a soffit lining, provided the appropriate concrete cover is guaranteed. A slab depth of 125 mm is necessary for an F 180 A rating. Temporary supports during erection are not normally required.

Metals

Large-format, thin-wall, profiled metal sheets have been produced industrially and used for building applications since the middle of the 19th century. Initially, presses stamped the flat sheet metal into corrugated profiles, which substantially increased their bending capacity perpendicular to the plane of the corrugations. Galvanising protects sheet steel against premature corrosion. Profiled sheets are used in the building envelope, as a loadbearing deck in the roof, as a roof covering or as planar wall elements. Since the mid-20th century the range of single-skin profiled sheets has been extended by trapezoidal, standing seam and tray profiles. Today, rolling plants make it possible to produce sheets from coated steel or aluminium in a continuous process. Since 1987 DIN 18807 has regulated the use and application of trapezoidal profiles in buildings. They are therefore acknowledged as standard components in the building industry. Trapezoidal profile sheets are regarded as very economic when used as a

1	Fastener along longitudinal edge (pop rivet, screw)	4	Web
2	Edge reinforcement	5	Top flange
3	Longitudinal edge bent up	6	Bottom flange
		7	Side lap
		8	Web rib
		9	Bottom flange rib

1 Fastener along longitudinal edge (pop rivet, screw)
2 Edge reinforcement
3 Longitudinal edge bent up
4 Web
5 Top flange
6 Bottom flange
7 Side lap
8 Web rib
9 Bottom flange rib
10 Top flange rib
11 Flat longitudinal edge
12 End lap
13 Beam, purlin, rail
14 Fastener at support
(screw, powder-actuated fastener)
15 Profiled sheet
16 Fastener at end lap (pop rivet, screw)

B 2.11

B 2.12

B 2.11 Metal trapezoidal profile sheet, details and terminology to DIN 18807
B 2.12 Trapezoidal profile sheet, 200 × 375, with transverse ribs in top flange
B 2.13 Trapezoidal profile sheet, positive installation
B 2.14 Trapezoidal profile sheet, negative installation
B 2.15 Surface textures
a, b Lines
c Hammered
B 2.16 Loading table for steel trapezoidal profile sheets for a uniformly distributed load q [kN/m²]

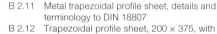

B 2.13

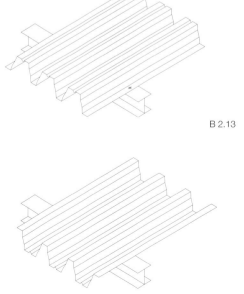

B 2.14

a

b

c

B 2.15

loadbearing deck for flat or pitched roofs, but corrugated sheets are hardly used any more. DIN 18807 has been accompanied by the product standard DIN EN 14782 since 2006.

Single-skin profiled sheets
Profiles are always designated in terms of their profile depth h and rib width b. Sheet lengths and widths specify the dimensions of individual profiled sheets. They are available in lengths up to approx. 24 m, and the widths of individual sheets are generally between 750 and 1075 mm. Besides the type of profile and the material, the load-carrying capacity depends on the profile depth h and the sheet gauge t. Profile depths are available in various gradations between 10 and 200 mm, the sheet gauges range from 0.63 to 1.5 mm. The German Trade Association for Lightweight Metal Building Systems (IFBS) provides up-to-date information on suppliers and contractors plus technical advice.
In a similar way to steel beams, we refer to top and bottom flanges when speaking about profiled sheets (Fig. B 2.11). These are connected by webs. Both flanges and webs can have ribs, corrugations or fins stamped or formed in them, which help to prevent local buckling and hence increase the load-carrying capacity substantially. They generally run in the longitudinal direction of the flanges and webs, but some newer profiles have such reinforcing elements in the transverse direction as well (Fig. B 2.12). Laying the sheets with the wider flange uppermost, the narrower flange at the bottom, is known as a positive installation (Fig. B 2.13), with the sheets reversed a negative installation (Fig. B 2.14).

Corrugated and trapezoidal profile sheets
Whereas corrugated sheets are used only for roofs with short spans, e.g. cycle sheds or canopies, trapezoidal profile sheets can achieve spans of up to 10 m in principle, with light loads and an appropriate choice of profile, even 14 m [14]. Exploiting the continuity effect over two or more spans increases the load-carrying capacity compared to that of a simply supported sheet. This advantage is easy to use because the overlaps of the sheets at the supports in the longitudinal direction are rigid and structurally effective. The individual sheets are connected by pop rivets, self-drilling screws or by bending.

Spans and load-carrying capacity
The manufacturers publish extensive tables specifying the load-carrying capacity of their profiles depending on cross-section, gauge, span, structural system and deflection limits. The values given are valid for uniformly distributed loads and pure bending. The figures do not usually cover other types of loading, e.g. local point loads, knife-edge loads or additional shear loads. Further project-specific analyses will be required in such cases. Owing to the low self-weight of thin-wall metal sheets, fixings to prevent wind uplift will be required in every case. In order to guarantee the serviceability of

roofs, the following deflection limits are specified in DIN 18807-3 for steel trapezoidal profile sheets:
• l/150 when used as a (weatherproof) covering or as the lower loadbearing deck in the case of a double-skin roof
• l/300 when used below a waterproofing material (warm roof)

Figs. B 2.16 and 2.17 (p. 40) list the design values for a number of popular profiles. These tables provide merely reference figures and are based mainly on data provided by manufacturers. They are in no way a substitute for a proper structural analysis.
The failure of trapezoidal profile sheet constructions because the yield strength in the effective flange cross-section has been exceeded, or by buckling of flanges and webs in compression, or by deformation of the webs at the supports, must be ruled out by adhering to the details given in the approval documents or by carrying out appropriate analyses.

Dimensions [mm]	Self-weight g [kN/m²]	Nom. mat. thk. d [mm]	Structural system	Span [m]										
				1.0	2.0	3.0	4.0	5.0	6.0	7.0	8.0	9.0	10.0	11.0
Trapezoidal profile (l/150) positive installation				**Load-carrying capacity q [kN/m²]**										
35/207	0.072	0.75	⌐	8.98	1.74	0.52	0.22	0.11						
	0.096	1.00		14.31	2.63	0.78	0.32	0.17						
	0.144	1.50		26.99	4.15	1.23	0.52	0.27						
		0.75		7.55	2.25	1.00	0.54	0.31						
		1.00		12.28	3.59	1.60	0.79	0.40						
		1.50		24.74	6.57	2.92	1.27	0.64						
		0.75		8.36	2.67	1.05	0.50	0.32						
		1.00		13.49	4.33	1.47	0.62	0.31						
		1.50		25.87	7.38	2.34	0.99	0.50						
135/310	0.097	0.75				2.88	2.13	1.75	1.32	0.96	0.65			
	0.130	1.00				5.37	3.97	2.82	1.92	1.21	0.81			
	0.195	1.50				14.28	7.91	4.79	2.94	1.85	1.24			
		0.75				3.18	2.14	1.77	1.35	1.02	0.80			
		1.00				5.36	3.97	2.82	2.21	1.63	1.25			
		1.50				14.26	8.32	5.42	3.89	2.86	2.13			
		0.75				3.60	2.47	1.88	1.39	1.05	0.85			
		1.00				6.10	4.02	2.93	2.23	1.73	1.42			
		1.50				14.28	8.30	5.53	4.00	3.08	2.29			
160/250	0.121	0.75					3.44	2.54	1.77	1.30	0.96	0.68	0.49	
	0.161	1.00					6.70	4.22	2.93	1.96	1.31	0.93	0.68	
	0.241	1.50					10.18	6.69	4.58	2.98	2.00	1.41	1.04	
		0.75					3.45	2.51	1.83	1.39	1.09	0.85	0.59	
		1.00					6.47	4.33	3.12	2.36	1.82	1.41	0.96	
		1.50					9.88	6.91	4.95	3.74	2.86	2.28	1.77	
		0.75					3.45	2.57	1.84	1.41	1.12	0.90	0.73	
		1.00					6.70	4.34	3.11	2.38	1.85	1.47	1.17	
		1.50					10.18	6.90	4.95	3.77	2.90	2.34	2.18	
200/375	0.118	0.75						2.12	1.76	1.51	1.22	0.96	0.74	0.56
	0.158	1.00						3.69	3.07	2.53	1.93	1.41	1.03	0.77
	0.237	1.50						8.84	6.19	4.55	3.16	2.22	1.62	1.22
		0.75						2.12	1.65	1.31	1.06	0.88	0.72	0.61
		1.00						3.77	2.87	2.26	1.82	1.50	1.22	1.04
		1.50						7.38	5.53	4.28	3.41	2.75	2.25	1.89
		0.75						2.12	1.76	1.51	1.22	0.96	0.81	0.69
		1.00						3.92	4.76	2.63	2.02	1.59	1.38	1.17
		1.50						8.82	6.19	4.55	3.48	2.75	2.56	2.16

The figures apply to a uniformly distributed load q.

B 2.16

Dimensions [mm]	Self-weight g [kN/m²]	Nom. mat. thk. d [mm]	Structural system	Span [m]							
				1.0	2.0	3.0	4.0	5.0	6.0	7.0	8.0
Trapezoidal profile (l/150) positive installation				**Load-carrying capacity q [kN/m²]**							
100/600	0.089 0.118	0.75 1.00	▭					0.96 1.59	0.61 1.02	0.43 0.67	
		0.75 1.00	▭					1.05 1.81	0.67 1.16	0.47 0.80	
		0.75 1.00	▭					1.32 2.26	0.84 1.44	0.59 1.00	
145/600	0.085 0.114	0.75 1.00	▭				1.36 2.48	0.87 1.59	0.60 1.10	0.44 0.81	
		0.75 1.00	▭				1.74 2.88	1.21 1.96	0.90 1.42	0.67 10.40	
		0.75 1.00	▭				2.08 3.50	1.36 2.39	0.94 1.72	0.69 1.26	
160/600	0.106 0.140 0.207	0.75 1.00 1.50	▭					1.13 1.86 2.84	0.79 1.29 1.97	0.58 0.95 1.45	0.44 0.73 1.11
		0.75 1.00 1.50	▭					1.28 2.15 3.27	1.01 1.62 2.47	0.74 1.19 1.81	0.57 0.91 1.39
		0.75 1.00 1.50	▭					1.50 2.57 3.91	1.18 1.95 2.96	0.90 1.49 2.26	0.69 1.14 1.73
Composite sheet (l/100)											
70/1000	0.55/ 0.45		▭	2.88 2.88 2.88	1.94 1.94 1.94	1.54 1.54 1.54	1.31 1.31 1.31	1.16 1.16 1.16			
120/1000	0.55/ 0.45		▭	4.53 3.78 4.21	2.80 2.52 2.80	2.00 2.00 2.00	1.60 1.60 1.60	1.36 1.36 1.36			
160/1000	0.55/ 0.45		▭	5.06 3.94 4.35	3.40 2.61 2.87	2.28 2.06 2.27	1.75 1.75 1.75	1.45 1.45 1.45			

The figures apply to a uniformly distributed load q.

B 2.17

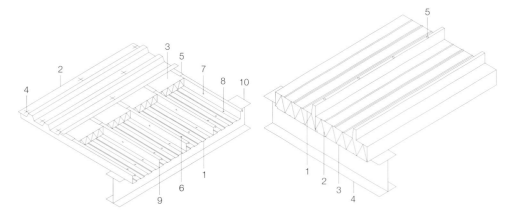

1	Tray profile sheet, continuously galvanised or plastic-coated	5	Separating strip if required
2	Outer skin, connected directly to the tray profile, continuously galvanised and plastic-coated	6	Fastener at support (screw or powder-actuated fastener)
3	Thermal insulation	7	Fastener for web
4	Fastener (outer skin to tray profile)	8	Structural fixing
		9	Sealing tape (side lap)
		10	Supporting construction (rafter, truss or column)

1	Outer facing (steel, aluminium, copper, etc.)
2	Core (PUR, mineral wool, PS)
3	Inner facing (steel, aluminium, copper, etc.)
4	Supporting construction
5	Fixing (self-drilling screw etc.)

B 2.18

B 2.19

Construction details

Special attention should be given to the design of supports, edges, joints and openings when using trapezoidal profile sheeting. DIN 18807 provides detailed information in this respect. For example, supports must generally be min. 80 mm deep, at least 100 mm on masonry. However, if the sheets are fixed immediately after laying, then the depth of the end support can be reduced to min. 40 mm for a supporting structure of steel or reinforced concrete, min. 60 mm for timber. Intermediate supports must always be at least 60 mm deep.

The practical overlap length for trapezoidal profile sheets with a waterproof roof covering is 50–150 mm in the direction of the span. In the case of joints designed to resist bending, this overlap length must be increased to suit the loading. Transverse joints are not permitted on flat roofs where the trapezoidal profile sheeting itself is used as the roof covering. The side laps between adjacent trapezoidal profile sheets must be connected together with pop rivets or self-drilling screws every max. 666 mm. Unsupported longitudinal edges must be strengthened with suitable sheet metal components with a gauge of at least 1 mm. The same applies to edges around openings, which for openings measuring 125 × 125 mm or larger must have a peripheral frame made from stiffening sheet metal components.

Cantilevering trapezoidal profile sheets are only possible in the direction of the span, i.e. in the direction of the ribs. Adequate transverse strengthening to the sheeting must be ensured at the unsupported end of the cantilever. This can be achieved, for example, by sheet metal angles or special stiffened sheets, and there must be a tension-resistant connection between every rib and such components. The other end of each cantilevering sheet must be secured against uplift immediately during laying.

Plate effect

Loadbearing roof decks can accommodate horizontal forces, e.g. due to wind or traffic, to a limited extent as well as vertical loads and hence contribute to bracing a building or components at risk of buckling. To do this, the profiled sheets must be connected together and connected to all peripheral supports with shear-resistant fasteners. Bays that satisfy these conditions and are intended to be used for resisting and transferring horizontal loads are known as shear diaphragms. They must be identified on the drawings and, to be on the safe side, on the structure itself as well because they are vital to the stability of the building. The shear capacities of the different types of trapezoidal profile sheets are given in the data provided by the manufacturers for design purposes. Owing to the "folded" nature of trapezoidal profile sheets, their stiffening effect is much lower than that of flat sheet metal.

Support on four sides for shear diaphragms as required by DIN 18807-3 usually results in extra work, which is why new research findings on

B 2.17 Loading table for tray profiles and composite sheets made from sheet steel for a uniformly distributed load q [kN/m²]
B 2.18 Metal tray profiles, details and terminology to DIN 18807
B 2.19 Composite sheets, details and terminology
B 2.20 Structural solid timber products (a–c) and wood-based product (d)
    a Squared solid timber section
    b Trio beam
    c Glued laminated timber (glulam)
    d Structural veneer lumber (SVL)

a   b   c   d   B 2.20

shear diaphragms supported on two sides were published in 2006. The authors propose applying reduction factors to the shear capacities determined in accordance with the standard [15].

Tray profiles
Motivated by the energy crisis, manufacturers of building materials developed loadbearing sheets in the 1970s that could accommodate thermal insulation, the so-called tray profiles. These too have flanges and webs, but the bottom flanges are much wider than the top ones (Fig. B 2.18). The resulting "tray" is filled with insulating material. The wide bottom flange is stiffened by shallow ribs. Deeper webs are also provided with ribs to improve their stability. Popular types range from 90/500 mm to 160/600 mm, with sheet gauges between 0.75 and 1.50 mm. These days, tray profiles are used less on roofs and primarily as a horizontally supported, inner leaf to double-leaf walls.

Standing seam profiles
The industrially manufactured standing seam profiles are also classed as single-skin profiled sheets. Similar to corrugated sheets, their use as both loadbearing and waterproofing elements simultaneously is generally restricted to less important buildings. They are normally used as the upper covering to a multi-layer roof assembly, where they only have to withstand wind uplift. As with the tray profiles, the bottom flange is flat and the top edges of the webs are folded over so that they can interlock with or be fixed to neighbouring sheets. Clips clamped in place upon folding over the top edges attach the sheets to the loadbearing structure. The sheets (0.63 mm gauge) are available in widths of 305–600 mm and depths of 50–75 mm. A range of surface textures is available, e.g. embossed lines, ribs, hammered (Fig. B 2.15, p. 38).

Composite sheets
It was around 1960 that manufactures started to produce sheet metal panels with a stabilising core of a rigid foam material, i.e. sandwich elements with a shear-resistant bond between the metal facings and the foam core (Fig. B 2.19). These can be used not only for the loadbearing and waterproofing functions, but also for the insulation. They can be supplied as factory-made roofing elements for quick, easy and inexpensive installation – ceiling, insulation and roof covering in one. The two facings can be supplied flat or profiled. The profiling on the outside is frequently deep so that the joints between the sheets do not lie in the water run-off level, whereas the inside face tends to have a shallower profile for a cleaner look.
Bonding together the stabilising core and the facing materials results in a complex composite effect, which is why the internal forces and stresses in the various components can only be determined approximately or by using numerical methods. DIN EN 14509 "Self-supporting double-skin metal-faced insulating panels – Factory made products – Specifications" has regulated these products since 2008. In contrast to the product standard DIN EN 14782 "Self-supporting metal sheet for roofing, external cladding and internal lining – Product specification and requirements", it also contains design guidelines and together with the ECCS and CIB [16] recommendations represents the state of the art regarding design and construction [17]. The National Technical Approval documents normal up until now are also still available from most suppliers. Common applications for these composite sheets are spans of approx. 5 m. When using them as wall elements, spans of up to 11 m are possible. To what extent composite sheets can also be used as bracing components is the subject of ongoing research [18].
Owing to the high insulation values of composite sheets, great temperature differences can develop between the inner and outer faces, depending on the weather conditions. High restraint stresses can develop depending on boundary conditions such as the structural system or the bending stiffness of the facing material. In summer such stresses can be twice as high as those caused by the other, normal,

external loads [19]. These additional stresses must be taken into account in the structural calculations, likewise the shear creep in the core under long-term loading.
Cores of polyurethane rigid foam are very common, but cores of expanded or extruded polystyrene, phenolic foam, cellular glass and mineral wool are also available (see "Insulating materials", pp. 94). Fire protection and sound insulation are good reasons for using composite sheets with a mineral wool core.

Timber

Timber has a long tradition in the building industry. And since industrialisation the applications of this natural raw material have been expanded by numerous innovations, ranging from the connectors of engineered structures to the development of modern wood-based products such as OSB or fibreboards. Timber is a renewable material regarded as environmentally friendly, carbon-neutral and sustainable. It is anisotropic, i.e. it exhibits very different properties in the axial, radial and tangential directions. The hygroscopic effects in particular lead to great differences in the shrinkage and swelling behaviour in these three directions, but the differences between the mechanical properties are also considerable. Designing so that the primary load-carrying function is parallel to the grain is therefore desirable. One principal feature of timber is the fact that its strength and deformation behaviour depend on the moisture content and duration of the loading. DIN 1052, which can be regarded as a code of practice, and DIN EN 1995 (Eurocode 5) contain the regulations for the design and construction of timber structures.
Different species of wood are processed to form various semi-finished products for use as building materials (Fig. B 2.20, p. 41). Solid timber comes closest to the raw material, as debarked round timber or as sawn products in the form of squared sections, planks, boards and battens (up to 40 × 80 mm). Various grading classes are available depending on the quality. The bonding together of solid timber components produces two- and three-part beams

a

b

c

d   B 2.21

or the much more common glued laminated timber (glulam).

A diverse range of wood-based products is available in addition to solid timber (Fig. B 2.21). The aim here is to reduce the disadvantages of the natural material, e.g. shrinkage and swelling, but also to make use of low-quality timber or waste. Boards are bonded together to form solid timber and multi-ply boards, boards and battens to form laminboard, blockboard and battenboard, veneers to form laminated veneer lumber (LVL) or plywood, chips to form oriented strand board (OSB) and particleboards bonded with synthetic resin, cement or gypsum and finished with various facing materials. The fibres are used to produce wood fibreboards with various degrees of hardness, or rather density, such as hardboard and medium board, wood fibre insulating boards, wood-wool boards, also gypsum fibreboard or cement fibreboard. The standards here are DIN EN 13986 "Wood-based panels for use in construction – Characteristics, evaluation of conformity and marking" and DIN 4074 "Strength grading of wood".

Roof decking of boards and planks

Timber in the form of rough-sawn or planed boards or planks can be used as the loadbearing deck to a flat roof (Fig. B 2.22). Boards must have a nominal thickness of min. 24 mm [20] and should have tongue and groove joints to help ensure transverse distribution of the loads to a certain extent and to create a level surface for the waterproofing materials. Planed boards are normally 95–155 mm wide. When the thickness is 40 mm or more we generally speak of planks [21]. The decking is nailed or screwed to the loadbearing joists positioned at a spacing of 600–1200 mm. However, larger spacings are possible, depending on the thickness of the decking. For example, 60 mm thick planks can span about 3 m.

Roof decking of wood-based products

Wood-based products are frequently used as the loadbearing deck to a flat roof instead of boards or planks (Fig. B 2.23). They are usually less expensive and also offer building physics, constructional, architectural or structural advantages. For example, depending on the binder content, OSB or particleboards can also function as a vapour barrier And, in contrast to solid timber, plywood can carry loads in two directions, while offering different surface finishes. DIN 18334 "German construction contract procedures (VOB) – Part C: General technical specifications in construction contracts (ATV) – Carpentry and timber construction works" specifies the following minimum thicknesses:

- Plywood (DIN EN 636): 15 mm
- OSB (DIN EN 300): 18 mm
- Pressed particleboards (DIN EN 312): 19 mm

According to DIN 18334, roof deckings made from wood-based products for metal, bitumen, slate and fibre-cement roof finishes, and also deckings below roof waterproofing materials, must be at least 22 mm thick.

Joints between wood-based products in the loadbearing direction must be positioned over the supports. Where the decking supports the waterproofing materials directly, it is also advisable, with wood-based products as well, to use products with tongue and groove joints. Suitable expansion joints at a spacing to suit the material should be provided in order to accommodate change of length of the boards. Measures to prevent wind uplift are especially important for roof deckings, which means that the boards must be connected to the supporting construction in such a way that the uplift forces are resisted by the fasteners and can be transferred to the primary structure. Depending on the material, loads and structural system, roof deckings made from wood-based products can achieve spans of up to about 5 m (Fig. B 2.26).

Plate effect of roof deckings

The roof construction is frequently used as a plate in order to brace a building against horizontal (wind) loads or provide lateral restraint to members at risk of buckling. Planar elements such as roof deckings made from wood-based products are ideal for this. In this situation, however, a number of conditions must be complied with. For example, an adequate plate effect in the decking is only possible when the joints between the wood-based products themselves and between the wood-based products and the loadbearing construction can resist shear and transfer the forces. The joints between the boards must be staggered. Without a more detailed analysis, the deflection of the bracing construction should be limited to l/500.

A plate effect in a roof decking made of boards can be achieved by using a diagonal arrangement. Under certain boundary conditions a decking of boards or planks can provide lateral restraint for roof trusses. To do this, the bracing forces counteracting the lateral buckling must be transferred to supports in the form of wind girders or plates.

Ribbed-panel and hollow-box forms

In order to achieve economic on-site operations, the decking of wood-based boards is frequently combined in the factory with the supporting construction to form large-format prefabricated elements. Parallel loadbearing webs below a decking of, for example, laminated veneer lumber form a ribbed-panel element (Fig. B 2.24). The decking is permanently bonded to the ribs and forms a structurally effective flange which increases the stiffness of the element substantially. If a decking is added below such a ribbed-panel element, the result is a hollow-box element (Fig. B 2.25). Such a form of construction is also economic from the structural viewpoint because bending loads on the plate can be carried by axial forces in the decking, which acts as a flange. Shear forces are resisted mainly by the vertical webs. Horizontal loads can be accommodated by and transferred to the primary structure via the plate effect of the elements. The spaces between the ribs or the voids of the hollow-box elements can be used for routing services or installing insulation. Project-specific loadbearing, building physics or constructional requirements can be met by adjusting the depth and spacing of the ribs (webs), and the element dimensions can be varied to suit different geometrical requirements. Ribbed-panel or hollow-box elements are available in widths of up to about 2.5 m and lengths of up to 23 m; standard depths are 200–500 mm.

Solid timber systems

Ecological solid timber roofs with good building physics performance can be achieved with the help of edge-fixed timber elements or cross-laminated timber. The basic idea, to produce solid timber elements from small-format squared sections, enables a good utilisation of the tree trunk and makes such forms of construction economically competitive. Solid timber components also easy to connect.

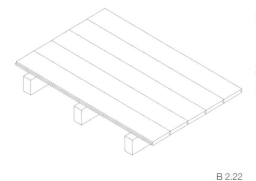

B 2.22

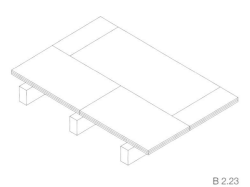

B 2.23

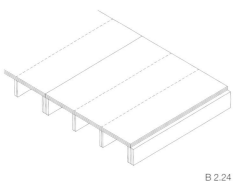

B 2.24

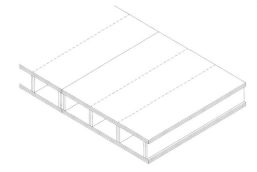

B 2.25

Material	Dimensions [mm]	Self-weight g_k [kN/m²]	Structural system	Snow load s_k [kN/m²]				
				0.75	1.00	1.25	1.50	2.00
Wood-based product				Span l [m]				
Lam. veneer lumber	21	0.5	▬	1.37	1.25	1.16	1.10	1.00
		0.8		1.27	1.22	1.16	1.10	1.00
	39	0.5		2.59	2.37	2.20	2.07	1.88
		0.8		2.40	2.30	2.19	2.07	1.88
	45	0.5		2.99	2.73	2.54	2.39	2.17
		0.8		2.77	2.65	2.53	2.39	2.17
	69	0.5		4.58	4.19	3.89	3.66	3.33
		0.8		4.25	4.07	3.88	3.66	3.33
	21	0.5	▬▬	1.86	1.70	1.58	1.49	1.35
		0.8		1.73	1.65	1.58	1.49	1.35
	39	0.5		3.51	3.21	2.98	2.81	2.55
		0.8		3.26	3.12	2.98	2.81	2.55
	45	0.5		4.05	3.71	3.44	3.24	2.94
		0.8		3.76	3.60	3.43	3.24	2.94
	69	0.5		6.21	5.69	5.28	4.97	4.51
		0.8		5.77	5.52	5.27	4.97	4.51
Ribbed-panel element								
Lam. veneer lumber	200 × 450	0.8	▬	9.40	9.00	8.57	8.07	7.31
		1.5		7.59	7.59	7.59	7.59	7.16
	240 × 600	0.8		10.26	9.82	9.35	8.80	7.97
		1.5		8.28	8.28	8.28	8.28	7.81
	400 × 450	0.8		16.91	16.18	15.42	14.51	13.16
		1.5		13.66	13.66	13.66	13.66	12.89
	500 × 600	0.8		19.22	18.39	17.52	16.49	14.94
		1.5		15.52	15.52	15.52	15.52	14.63
Hollow-box element								
Lam. veneer lumber	200 × 450	0.8	▬	12.99	12.42	11.83	11.13	10.07
		1.5		10.47	10.47	10.47	10.47	9.87
	240 × 600	0.8		14.38	13.75	13.09	12.30	11.13
		1.5		11.57	11.57	11.57	11.57	10.90
	400 × 450	0.8		20.90	20.00	19.05	17.93	16.24
		1.5		16.87	16.87	16.87	16.87	15.91
	400 × 600	0.8		20.36	19.47	18.55	17.44	15.79
		1.5		16.41	16.41	16.41	16.41	15.47

The above table contains reference values only and is essentially based on data supplied by manufacturers. The values are in no way intended to replace a proper structural analysis.

B 2.26

Material	Dimensions [mm]	Self-weight g_k [kN/m²]	Structural system	Snow load s_k [kN/m²]				
				1.00	1.50	2.00	3.00	4.00
Solid timber element				Span l [m]				
Cross-laminated timber	95	1.0	▬		3.50	3.50	3.00	3.00
		1.5			3.50	3.25	3.00	
	125	1.0			4.50	4.00	4.00	3.50
		1.5				4.00	4.00	3.50
	186	1.0			7.00	6.50	6.00	5.75
		1.5			6.50	6.50	6.00	5.50
	201	1.0				7.00	7.00	6.50
		1.5			7.00	7.00	6.50	6.50
Edge-fixed timber	8	1.0	▬	3.90	3.60	3.40	3.10	2.90
	14	1.0		6.90	6.40	6.00	5.40	5.00
	16	1.0		7.80	7.30	6.80	6.20	5.80
	22	1.0		10.80	10.00	9.40	8.50	7.90

The above table contains reference values only and is essentially based on data supplied by manufacturers. The values are in no way intended to replace a proper structural analysis.

B 2.27

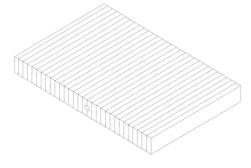

B 2.28

B 2.29

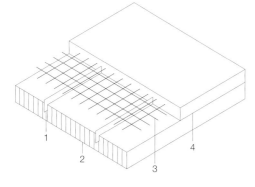

B 2.30

Edge-fixed timber elements
The edge-fixed timber elements used for roofs consist of conventional boards or planks with thicknesses between 24 and 60 mm which are placed upright alongside each other (Fig. B 2.28). The use of mechanical fasteners such as hardwood dowels, nails, screws or, less often, adhesive enables planar elements to be fabricated with depths between 80 and 260 mm depending on span and loading. These factory-prefabricated components are quickly erected, enable long spans and can be loaded immediately after installation. The overall depth of such a roof is much less than that of a traditional timber joist construction because the roof carries loads over its entire width. For example, the stiffness of a 120 mm deep edge-fixed timber element corresponds to that of a timber joist floor with 240 × 80 mm joists at a spacing of 625 mm, and the load-carrying capacity of such a floor exceeds that of a traditional timber joist construction [22]. The different properties of edge-fixed timber elements perpendicular and parallel to the boards must be taken into account. For in-plane loads acting perpendicular to the direction of the boards the structurally effective cross-section is made up of a multitude of individual board cross-sections with non-rigid connections. On the other hand, in-plane loads parallel to the direction of the boards must be resisted by the shear-resistant connections between the boards as they slide against each other. In addition, it is important to ensure shear-resistant connections between the individual timber elements. Rebated or tongue and groove joints are helpful here.

Cross-laminated timber elements
A roof employing cross-laminated timber elements (Figs. B 2.29 and 2.30) combines the advantages of glued laminated timber and those of plywood in one system and therefore expands the potential applications to large areas. Roofs with cross-laminated timber elements offer high dimensional stability and a good load-carrying capacity. In contrast to roofs with edge-fixed timber elements, roofs with cross-laminated timber elements can carry loads in two directions. In addition, the surface qualities can be adjusted to suit the respective usage by employing suitable facing plies; whereas the appearance is hardly important for an industrial building, various species of wood such as spruce, larch or Douglas fir can be chosen for residential applications. The sizes of individual elements are essentially governed by road transport limitations.

Timber-concrete composite systems
Besides pure timber constructions, flat roofs can also be built using various timber-concrete composite systems. Concrete components are positioned above or, less often, below timber components. Special connectors are installed to create a shear-resistant connection between the concrete and the timber and thus ensure a composite loadbearing action. This results in a two- to five-fold increase in the load-carrying capacity and stiffness compared to supporting beams without any composite action. Where a concrete slab is supported on timber beams or boards, the concrete essentially resists the compression, the timber the tension (Fig. B 2.31

and B 2.32). With the timber on top of the concrete, the tension must be resisted by steel reinforcement in the concrete and the concrete itself has hardly any loadbearing function.
In contrast to suspended floors, timber-concrete composite construction is not widely used for roofs because its particular advantages – sound insulation, fire protection, stiffness and vibration behaviour – are seldom critical for roofs. What can be relevant, however, is the building physics advantage of a larger thermal mass compared with a normal timber roof.

Glass

Glass has been used as a building material in the building envelope since Roman times. And as architects strive for ever greater dematerialisation and transparency, glass has experienced an incredible development over the last 20 years. New findings regarding its loadbearing and failure behaviour have pushed back the boundaries of feasibility and found their way into standards and technical recommendations. Combined with progress in material developments and glass production, the result has been ever more delicate designs and ever longer spans. Besides its traditional function in the building envelope, glass has in recent years been used increasingly as part of the primary structure, e.g. as columns or beams (Fig. B 2.36).

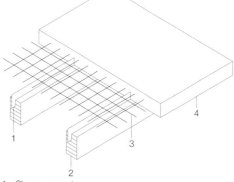

1   Shear connector
2   Solid timber element
3   Reinforcement
4   Concrete topping

B 2.31

1   Shear connector
2   Timber joist
3   Reinforcement
4   Concrete slab

B 2.32

B 2.28   Edge-fixed element
B 2.29   Cross-laminated timber
B 2.30   Cross-laminated timber, section through system
B 2.31   Timber-concrete composite system, in situ concrete topping on edge-fixed timber element
B 2.32   Timber-concrete composite system, concrete slab on timber joists
B 2.33   The principle of the production of toughened glass
B 2.34   The mechanical properties of soda-lime-silica glass according to DIN EN 572-1
B 2.35   Comparison of the stress-strain diagrams for glass and steel. Subjected to tension (F), steel exhibits plasticity after exceeding the elastic limit and is hence highly ductile (f) up to the point of failure. But glass exhibits a linear elastic behaviour up to the point of failure, and does not exhibit any plastic material capacity.
B 2.36   Dining hall, Dresden TU (D), 2007; architects: Maedebach, Redeleit & Partner; structural engineers: Leonhardt, Andrä & Partner

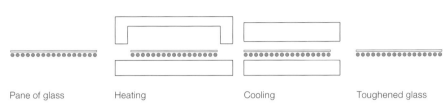

| Pane of glass | Heating | Cooling | Toughened glass |

B 2.33

Mechanical properties of soda-lime-silica glass to DIN EN 572-1

Density	25.0 KN/m^3
Modulus of elasticity	$70,000 \text{ N/mm}^2$
Poisson's ratio	0.23
Coefficient of thermal expansion	$9.0 \times 10^{-6} \text{ 1/K}$
Thermal fatigue resistance	40 K

B 2.34

Production

The production of the various types of glass used in the building industry takes place in two stages: the manufacture of the basic products followed by further processing to form the final products installed on the building site. The different methods of production lead to different mechanical and optical properties, also different maximum sizes.

Float glass

These days, almost all sheet glass is produced by the float method. Soda-lime-silica glass is the most common type produced (Fig. 2.34). In the production process the raw materials are melted at a temperature of approx. 1600 °C and the resulting dough-like fluid mixture is poured in a continuous process on to a 70 m long bath of molten tin at a temperature of about 1100 °C, where it spreads out to form a layer of constant thickness as it cools. The surface tensions of glass and tin result in a very smooth, consistent surface. The hot glass, at a temperature of approx. 600 °C, is drawn off continuously from the cooler end of the float bath and allowed to cool further while avoiding inducing any stresses into the material. In order to minimise damage along the edges caused by the production process, the ribbon of glass is trimmed to a width of 3.21 m and cut into panes 6 m long. The panes of float glass produced in this way are typically between 1.5 and 19 mm thick.

Rolled glass (cast glass)

The molten glass is poured on to a table and subsequently rolled to form sheet glass. Colourless or coloured textured glasses and wired glass are produced in this way. The manufacturing process causes significant microcracks in and damage to the surface, which is why the strength of this type of glass is much lower than that of float glass. The tolerances are up to 10 % of the material thickness. The maximum pane size is 2.5 × 4.5 m.
The production of wired glass involves rolling a fine wire mesh into the glass which holds fragments of glass in place upon breakage.

Microstructure and failure behaviour

Owing to its high atomic bonding forces, glass with an undisrupted microstructure and a perfectly smooth surface has a very high mechanical strength. However, this strength cannot be achieved in practice. Disruptions to the microstructure within the material and cracks or scratches on the surface result in notch effects with extremely high peak stresses when subjected to mechanical actions. As with all brittle materials, such stresses cannot be dissipated by way of plastic deformation. And, as such damage can never be completely avoided in practice, only a fraction of the true material strength can be exploited in a building component. At its tip, a crack begins to grow as soon as a critical tensile stress has been exceeded. Glass has no plastic loading reserves and hence no residual loadbearing capacity, and so such stresses finally lead to abrupt failure of the glass component (Fig. B 2.35).
The mechanical strength of glass is therefore not a precisely defined material parameter, but rather a variable that depends on the degree of damage to the surface (including edges and drilled holes). The probability that a certain mechanical action will increase local stresses at a critical flaw (crack depth) can only be catered for by employing statistical methods and the design is based on such assumptions. Only after the basic glass product has been further processed to form toughened, laminated and insulating glasses can it be used as a loadbearing component.

Sheet glass treatments

The basic glass products are further processed in several stages to form the final glass products. After cutting the panes to size, finishing the edges and drilling any holes necessary, various methods can be used to optimise the mechanical properties of the glass panes.

Toughened safety glass

The production of thermally toughened glass, so-called toughened safety glass, involves heating the finished panes of float glass (including drilled holes etc.) in a horizontal tempering furnace up to a temperature of about 650 °C and afterwards cooling them rapidly with cold air (Fig. B 2.33).
The inner core of the pane cools at a slower rate, but the contraction associated with this is prevented by the surfaces of the pane, which have already solidified. This mutual restraint generates tensile stresses in the core, compressive stresses at the surfaces. These prestresses are in equilibrium over the thickness of the material.
The compressive prestress closes the microcracks in the surface. And superimposing the prestress on the bending stress results in the true stress distribution in the pane of glass (Fig. 2.37). The microcracks remain closed until the tensile bending stresses, which depend on the load, exceed the prestress.
The glass passes over ceramic rollers in the tempering furnace. As the panes have been heated to their softening point, the glass sags

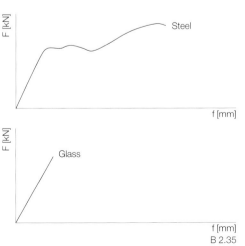

B 2.35

B 2.36

45

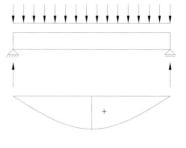

| Prestress | Bending stress | Resultant stress |

B 2.37

Type of glass	Permissible tensile bending stress [N/mm²] for…	
	Overhead glazing	Vertical glazing
Tough. safety glass (float glass)	50	50
Tough. safety glass (rolled glass)	37	37
Enamelled tough. safety glass (from float glass) [1]	30	30
Float glass	12	18
Rolled glass	8	10
Lam. safety glass (float glass)	15 (25) [2]	22.5

[1] Enamel on the tension side
[2] Only permissible for the lower pane of overhead glazing for the loading case "failure of the upper pane"

B 2.38

slightly between the rollers and this gives rise to minor deviations from the ideal flat geometry, so-called roller waves, which, owing to the thermal toughening process, cannot be completely avoided. Glass can also be toughened by a chemical process involving the exchange of ions in a calcium salt solution. However, this is an elaborate, costly process and so chemical toughening is rarely used for architectural applications.

Heat-strengthened glass
Heat-strengthened glass is essentially produced in the same way as toughened safety glass. However, the cooling process is much slower, which means that the induced prestresses are much lower.
The advantage of heat-strengthened glass lies in its failure behaviour. In contrast to a pane of toughened safety glass with its fracture pattern of small fragments, a pane of heat-strengthened glass breaks into larger pieces. Whether a glass product is classed as heat-strengthened or toughened safety glass depends on its fracture pattern, i.e. on the number of fragments within a defined area (Fig. B 2.39).

Laminated safety glass
Laminated safety glass consists of two or three plies of float, toughened safety or heat-strengthened glass alternating with layers of elastic and highly tear-resistant sheets of polyvinyl butyral (PVB), which are bonded together in a rolling process (Fig. B 2.43). The thickness of the PVB material is always a multiple of 0.76 mm and is defined according to the anticipated actions. When the glass fails due to a mechanical overload, the plastic interlayer binds the fragments together. In contrast to normal float glass, laminated safety glass is therefore relatively safe because dangerous fragments are held in place once the pane is broken. Overhead glazing must therefore always be made from laminated safety glass.

a

b

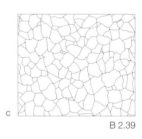

c

B 2.39

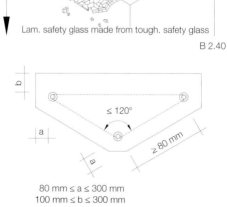

Lam. safety glass made from float glass

Lam. safety glass made from heat-strength. glass

Lam. safety glass made from tough. safety glass

Better loadbearing capacity

Better residual loadbearing capacity

B 2.40

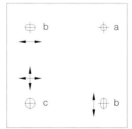

B 2.41

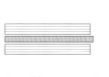

≤ 120°

≥ 80 mm

80 mm ≤ a ≤ 300 mm
100 mm ≤ b ≤ 300 mm

B 2.42

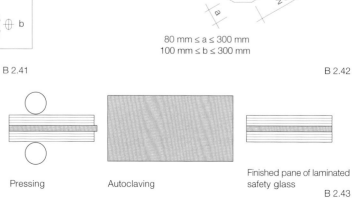

Assembling panes and PVB interlayers — Pressing — Autoclaving — Finished pane of laminated safety glass

B 2.43

B 2.37 Distribution of the thermal prestress over the section and the superimposition of prestress and bending stress
B 2.38 Permissible tensile bending stresses for various types of glass
B 2.39 Schematic presentation of the fracture patterns of…
a Float glass
b Heat-strengthened glass
c Toughened safety glass
B 2.40 Relationship between loadbearing capacity and residual loadbearing capacity
B 2.41 Restrained and unrestrained supports for glass
a Restrained support
b Vertical support for accommodating displacements in the horizontal direction
c Unrestrained support with allowance for displacement in both directions
B 2.42 Angle and spacing definitions for glass supported by drilled point fixings
B 2.43 The principle of the manufacture of laminated safety glass with a PVB interlayer

the vertical < 10°
- Overhead glazing: inclination with respect to the vertical > 10°

Linear supports

The design and construction of overhead glazing on linear supports (see "Mounting", p. 110) is dealt with in the "Technical Rules for the Use of Glazing on Linear Supports" (TRLV) published by the Deutsche Institut für Bautechnik (DIBt, German Institute of Building Technology):

Wired glass or laminated safety glass made from float or heat-strengthened glass must be used for single glazing or the lower pane of insulating glass.

The deflection of the support members may not exceed 1/200 of the length of the pane to be supported, and in no case should exceed 15 mm.

Panes of laminated safety glass made from float and/or heat-strengthened glass spanning > 1.20 m must have linear supports on all sides. The aspect ratio of such panes may not exceed 3:1. However, the maximum pane size is not defined and depends on the supplier.

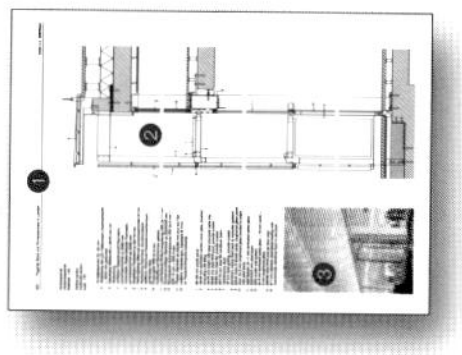

Wired glass is only permissible up to a span of 0.7 m in the main direction.

If the planned loadbearing components of glass deviate from the recommended values, an individual Approval must be applied for.

The permissible bending stresses required for a stability analysis can be taken from Fig. B 2.38.

Point supports

A statically determinate form of support should generally be chosen for point-supported overhead glazing (see "Mounting", p. 110). There should be one fixed (restrained) point and the other supports should be unrestrained in order to guarantee that the pane is properly supported without inducing any restraint stresses (Fig. B 2.41). Alternatively, each point support itself can be designed as a rod with pinned ends [23]. The main points relevant for overhead glazing in the "Technical Rules for the Design and Construction of Point-Supported Glazing" (TRPV) are summarised below:

- Only panes of laminated safety glass made from heat-strengthened or toughened safety glass may be used. Laminated safety glass made from panes of heat-strengthened glass with the same thickness (at least 2 No. 6 mm) should be used for single glazing.
- The top edge of the glazing may not be more than 20 m above ground level. The size of the glass panes may not exceed 2500 × 3000 mm.
- The deflection of the glazing should be limited to 1/100 of the relevant span.
- Drilled holes are to be positioned so that there is a clear width of glass of min. 80 mm between the hole and an edge or another drilled hole. Furthermore, at corners this distance must be min. 80 mm to one edge and min. 100 mm to the other. An unsupported glass edge may cantilever no more than 300 mm beyond the area bounded by the glass fixings (Fig. B 2.42).
- Every pane of laminated safety glass supported exclusively by point fixings shall require at least three such fixings. The largest enclosed angle may not exceed 120° (Fig. B 2.42).
- Disc fixings must have a circular disc with a minimum diameter of 50 mm on both sides.

References

[1] Deutscher Ausschuss für Stahlbeton (DAfStb) (pub.): Erläuterungen zur DAfStb-Richtlinie »Wasserundurchlässige Bauwerke aus Beton«. Berlin, 2006, p. 18
[2] Lohmeyer, Gottfried; Ebeling, Karsten: Schäden an wasserundurchlässigen Wannen und Flachdächern aus Beton. Schadenfreies Bauen, vol. 2. Stuttgart, 2007, p. 145
[3] ibid.
[4] Lohmeyer, Gottfried: Schäden an Flachdächern und Wannen aus wasserundurchlässigem Beton. Schadenfreies Bauen, vol. 2. Stuttgart, 1996, p. 107
[5] Lohmeyer, Gottfried; Ebeling, Karsten: Weiße Wannen einfach und sicher. Konstruktion und Ausführung wasserundurchlässiger Bauwerke aus Beton. Düsseldorf, 2009, p. 31
[6] Hierlein, Elisabeth: Betonfertigteile im Geschoss- und Hallenbau. Grundlagen für die Planung. Pub. by Fachvereinigung Deutscher Betonfertigteilbau e.V. (FDB). Düsseldorf, 2009, p. 37
[7] Bindseil, Peter: Stahlbetonfertigteile. Konstruktion – Berechnung – Ausführung. Cologne, 2007, p. 7
[8] Fachverband deutscher Fertigteilbau: http://www.fdb-wissensdatenbank.de, Position as of 15 Feb 2010
[9] ibid. [7]
[10] ibid. [7], p. 133
[11] ibid. [5], p. 69
[12] ibid. [7], p. 134
[13] von Busse, Hans-Busso et al.: Atlas Flache Dächer. Munich/Basel, 1994, p. 63
[14] Pöter, Hans: Metallleichtbaukonstruktionen: Früher und heute. In: Stahlbau 05/2009, pp. 288ff.
[15] Dürr, Markus; Kathage, Karsten; Saal, Helmut: Schubsteifigkeit zweiseitig gelagerter Stahltrapezbleche. In: Stahlbau 04/2006, pp. 280ff.
[16] European Convention for Constructional Steelwork (ECCS): European Recommendations for Sandwich Panels 2000. CIB Publication 257, Brussels, 2001
[17] Böttcher, Marc: Dach- und Fassadenelemente aus Stahl. Erfolgreich Planen und Konstruieren. Pub. by Stahl-Informations-Zentrum. Düsseldorf, 2007
[18] Berner, Klaus: Selbsttragende und aussteifende Sandwichbauteile. Möglichkeiten für kleinere und mittlere Gebäude. In: Stahlbau 05/2009, pp. 298ff.
[19] Bauen mit Stahl (pub.): Stahlbau Arbeitshilfe 46. Sandwichelemente. Düsseldorf, 2000
[20] DIN 1052 Design of timber structures – General rules and rules for buildings. Dec 2008, p. 28
[21] DIN 4074 Strength grading of wood. Dec 2008
[22] Winter, Stefan; Schopbach, Holger: Hoch gestapelt – Brettstapeldecken in der Quasi-Balloon-Bauweise. In: Holzbau – die neue quadriga, 01/2004
[23] Weller, Bernhard et al.: Glass in Building. Munich, 2009

Part C Building physics

Fig. C   Velodrome, Berlin (D), 1997, Dominique Perrault

Thermal insulation

Christian Bludau, Hartwig M. Künzel

C 1.1

Thermal insulation is crucial for guaranteeing a hygienic interior climate, protecting the building fabric against moisture and reducing energy consumption. And it is primarily energy-efficient building design aspects that are crucial to cutting the energy needed for heating and cooling buildings. The aim is not necessarily just to maximise the use of renewable energy sources, but to minimise the energy consumption as a whole using technical measures. In this context, thermal insulation measures and – with respect to summertime thermal performance – shading measures for transparent components are the most important issues.

The principles of heat transfer

Where a building component separates areas with different air temperatures, a heat flow towards the lower temperature is always established depending on the in situ boundary conditions. Heat transfer can take place in the following forms (Figs. C 1.3 and 1.4):
• Conduction
• Convection
• Radiation

The heat flow that actually takes place is frequently a combination of the above forms of heat transfer.
Fig. C 1.5 lists the most important terms and symbols used in this chapter. DIN EN ISO 7345 specifies further physical variables, symbols and units [1].

Conduction
Heat transfer by way of conduction is mainly found in solid bodies, but is also possible in stationary liquids and gases. The heat flow Φ_{cd} by way of conduction is calculated as follows:

$$\Phi_{cd} = -A \cdot \lambda \cdot \frac{\Delta\theta}{d} \quad [W]$$

where:
A   area [m²]
λ thermal conductivity [W/mK]
θ temperature [°C]
d   thickness [m]

The thermal conductivity of a material depends primarily on its density and moisture content. Thermal conductivity decreases with decreasing density. Stationary air has a thermal conductivity of only about 0.025 W/mK, and that is why highly porous materials – whose pores are filled with stationary air – have a low thermal conductivity. (Fig. C 1.6, p. 52). As the moisture content rises, so the density increases and hence the thermal conductivity, too (Fig. C 1.7, p. 52) [2]. Thermal conductivity also depends on temperature. However, for building materials this influence amounts to < 0.1 % per degree temperature rise, although for insulating materials with thermal conductivities < 0.1 W/mK this can increase to 0.4 % per degree. This effect can be neglected in building design because of the low temperature differences involved (Fig. C 1.8, p. 52).
By including the area affected by the heat flow in the equation, we can calculate the heat flow rate q:

$$q = -\frac{\Phi_{cd}}{A} = -\lambda \cdot \frac{\Delta\theta}{d} = -\frac{\Delta\theta}{d/\lambda} = -\frac{\Delta\theta}{R} \quad [W/m^2]$$

And from this we can calculate the thermal resistance R of a layer of material with thickness d:

$$R = \frac{d}{\lambda} \quad [m^2K/W]$$

Convection
Convection is the heat transfer that takes place as a result of the flow (movement) of gases and liquids. As these transfer processes are relatively complicated and difficult to represent mathematically, a heat transfer coefficient h_c for convection has been introduced for practical applications. The heat flow is then calculated as follows:

$$\Phi = h_c \cdot A \cdot (\theta_L - \theta_O) \quad [W]$$

where:
h_c heat transfer coefficient [W/m²K]
θ_L air temperature [°C]
θ_O surface temperature [°C]

C 1.1   Thermographic image of a flat roof with a dark roof covering and a light-coloured test area
C 1.2   Thermal conductivities of common building materials
C 1.3   Heat transfer processes in a roof construction in winter for an outward heat flow
C 1.4   Schematic presentation of the three heat transfer processes
C 1.5   Symbols and units for building physics variables

	Thermal conductivity λ [W/mK]
Metals	approx. 50–400
Concrete	1.5–2.5
Timber	0.13–0.2
Insulating materials	0.02–0.1
Stationary air	approx. 0.025

C 1.2

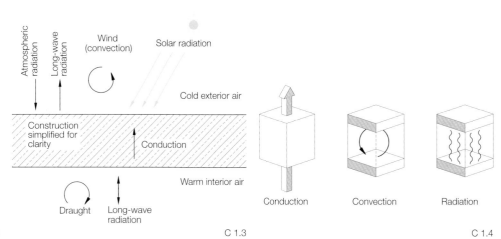

C 1.3   C 1.4

Symbol	Physical variable	Unit of measurement
A	Area	m²
c	Specific heat capacity	J/kgK
d	Thickness	m
g	Total energy transmittance of glazing	–
f_{Rsi}	Temperature factor for inner surface	–
h	Heat transfer coefficients	W/m²K
I	Intensity, energy	W/m²
l	Length	m
n	Air change rate	1/h
q	Heat flow rate	W/m²
R	Thermal resistance	m²K/W
t	Time	s
T	Temperature	K
U	Thermal transmittance	W/m²K
V	Volume	m³
α (alpha)	Radiation absorptance	–
ε (epsilon)	Radiation emissivity	–
θ (theta)	Celsius temperature	°C
λ (lambda)	Thermal conductivity, thermal conductivity design value Wavelength	W/mK m
ρ (rho)	Density Radiation reflectance	kg/m³ –
τ (tau)	Radiation transmittance	–
Φ (phi)	Heat flow	W
χ (chi)	Point thermal transmittance	W/K
Ψ (psi)	Linear thermal transmittance (thermal bridge loss coefficient)	W/mK

C 1.5

The heat transfer coefficient h_c is not a material parameter; rather, it represents the influence of several factors such as temperature, flow velocity, surface characteristics and surface geometry. The incident flow at the surface is one of the main factors here, and we distinguish between the following effects [3]:
• Free (or natural) convection due to temperature differences and the resulting density differences, e.g. convection vortex in enclosed air spaces (double glazing) with boundary surfaces at different temperatures.
• Forced convection due to differences in the total pressure, e.g. wind-induced air flows through a building or dissipation of heat in a ventilated flat roof (Fig. C 1.10, p. 52).

Radiation
The energy input due to radiation is much more important for roofs than is the case with, for example, opaque wall constructions.
Heat transfer by way of radiation takes place in the form of electromagnetic waves and requires a radiation-permeable medium (vacuum, gas). Radiation is characterised by the wavelength λ specified in μm or nm. A body radiates heat (energy) within a certain range of wavelengths (spectrum) depending on its temperature. The higher this temperature, the shorter the wavelength of the radiation emitted. The spectral ranges important for flat roofs are as follows:
• Short-wave radiation:
  UV (ultraviolet: $\lambda < 0.380$ μm)
  VIS (visible light:
  0.380 μm $\leq \lambda \leq 0.780$ μm)
• Long-wave radiation:
  IR (infrared: $\lambda > 0.780$ μm)

The most important source of radiation is the sun, the spectrum of which ranges from the short-wave X-ray and ultraviolet radiation via the range of visible light to the long-wave infrared and radio waves (Fig. 1.9, p. 52).
The solar radiation incident on the surface of a building component is split into various parts with respect to its intensity: a portion reflected from the surface, a portion absorbed by the surface and, for transparent building components, a portion transmitted through the surface. Only the absorbed radiation portion is converted

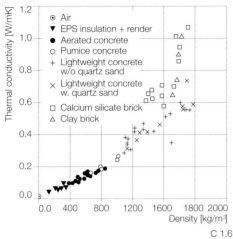

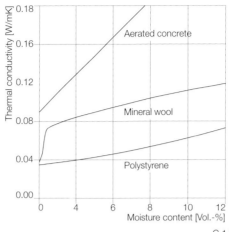

C 1.6

C 1.7

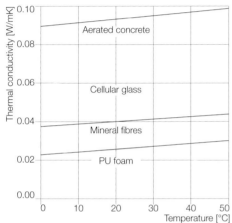

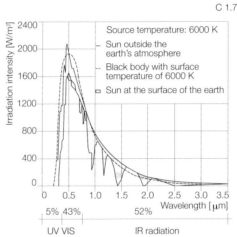

C 1.8

C 1.9

Type of convection	Thermal transmittance [W/m²K]
Free (natural)	3–10
Forced	10–100

C 1.10

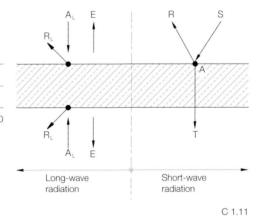

Long-wave radiation   Short-wave radiation

C 1.11

C 1.6   Relationship between thermal conductivity and density
C 1.7   Relationship between thermal conductivity and moisture content
C 1.8   Relationship between thermal conductivity and temperature
C 1.9   The solar spectrum
C 1.10   Thermal transmittance for free (natural) and forced convection of air
C 1.11   The processes at the internal and external surfaces in the short- and long-wave ranges using the example of a transparent building component
         S   solar irradiation
         R   reflected component
         A   absorbed component
         T   transmitted component
         E   long-wave emissions
         A_L absorbed long-wave irradiation

C 1.12   Daily progressions of direct and diffuse solar irradiation intensities on building component surfaces with different orientations at different times of the year
C 1.13   Emission coefficients for a selection of common building materials
C 1.14   Schematic presentation of the temperature distribution in summer and winter over the cross-section of an unventilated reinforced concrete roof
         s_B thickness of reinforced concrete layer
         s_D thickness of thermal insulation layer
         s_G limit value at a certain layer depth in which the heat wave "suffocates"
         θ_o temperature at upper surface of roof
         θ_T temperature at interface between layer of thermal insulation and layer of reinforced concrete
         θ_u temperature at underside of roof

into heat at the surface of the building component. Fig. C 1.11 illustrates these processes. In the short-wave range, the incident solar radiation is partly reflected from, partly absorbed by the surface, and the remaining radiation passes through the component (transmission). In the long-wave range, the same effect is evident on both sides of the component: long-wave radiation strikes the surface and is partly absorbed. At the same time, the building component radiates energy (emission) in the long-wave range depending on its temperature. The following general formula applies for the incident radiation:

$$\Phi_e = \Phi_e \cdot \rho_e + \Phi_e \cdot \alpha_e + \Phi_e \cdot \tau_e \ [W]$$

which can also be expressed as:
$$\rho_e + \alpha_e + \tau_e = 1$$

where:
Φ_e intensity of irradiation, e.g. sun [W]
ρ_e direct reflectance [-]
α_e direct absorptance [-]
τ_e direct transmittance [-]

The radiometric variables depend on the wavelength and the angle of incidence. As the angle of incidence decreases, e.g. on glazing, so α_e and τ_e tend towards 0, whereas ρ_e tends towards 1. This has an effect on the transmission of light and energy through glazing. DIN EN 410 covers the calculation of the photometric and radiometric variables for glazing.

Short-wave radiation
The intensity I of the solar spectrum incident on the surfaces of a building component is calculated from the heat flow divided by the area:

$$I = \frac{\Phi_e}{A} \ [W/m^2]$$

The portion of this heat flow absorbed by the surface is determined by the absorptance:

$$I_s = I \cdot \alpha_s \ [W/m^2]$$

Fig. C 1.12 shows the orders of magnitude of the maximum irradiation intensity of the sun on surfaces depending on orientation and time of year. For the short-wave spectral range, the absorptance can be assigned roughly depending on the colour of the surface:
• 0.1–0.3: white and very light-coloured surfaces
• 0.3–0.6: light-coloured to grey surfaces
• 0.6–0.9: dark surfaces
So-called ideal black surfaces exhibit an absorptance of about 1.0. Fig. C 1.15 (p. 54) shows how the absorptance affects the surface temperature.

Long-wave radiation

Every body emits electromagnetic radiation because of the thermal movement of its constituent molecules. At the ambient temperature this radiation is in the long-wave or infrared range. The long-wave radiation emission I_{lr} of warm surfaces (e.g. radiator, wall) is calculated according to the Stefan-Boltzmann law and is proportional to the long-wave emissivity ε of the surface:

$$I_{lr} = \varepsilon \cdot I_e \cdot \sigma \cdot T_0^4 \ [W/m^2]$$

where:

I_{lr} long-wave irradiance of a real body [W/m²]
ε emissivity [-]
I_e energy emitted by a black body [W/m²]
σ Stefan-Boltzmann constant, 5.67·10⁻⁸ [W/m²K⁴]
T_0 surface temperature of body [K]

The emissivity of the surface depends mainly on the wavelength. The visibly perceived colour plays only a subordinate role because it depends on the optical properties in the short-wave spectral range. Non-metallic materials generally exhibit an emissivity of about 0.9, whereas for metallic materials it ranges from 0.04 for bright or polished surfaces to 0.9 for rough or oxidised surfaces (Fig. C 1.13). The emissivity and absorptance values of a surface are usually identical.

Radiation balance

A radiation balance includes all the radiation effects taking place on a roof surface:

$$I_{tot} = I_s \cdot \alpha_s + \varepsilon \cdot I_i - I_e \ \ [W/m^2]$$

where:

I_{tot} net radiation at the component surface [W/m²]
I_s short-wave solar radiation [W/m²]
α_s short-wave absorbtance of component surface [-]
ε long-wave emissivity and absorptance of component surface [-]
I_i long-wave atmospheric radiation [W/m²]
I_e long-wave radiation emitted from component surface [W/m²]

A positive value of I_{tot} leads to a rise in temperature at the component surface, a negative value results in a cooling of the surface. If the surface temperature is lower than the dew point temperature, condensation can collect on the roof surface, but it can also result in condensation within the construction (interstitial condensation), depending on the design of the roof (see "Moisture loads", pp. 64–65).

Thermal actions on flat roofs

Due to its exposed position, a flat roof is subjected to climatic effects, primarily solar irradiation, to a greater extent than other building components. That can lead to high surface temperatures and damage caused by those high temperatures. Fig. C 1.14 shows in schematic form the seasonal temperature fluctuations that lead to high temperature gradients between the inner and outer surfaces [4]. Considerable temperature fluctuations occur on the surface of the roof over the course of a day and a year.

The surface temperatures also depend on the type of roof covering. Roofs with a covering of flexible sheeting or sheet metal give rise to the highest temperatures, especially with dark colours. Fig. C 1.15 (p. 54) shows the surface temperatures measured on flat roofs with black and white flexible sheeting on a warm summer's day. The temperature rise of the black surface is much greater than that of the white one. And where winter temperatures of -20 °C and lower can be expected that means dark-coloured flexible sheeting has to withstand temperature ranges of up to 100 K! On roofs with a low level of thermal insulation, light-coloured surfaces can make a significant contribution to summertime thermal performance; and light-coloured roof surfaces are very important in hotter climates because they can help to save cooling energy. So light-coloured, reflective surfaces are therefore preferred for flat roofs in the countries of, for example, Southern Europe, whereas the energy-saving advantages of reflective flexible sheeting are minimal in Central Europe, where the risk of moisture damage increases due to the reduced drying-out potential in summer (see "Moisture control design using steady-state vapour diffusion calculations", pp. 69–70) [5]. Temperature differences can lead to high thermally induced changes of length in the materials affected. With sheet metal roof coverings, for instance, it is absolutely essential to include a separating layer between the metal and the material underneath because the coefficients of expansion of the metal and the other material are so different. Significant changes in length in the loadbearing structure can also occur in roof designs with internal insulation. Monolithic concrete slabs in particular should therefore employ an impermeable concrete mix and should be supported on sliding bearings on the walls. Furthermore, the top surface of the concrete should be covered with a layer of gravel to attenuate the temperature differences. Temperature fluctuations at the level of the flexible sheeting are much lower on roofs with a further layer of finishes. Fig. C 1.16 (p. 54) shows the temperatures measured at the waterproofing layer on a gravel-covered roof and a green roof (depth of gravel/substrate: 130 mm) compared with the outside air temperature on a summer's day, and Fig. C 1.17 (p. 54) shows the temperatures on a cold winter's day. In winter planting can keep the temperature

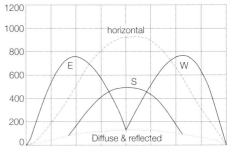

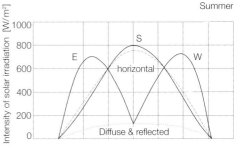

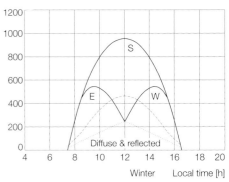

C 1.12

Material	Emission coefficient
Aluminium, polished	0.04–0.06
Aluminium, rough	0.07–0.09
Steel, polished	0.14–0.32
Steel, oxidised	0.80–0.90
Iron, polished	0.20–0.30
Iron, oxidised	0.60–0.90
Concrete	0.92–0.94
Timber	0.80–0.90
Flexible bitumen sheeting	0.90–0.93

C 1.13

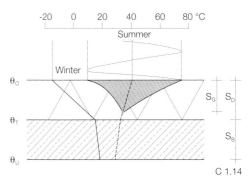

C 1.14

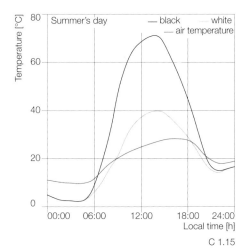

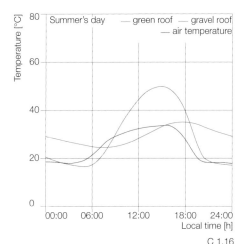

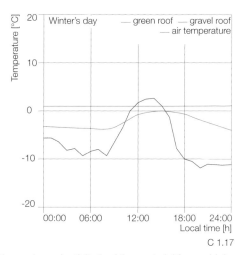

C 1.15

C 1.16

C 1.17

around freezing for a long time because the absorbed water freezes. And in those regions where a thermally insulating layer of snow can lie on the surface of the roof for a long period, the fluctuations at the surface in the winter can be much smaller.

Where the insulation is above the loadbearing structure, the fluctuations that occur are attenuated by the insulation before they have an effect on the roof itself and then also the interior. But if the roof insulation is instead positioned below the loadbearing structure (e.g. internal insulation), the loadbearing structure must be able to accommodate larger temperature fluctuations.

The night-time situation
Radiation exchange processes on the roof surface always take place in parallel, and this has an influence on the thermal and moisture-related behaviour of the roof. During the day, the solar irradiation – depending on the orientation of the surface, the intensity of the solar irradiation and the solar absorptance of the surface – outweighs the long-wave radiation emissions from the surface. The roof heats up. During the night, on the other hand, in the absence of solar irradiation, the long-wave radiation emissions can lead to a significant overcooling of the roof surface amounting to 10 K below the outside air temperature. This effect can be aggravated by the formation of a cold-air pool, e.g. behind a parapet (Fig. C 1.18) [6]. Heavier cool air collects near the parapet and acts as an insulating layer

between the roof and the ambient air. This can cause further overcooling amounting to as much as 3 K, which means that on cloudless nights an overcooling effect of max. 13 K below the outside air temperature is possible. In exposed positions at high altitudes this can mean that on cool nights in July, when the outside air is only a few degrees above zero, ice forms on the roof because the surface temperature reaches -8 °C!

Heat transmission in building components

A temperature gradient becomes established through the cross-section of a building component because of the different temperature conditions on the inner and outer surfaces of the component.

Heat flows through building components
If we assume a steady-state, non-variable condition for the temperature distribution, then this gradient does not undergo any change over time (Fig. C 1.19). As in the steady-state condition no storage of thermal energy takes place, the outward heat flow rate q_i must be equal to the inward heat flow rate q_e. The following therefore applies:

$$q_i = q_e = q \ [W/m^2]$$

The heat flowing through the component by way of conduction is directly proportional to the

thermal conductivity λ of the material from which the component is made, and inversely proportional to the thickness d of the component. From this we can derive the following formula:

$$q = \frac{\theta_{si} - \theta_{se}}{\dfrac{d}{\lambda}} \ [W/m^2]$$

where:
θ_{si} temperature of inside surface [°C]
θ_{se} temperature of outside surface [°C]
d   thickness [m]
λ thermal conductivity [W/mK]

Consequently, we must know the thermal conductivity λ of a material in order to calculate the heat flow through it. In order to determine this parameter, a sample of the material with certain dimensions is tested for its thermal conductivity under defined boundary conditions over a defined period of time (see "Conduction", p. 50).
The heat transfer between interior air, exterior air and the component takes place primarily by way of convection and radiation. The surface resistances relevant here depend on the flow conditions in the adjoining boundary air layer. The heat transfer coefficients and the surface resistances are calculated using the following formula:

$$q_{tr} = h \cdot \Delta\theta = \frac{\Delta\theta}{1/h} = \frac{\Delta\theta}{R_s} \ [W/m^2]$$

where:

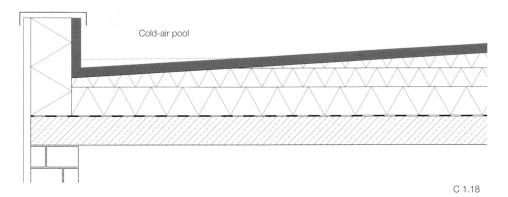

Cold-air pool

C 1.18

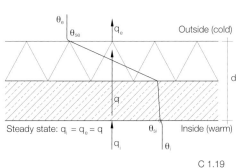

Steady state: $q_i = q_e = q$

C 1.19

Type of heat transfer	External surface resistance R_{se} [m²K/W]	Internal surface resistance R_{si} [m²K/W]
DIN EN ISO 6946 for thermal performance calculations		
Upward heat flow	0.1	
Horizontal heat flow	0.13	0.04
Downward heat flow	0.17	
DIN 4108-3 for calculations for avoiding condensation		
Horizontal and upward heat flows, also pitched roofs	0.13	0.04 (0.08)[1]
Downward heat flow	0.17	

[1] R_{se} = 0.08 m²K/W when the outside surface is in contact with a ventilated layer of air, e.g. in ventilated roofs.

C 1.20

C 1.21

q_{tr} heat flow rate [W/m²]
h heat transfer coefficient [W/m²K]
R_s surface resistance [m²K/W]

Fig. C 1.20 lists the internal and external surface resistances for planar components. The greater air velocities outside the building lead to external surface resistances that are lower than the internal surface resistances. The standardised heat transfer coefficients take into account the convection and radiation components.
DIN EN ISO 6946 should be used to calculate the surface resistances of all kinds of materials, also those with more complex geometries (see p. 58).

Assuming an internal surface resistance $R_{si} = 1/h_i$ and an external surface resistance $R_{se} = 1/h_e$, the heat flow for a multi-layer component with n layers is as follows:

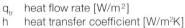

$$q = \frac{\theta_i - \theta_e}{\frac{1}{h_i} + \sum_n \left(\frac{d}{\lambda}\right)_n + \frac{1}{h_e}} = \frac{\theta_i - \theta_e}{R_{si} + \sum_n \left(\frac{d}{\lambda}\right)_n + R_{se}} \quad [W/m^2]$$

where:
h_i internal heat transfer coefficient [W/m²K]
h_e external heat transfer coefficient [W/m²K]
R_{si} internal surface resistance [m²K/W]
R_{se} external surface resistance [m²K/W]
θ_e temperature of exterior air [°C]
θ_i temperature of interior air [°C]
The denominator in the equation contains the parameters that link the heat flow rate with the temperature difference driving the flow. These parameters are designated resistances. The total thermal resistance R_T is the sum of the internal and external surface resistances (R_{se}, R_{si}) plus the thermal resistances of the individual layers of the component.

Thermal transmittance
The thermal transmittance U (often simply referred to as U-value) is the critical factor for the transmission heat losses through a component and the inverse of the total thermal resistance R_T. Fig. C 1.22 (p. 56) shows the relationship between U-value, total thermal resistance and insulation thickness, and Fig. C 1.23 (p. 56) shows the U-value depending on the insulation thickness for the thermal conductivity classes (WLS) normally used these days.
The influence of effects such as thermal bridges and the energy infiltration due to water in an inverted roof can be taken into account by a surcharge added to the U-value (see p. 56).
The relationship between the thermal transmittance U of a building component and the heat flow rate q can be expressed as follows:

$$q = U \cdot (\theta_i - \theta_e) \quad [W/m^2]$$

The heat flow Φ through a defined area A of the component is then calculated as follows:

$$\Phi = q \cdot A = U \cdot (\theta_i - \theta_e) \cdot A \quad [W]$$

A heat flow becomes established in the direction of the lower temperature. With steady-state boundary conditions, this heat flow remains constant. The heat transfer can take place in different ways. In solids the heat flows mainly by way of conduction, in gases and liquids mainly by way of convection, and in radiation-permeable materials by means of radiation as well. The thermal conductivity values given in publications and the product catalogues of building materials suppliers already take into account the different transmission mechanisms.

Layers of air
In a layer of air the heat transfer is made up of conduction, convection and long-wave radiation. For simplicity, a resultant total thermal resistance is used. In doing so, the thermal resistances of layers of air (Fig. C 1.24, p. 56) according to DIN EN ISO 6946 in the component cross-section can be considered to correspond to the homogeneous layers perpendicular to the heat flow if the following conditions are satisfied:
- Boundary surfaces are parallel with ε values > 0.8, a requirement that is satisfied by most building materials
- The layer thickness is < 0.1 times the thickness of one of the other adjacent layers
- Max. layer thickness = 0.3 m
- No exchange of air with the interior

For all other cases the resistance must be calculated according to DIN EN ISO 6946 Annex B.

C 1.15   Surface and air temperatures measured on a black and a white roof on a summer's day. The significant overcooling below the outside temperature for both types of flexible sheeting is clearly apparent.
C 1.16   Roof and air temperatures measured on a summer's day. The temperature rise of the gravel roof, related to the exterior air temperature, is delayed due to the high thermal mass of the gravel finish. The temperature below the green roof is plainly attenuated; over the course of the day it tracks the exterior air temperature but with a considerable delay.
C 1.17   Roof and air temperatures measured on a winter's day. The layer of gravel is frozen right through, the surface temperature does not reach even the minimum temperature of the air. The green roof functions simultaneously as insulator and storage medium; its temperature remains constant at 1 °C despite the low temperature of the outside air.
C 1.18   A cold-air pool behind a tall peripheral parapet (excessive cooling)
C 1.19   Heat transfer by way of conduction and convection, illustrated using the example of the temperature gradient over the cross-section of a simplified roof construction for a steady-state temperature condition.
C 1.20   Surface resistances R_{si} and R_{se}. The value for the upward heat flow is used for flat roofs.
C 1.21   Roof finished with white synthetic waterproofing material

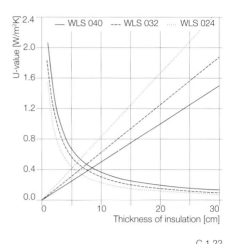

C 1.22

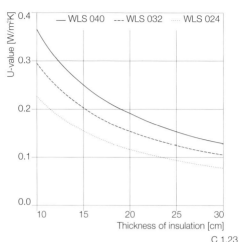

C 1.23

Thickness of air layer [mm]	Direction of heat flow		
	upwards	horizontal	downwards
0	0.00	0.00	0.00
5	0.11	0.11	0.11
7	0.13	0.13	0.13
10	0.15	0.15	0.15
15	0.16	0.17	0.17
25	0.16	0.18	0.19
50	0.16	0.18	0.21
100	0.16	0.18	0.22
300	0.16	0.18	0.23

Intermediate values may be obtained by linear interpolation.

C 1.24

Thermal bridges

Thermal bridges are local, limited areas of building components where there is an increased heat flow from inside to outside (heat loss) and hence a lower interior surface temperature compared with the adjacent areas. This can lead to a higher relative surface humidity, even condensation in these areas. The relative humidity at the surface should not exceed 80 % if the risk of mould growth is to be avoided. Further information on the calculation of the surface temperature in order to avoid critical surface humidity levels can be found in DIN 4108-2 and DIN EN ISO 13788.

In principle, we can distinguish between the effects of material and geometrical thermal bridges. In practice thermal bridges are usually a combination of these two types.

A material thermal bridge occurs when the thermal conductivities of the adjacent materials are different, e.g. a loadbearing column within a wall panel (Fig. C 1.25).

With a geometrical thermal bridge, e.g. an external wall corner, the heat outflow surface area on the cold surface is larger than the corresponding heat inflow surface area on the warm surface. The result is a greater loss of energy in this area (Fig. C 1.26).

Thermal bridge calculation

The effects of thermal bridges are determined these days with the help of multi-dimensional numerical methods. These calculations also form the basis of the information given in the catalogues of thermal bridge details [7], where usually the minimum interior surface temperature and the additional heat flow towards the undisturbed area are given. A temperature factor f_{Rsi} (dimensionless surface temperature) is specified for the point of the minimum surface temperature from which the surface temperatures can be determined assuming arbitrary steady-state temperature conditions (θ_i, θ_e):

$$f_{Rsi} = \frac{\theta_{si} - \theta_e}{\theta_i - \theta_e} \quad [-]$$

In order to avoid mould growth, DIN 4108-2 specifies a minimum temperature factor of $f_{Rsi} \geq 0.7$. We distinguish between linear- and point-type thermal bridges; for the former, an increased heat loss compared with the undisturbed area is considered by using a linear thermal transmittance value Ψ, and for the latter a point thermal transmittance value χ. The calculation is carried out using the following equations:

Linear-type thermal bridge:

$$\Phi_{T,l} = (U \cdot A + \Psi \cdot l) \cdot (\theta_i - \theta_e) \ [W]$$

Point-type thermal bridge:

$$\Phi_{T,p} = (U \cdot A + \chi) \cdot (\theta_i - \theta_e) \ [W]$$

where:

$\Phi_{T,l}$ heat flow through linear-type thermal bridge [W]

$\Phi_{T,p}$ heat flow through point-type thermal bridge [W]
U   thermal transmittance of component [W/m²K]
A   area of component [m²]
Ψ linear thermal transmittance value [W/mK]
l   length of linear thermal bridge [m]
χ point thermal transmittance value [W/K]
θ surface temperature [°C]

DIN 4108-6 recommends determining the thermal bridge loss coefficients according to DIN EN ISO 10211 or the use of a catalogue of thermal bridge details.

Linear-type thermal bridges are regarded as two-dimensional heat-flow and temperature fields, e.g. an external wall corner extending the full height of the room. Point-type thermal bridges with three-dimensional fields ensue, for example, at the corners of rooms or at pipe penetrations. Such thermal bridges are overestimated if calculated according to the Ψ values, this may result in negative χ values [8]. According to the new thermal performance requirements, it is primarily linear-type thermal bridges that must be considered because they constitute a not inconsiderable part of the thermal envelope and therefore make a significant contribution to the total transmission heat losses.

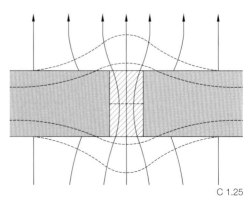

C 1.25

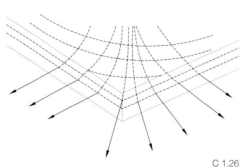

C 1.26

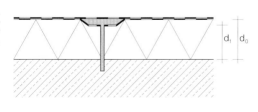

C 1.27

Thermal resistance on room side of waterproofing as proportion of total thermal resistance [%]	Surcharge ΔU [W/m²K]
< 10	0.05
10–50	0.03
> 50	0

C 1.28

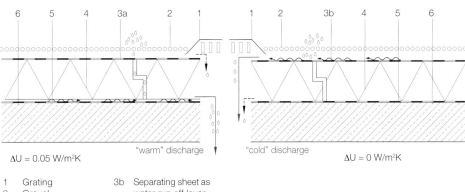

ΔU = 0.05 W/m²K

"warm" discharge

"cold" discharge

ΔU = 0 W/m²K

a

1   Grating
2   Gravel
3a   Sheeting to prevent loss of gravel
3b   Separating sheet as water run-off layer
4   Insulation (XPS)
5   Waterproofing
6   Loadbearing structure

b

C 1.29

Surcharges on the U-value

Effects such as conduction via penetrations, cool water seeping below insulating materials, etc. can be included in the calculations by way of U-value surcharges. The influence with respect to transmission heat losses via a thermal bridge can be taken into account by way of a specific global thermal bridges surcharge ΔU_{WB}. In doing so, the Ψ values for the relevant thermal bridges can be taken from a catalogue of thermal bridge details and multiplied by the lengths of the individual thermal bridges and the temperature factor f_{Rsi}, added together, and divided by the thermal envelope [9]. For linear-type thermal bridges this results in the following:

$$\Delta U_{TB} = \frac{\sum (\Psi \cdot l \cdot f_{Rsi})}{A} \ [W/m^2K]$$

For point-type thermal bridges (e.g. anchors) the surcharge is calculated accordingly:

$$\Delta U_{TB} = \frac{\sum (X \cdot n \cdot f_{Rsi})}{A} \ [W/m^2K]$$

DIN EN ISO 6946 specifies a similar correction for mechanical fasteners. According to that, ΔU_f is calculated as follows:

$$\Delta U_f = \alpha \cdot \frac{\lambda_f \cdot A_f \cdot n_f}{d_o} \cdot \left(\frac{R_1}{R_{T,h}}\right)^2 \ [W/m^2K]$$

where:
ΔU_f surcharge on U-value [W/m²K]

C 1.22   Relationship between the U-value, thermal resistance and thickness of a layer of insulation
C 1.23   Relationship between the U-value and the thickness of a layer of insulation for three thermal conductivity classes (WLS)
C 1.24   Thermal resistances of stationary layers of air to DIN EN ISO 6946
C 1.25   Material thermal bridge (e.g. loadbearing column within a wall panel)
C 1.26   Geometrical thermal bridge (e.g. roof slab)
C 1.27   Mechanical fastener fitted in a recess
C 1.28   Surcharges for inverted roofs to DIN 4108-2
     d_0 Thickness of layer of insulation fixed by fastener
     d_1 Thickness of layer of insulation penetrated by fastener
C 1.29   Surcharge ΔU added to thermal resistance
     a   Water draining below the insulation
     b   Water draining above the insulation

λ_f thermal conductivity of fastener [W/mK]
A_f cross-sectional area of one fastener [m²]
n_f number of fasteners per m² [1/m²]
d_0 thickness of layer of insulation containing the fastener [m]
R_1 thermal resistance of layer of insulation penetrated by fastener [m²K/W]
$R_{T,h}$ thermal resistance of component ignoring thermal bridges [m²K/W]
α conversion coefficient [-]

Coefficient α is 0.8 for complete penetration and $0.8 \cdot d_1/d_0$ when the fastener is fixed in a recess (Fig. C 1.27). The thickness d_1 of the insulation penetrated by the fastener can be greater than the actual thickness d_0 of the layer of insulation if the fastener is installed at an angle.

Inverted roofs
On an inverted (or upside-down) roof, the design is such that water can seep below the thermal insulation. As this precipitation or meltwater drains away, it cools the water run-off level and the construction below, and this energy loss has to be taken into account by a surcharge to the U-value. The surcharge ΔU depends on the value of the thermal resistance below the waterproofing as a percentage of total thermal resistance of the roof construction. Fig. C 1.28 shows appropriate surcharges. If the proportion of the insulation below the waterproofing amounts to 50 % or more, the U-value does not have to be increased. On an inverted roof with a 200 mm deep concrete slab, for example, the percentage proportion of the insulating value of the loadbearing structure with respect to XPS insulation 200 mm thick is just under 2 %. In this case the U-value of the construction would have to be å roofs with a lightweight supporting construction and a mass per unit area < 250 kg/m², the thermal resistance R below the waterproofing must be at least 0.15 m²K/W. This requirement is intended to prevent the occurrence of condensation on the soffit in the event of sudden rainfall. Condensation can form when during rainfall cold water flowing beneath the insulation on top of the waterproofing causes a severe cooling of the soffit.
The surcharge on the U-value can be omitted when a separating layer for draining the water

is used in combination with an extruded polystyrene rigid foam (XPS) to DIN EN 13164. If the combination of products in the total system, consisting of gravel, drainage mat (= separating layer) and thermal insulation, is covered by a National Technical Approval, there is no need to consider a surcharge. It is then guaranteed that most of the water can drain away on top of the upper separating layer (Fig. C 1.29).
DIN EN ISO 6946 Annex D4 specifies another correction method for inverted roofs. This method takes into account the rainwater draining between the insulation and the waterproofing according to the following equation:

$$\Delta U_r = p \cdot f \cdot x \cdot \left(\frac{R_1}{R_T}\right)^2 \ [W/m^2K]$$

where:
U_r surcharge on U-value [W/m²K]
p   average amount of precipitation during the heating period [mm/day]
f   drainage factor
x   factor for increasing the heat loss due to precipitation [W·day/m²K·mm]
R_1 thermal resistance of insulation above the waterproofing [m²K/W]
R_T thermal resistance of construction prior to correction [m²K/W]

The result is correction values for the U-value between approx. 0.01 and 0.08 W/m²K which depend on the amount of precipitation and the edges of, or rather the joints between, the thermal insulation boards. The method should be used according to the details given in the technical specifications for the insulating materials. Research into the moisture absorption of insulation made from the extruded polystyrene rigid foam typically used in inverted roofs has revealed that the insulation absorbs a not inconsiderable amount of moisture by way of diffusion, even though it does not exhibit any water absorption by way of capillary action. On green roofs, where high relative humidity levels always prevail in the region of the insulation, the heat transfer for a layer of insulation 100 mm thick can increase by up to 25 % over a period of 30 years and hence reduce the U-value of the insulation [10].

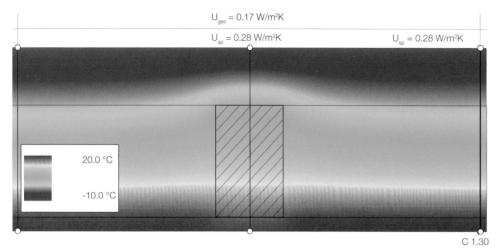

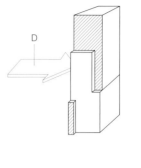

$U_{ges} = 0.17 \text{ W/m}^2\text{K}$

$U_{sp} = 0.28 \text{ W/m}^2\text{K}$ $U_{sp} = 0.28 \text{ W/m}^2\text{K}$

20.0 °C

-10.0 °C

C 1.30

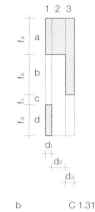

a   b   C 1.31

Standardised methods of calculation

Important methods of calculating thermal performance are:
• Thermal resistance of components to DIN EN ISO 6946
• Minimum wintertime thermal performance to DIN 4108-2
• Energy-related thermal performance according to the Energy Conservation Act (EnEV, *Energieeinsparverordnung*)

Thermal resistance of building components

DIN EN ISO 6946 describes how to calculate the thermal resistance R of inhomogeneous components by using a simplified combined method that considers not only the conduction perpendicular to the surface, but also a transverse thermal distribution. This method is based on the principle of combining resistances in parallel and in series. The prerequisite for the applicability of the method is a strictly orthogonal component make-up, i.e. the boundaries between the constituent parts may be only perpendicular to and parallel with the component surfaces.

First of all, the component is divided into its constituent layers. The thermal resistances of the ensuing segments are then calculated perpendicular to the component surface in the direction of the anticipated heat flow. Afterwards, these resistances are placed in parallel. This means that all heat flows transverse to the principal heat flow direction remain unconsidered, i.e. an additional heat loss via constituent layers with a higher thermal conductivity is ruled out and the result is the upper resistance limit for the thermal resistance.

The opposite procedure, i.e. combining all the thermal resistances in parallel first and adding these together in the principal heat flow direction perpendicular to the component surface, results in the lower resistance limit for the thermal resistance. This calculation includes an ideal heat exchange at the boundary layers parallel to the component surface so that more thermal energy is available for component layers with a greater thermal conductivity.

In order to calculate the thermal transmittance U, an inhomogeneous component is divided into

m (m = a, b, c, …) segments perpendicular to the surface (Fig. C 1.31a) and j (j = 1, 2, 3, …) layers parallel to the surface (Fig. C 1.31b). The component is therefore divided into mj parts that are thermally homogeneous.

The upper resistance limit of the thermal resistance has to be determined assuming a one-dimensional heat flow perpendicular to the surfaces of the component. This is calculated using the following equation:

$$\frac{1}{R'_T} = \frac{f_a}{R_{Ta}} + \frac{f_b}{R_{Tb}} + \frac{f_c}{R_{Tc}} + \dots \text{ [m}^2\text{K/W]}$$

The lower resistance limit has to be determined assuming that all plane surfaces parallel to the surfaces of the component are isothermal. A thermal resistance R_j for every thermally inhomogeneous layer must be calculated using the following equation:

$$\frac{1}{R_j} = \frac{f_a}{R_{aj}} + \frac{f_b}{R_{bj}} + \frac{f_c}{R_{cj}} + \dots \text{ [m}^2\text{K/W]}$$

The lower resistance limit is then as follows:

$$R''_T = R_1 + R_2 + R_3 + \dots + R_j \text{ [m}^2\text{K/W]}$$

The thermal transmittance U is formed by the arithmetic mean of the R'_T and R''_T values:

$$R_t = \frac{R'_T + R''_T}{2} \quad \text{[m}^2\text{K/W]}$$

$$U = \frac{1}{R_{si} + R_T + R_{se}} \quad \text{[W/m}^2\text{K]}$$

This method may be only be used up to an $R'_T : R''_T$ ratio of 1.5; a higher ratio indicates a thermal bridge. In this case a two- or three-dimensional thermal bridges program must be used to calculate the U-value. More accurate results can be obtained – especially for components made up of constituent parts with very different thermal conductivities – by using numerical methods (to DIN EN ISO 10211). According to this method, a timber flat roof (Fig. C 1.32) with additional insulation, a joist spacing of 700 mm and material data according to Fig. C 1.33 has a U-value 0.17 W/m²K for the entire construction (Fig. C 1.30). If we cut through

a bay between joists only, without taking the transverse thermal distribution into account, combining the resistances in series results in a U-value of 0.15 W/m²K, compared with a U-value at the joists of 0.28 W/m²K.

Thermal bridge programs can be used to determine more accurately the thermal parameters of more complex, also two- or three-dimensional, components. Fig. C 1.30 shows the result of a calculation for a timber flat roof carried out with just such a program. Here, the resulting total U-value U_{tot} for the entire component is 0.17 W/m²K, for the section through the joists the U_{sp} value is 0.28 W/m²K, and for the section through the roof between the joists the U_g value is 0.14 W/m²K. The small deviations between individual sections can be attributed to the simplified method of DIN EN ISO 6946 compared with numerical methods.

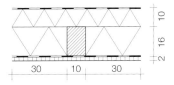

C 1.32

Material	Layer thickness d [cm]	Thermal conductivity λ [W/mK]
Insulation above loadbearing members	10	0.04
Insulation between loadbearing members	16	0.04
Loadbearing members	16	0.13
OSB	2	0.13

C 1.33

Component/system	Property	Value of reference design
Roof, topmost suspended floor, walls below roof slope	Thermal transmittance	$U = 0.20$ W/m²K
Velux-type roof windows	Thermal transmittance	$U_w = 1.40$ W/m²K
	Total energy transmittance of glazing	$g_\perp = 0.60$
Rooflights	Thermal transmittance	$U_w = 2.70$ W/m²K
	Total energy transmittance of glazing	$g_\perp = 0.64$
Aforementioned components	Thermal bridges surcharge	$\Delta U_{WB} = 0.05$ W/m² K

C 1.34

Minimum wintertime thermal performance

This method comprises the measures for reducing heat transfer through the enclosing surfaces of a building and through the surfaces separating areas of different temperatures. The minimum requirement for thermal performance is intended to guarantee a hygienic interior climate in winter and protect the building fabric permanently against climate-related moisture effects. Wintertime thermal performance ensures that an adequately high surface temperature prevails at the inner surfaces of external components during the heating period. This prevents the temperature dropping below the dew point and an excessive surface humidity ($\geq 80\%$ relative humidity), which can lead to condensation on the surface or mould growth. In addition, wintertime thermal performance is intended to define component designs that minimise transmission heat losses and hence comply with the limit values specified in the Energy Conservation Act. DIN 4108-2 assumes that the rooms are adequately heated and ventilated appropriate to their usage. In summer the heating of habitable rooms as a result of solar heat gains should be prevented and comfortable interior conditions created. Mechanical cooling systems should be avoided wherever possible. Minimum wintertime thermal performance is defined in the standard as a component thermal resistance of min. 1.2 m²K/W for roofs, although lightweight forms of construction with a mass per unit area < 100 kg/m² may have a thermal resistance of min. 1.75 m²K/W. Surcharges for the insulation on inverted roofs may be added to these figures. According to DIN 4108, for a conventional flat roof with thermal insulation ($\lambda = 0.04$ W/mK) on top of a 200 mm deep concrete slab ($\lambda = 2.1$ W/mK) this results in an insulation thickness requirement of approx. 40 mm. In this case the insulation is only that required for hygiene reasons. The requirements of the Energy Conservation Act must also be complied with.

Energy-related thermal performance according to Energy Conservation Act

Germany's Energy Conservation Act (EnEV, *Energieeinsparverordnung*) is a combination of legislation covering thermal insulation and heating installations. It regulates the total energy requirement of a building for covering the heat losses through the building envelope (thermal insulation provisions) and for operating the heating plant, the hot-water systems and, if applicable, mechanical ventilation systems (heating installation provisions). According to this legislation a building must be designed in such a way that its annual primary energy requirement does not exceed the requirement of a reference building with the same geometry, usable floor area and orientation. The calculations relate to the entire building, but only the information relevant to roofs is discussed here. Fig. C 1.34 lists the values for the roof of the reference design for both residential and non-residential buildings, the only function of which is to determine the maximum permissible annual primary energy requirement. These values do not represent individual requirements, but are instead used for the total calculation. The reference design includes a thermal bridge surcharge ΔU_{TB} of 0.05 W/m²K. The reference building prescribes a U-value of 0.20 W/m²K for the roof. For a conventional flat roof with a 200 mm deep concrete slab ($\lambda = 2.1$ W/mK) and thermal insulation ($\lambda = 0.04$ W/mK), the Energy Conservation Act specifies an insulation thickness of approx. 200 mm.

Recommended thicknesses for layers of insulation

The increasingly stringent requirements regarding thermal performance mean that the prescribed thicknesses for layers of insulation are also increasing considerably. New buildings not only have to comply with the Energy Conservation Act (EnEV) valid at the time of their erection, but wherever possible should also take into account the fact that legislation will become even more stringent in the future. Such an approach is also advisable for refurbishment projects. Following an assessment, the existing insulation, if any, can be retained and a further layer of insulation added (see "Refurbishment", p. 117).

C 1.30 Output of U-value results calculated with the help of a thermal bridges program
C 1.31 Thermally inhomogeneous component taken from DIN EN ISO 6946:
a Segments m = a, b, c, d
b Layers j = 1, 2, 3...
C 1.32 Dimensions [cm] for a flat roof with insulation above and between the loadbearing members. The values given in Fig. C 1.33 have been assumed for the materials.
C 1.33 Material values and layer thicknesses using the example of a flat roof with additional insulation
C 1.34 Prescribed values for a reference building related to an EnEV-compliant roof
C 1.35 Aerial photograph of terraces of bungalows in Munich in winter

C 1.35

	Prescribed U-value [W/m²K]	Insulation thickness for thermal conductivity class WLS 040 [cm]
DIN 4108-2	8.3	4
EnEV 2001	0.25	15
EnEV 2007	0.25	15
EnEV 2009	0.2	19
EnEV 2012[1]	0.19–0.17	20–22
Energy-saving optimum	approx. 0.13	30

[1] Anticipated values

C 1.36

C 1.36 Minimum thicknesses of thermal insulation required for a flat roof with a 200 mm deep loadbearing concrete slab

C 1.37 Permissible values for the proportion of window area S related to the floor area, below which verification of summertime thermal performance is not required.

New-build

The insulation layer thicknesses specified in DIN 4108-2 for avoiding mould growth have been superseded by those in the EnEV and are now used mainly for assessing damage in existing buildings. Currently, EnEV 2009 specifies a maximum U-value of 0.20 W/m²K for the roof. It is expected that the maximum total energy requirement will be reduced by a further 30 % in the next edition of the EnEV, which is planned for 2012. That means that the required maximum U-value for roof design will probably be 0.17–0.19 W/m²K. Fig. C 1.36 lists the U-values prescribed by standards and legislation, and the corresponding insulation layer thicknesses. At the moment a theoretical insulation layer thickness of 190 mm is possible, but upon publication of the revised EnEV this value will certainly increase to 200–220 mm [11].

Refurbishment

Many older flat roofs do not come even close to complying with today's standards of insulation. Even those roofs that satisfied the requirements when they were built cannot achieve the latest standards. This is clearly shown by the increase in the prescibed U-value since the Energy Conservation Act was introduced (Fig. C 1.36). In addition, damage to or saturation of the insulation weakens the insulating effect of an insulating material. Good thermal insulation is the prime requirement for creating a comfortable internal climate in both summer and winter, and for minimising energy consumption. The most important factor when assessing the thermal performance of a flat roof is its thermal transmittance (U-value). When a roof is refurbished or rebuilt (e.g. new covering externally or new lining internally), the maximum U-value of 0.20 W/m²K specified in the EnEV may not be exceeded. Consequently, the amount of thermal insulation must be increased or new materials installed.

The following example shows the thickness of insulation required for a refurbishment project: An existing flat roof consisting of a 160 mm deep reinforced concrete slab and 80 mm insulation with a λ value of 0.04 W/mK is to be refurbished. The existing U-value of the roof is calculated to be 0.45 W/m²K.

If only the insulation is to be renewed and increased (i.e. the existing roof and insulation are still intact), according to the current EnEV a total insulation thickness of min. 190 mm (thermal conductivity class WLS 040) must be achieved. That means an additional layer of insulation at least 110 mm thick must be installed in the roof in order to comply with the stipulations of the EnEV.

If the insulation is damaged or the roof is no longer watertight, the roof must be rebuilt. Insulating materials with a thermal conductivity of 0.032 W/mK are available. The installation of new insulation means that a layer min. 160 mm thick will be required for material with a thermal conductivity of 0.032 W/mK in order to comply with the EnEV.

But as not all insulating materials are available in all thicknesses, the calculated value will have to be rounded up to the next available thickness.

Summertime thermal performance

Summertime thermal performance means creating a comfortable internal climate for all users by preventing an excessive temperature rise in the room(s) below a flat roof.

DIN 4108-2 contains the requirements regarding summertime thermal performance. The aim is, as part of the effort to save energy, to exploit the design of the building to limit solar heat gains and to avoid, or least minimise, the use of energy for mechanical cooling and/or ventilation measures. It is assumed that complying with the requirements will ensure that the maximum reasonable internal air temperature will be exceeded only occasionally. To achieve this, the following factors must be designed accordingly:
- Thermal insulation for reducing transmission heat flows
- Shading measures to reduce solar heat gains via the windows
- Using the thermal mass of internal components
- Ventilation, at night in particular

Thermal insulation

In summer the exterior air temperature – depending on the orientation of the external components and the time of year – is on average higher than the interior air temperature because of the high amount of solar irradiation. Although these transmission heat gains from outside normally account for only a small proportion compared to the other heat flows, especially in hot climates, the radiation energy passing through glazing still dominates in many cases. The components should be designed to prevent unnecessary heat flows into the interior via the roof. This is especially important for the extreme irradiation values that occur during the summer and for flat roofs that are exposed to high temperatures. As a rule, this condition is satisfied by the wintertime thermal performance requirements.

Shading

Especially with glass roofs, but also with other flat roofs with large areas of glass, solar heat gains due to radiation can be considerable. In such instances the summertime thermal performance requirements must be complied with, which for large areas of glazing can usually be achieved by providing some form of sunshading. Small windows are better for complying with summertime thermal performance requirements, although they do reduce the amount of daylight in the building. Large windows lead to problems in summer, but do ensure useful passive solar heat gains in winter. One compromise is to use large windows fitted with external sunshades.

DIN 4108-2 prescribes a simplified method that uses a solar gain index determined for each room depending on the glazing. This index may not exceed a maximum permissible value that depends on the climate and the constructional parameters.

Verification is not required when the permissible values for the area of window in each facade A_W/A_{facade} (including the roof surface) are not exceeded (Fig. C 1.37). Also excluded from this verification requirement, according to the standard, are detached and semi-detached houses whose windows facing south, east or west are fitted with permanent external sunshades with a shading coefficient $F_C \leq 0.3$ (e.g. roller shutters).

The solar gain index S is calculated according to DIN 4108-2 as follows:

Inclination of windows to the horizontal	Orientation of windows[1]	Window area as percentage of floor area[2] f_{AG} [%]
> 60°–90°	North-west via south to north-east	10
	All other northward orientations	15
0°–60°	All orientations	7

The window area proportions given are based on the climate values of climate region B to draft standard DIN V 4108-6.
[1] If the room under consideration has windows facing in several directions, the lower limit value for fAG shall govern.
[2] The window area proportion fAG results from the ratio of the window area to the floor area of the room or group of rooms under consideration. If the room or group of rooms under consideration has several facades or, for example, projecting bays, the fAG is to be calculated from the sum of all window areas relative to the floor area.

C 1.37

$$S = \frac{\sum (A_W \cdot g_{total})}{A_G} \quad [-]$$

where:
A_W window area (clear structural opening) [m²]
g_{total} total energy transmittance of glazing including sunshade, calculated according to the equation given below, DIN EN 13363-1, DIN EN 410 or the guaranteed data of suppliers [-]
A_G net floor area of room [m²]

The total solar gain index for a room is found by adding together the figures for each individual window. For simplicity, the following equation can be used to calculate the total energy transmittance g_{total}:

$$g_{total} = g \cdot F_c$$

where:
g total energy transmittance of glazing to DIN EN 410 [-]
F_c shading coefficient for permanent sunshade [-]

DIN 4108-6 specifies reference values for the total energy transmittance of transparent building components normally intended for a perpendicular angle of incidence. FC values for sunshades are given in DIN 4108-2. For canopies, loggias and awnings it must be ensured that maintaining a certain shading angle prevents direct sunlight from reaching the window. The calculated solar gain index S may not exceed a maximum value S_{perm}:

$$S_{perm} = \sum S_\chi$$

The proportional solar gain index S_χ in this equation depends on the climate region and can be obtained from DIN 4108-2 Tab. 9. Germany is divided into three "summer climate regions": cool summers (region A), mild summers (region B) and hot summers (region C), for which maximum interior temperatures of 25 °C, 26 °C and 27 °C respectively are specified. These limit values should not be exceeded for more than 10 % of the time the room is occupied (nor-

mally 24 hours per day for housing, 10 hours per day for offices).

References
[1] Richter, Ekkehard; Fischer, Heinz M.: Lehrbuch der Bauphysik. Schall – Wärme – Feuchte – Licht – Brand – Klima. Wiesbaden, 2008
[2] Künzel, Helmut: Wie ist der Feuchteeinfluss auf die Wärmeleitfähigkeit von Baustoffen unter heutigen Bedingungen zu bewerten? In: Bauphysik 11/1989, pp. 185ff.
[3] ibid. [1], pp. 128f.
[4] Sedlbauer, K.; Gottschling, H.: Sommerliche Temperaturbeanspruchung der Dachhaut bei belüfteten und nicht belüfteten Flachdächern. In: IBP-Mitteilung 357. Stuttgart, 1999
[5] Künzel, Hartwig M.; Sedlbauer, Klaus: Reflektierende Flachdächer – Sommerlicher Wärmeschutz kontra Feuchteschutz. In: IBP Mitteilung 482. Stuttgart, 2007
[6] Dederich Ludger: Flachdächer in Holzbauweise. Pub. by Holzabsatzfonds. Bonn, 2008. pp. 40ff.
[7] Hauser, Gerd; Stiegel, Horst: Wärmebrücken-Atlas für den Mauerwerksbau. Wiesbaden, 2001; DIN 4108 Beiblatt 2
[8] Hauser, Gerd; Stiegel, Horst: Pauschalierte Erfassung der Wirkung von Wärmebrücken. In: Bauphysik 17/1995, pp. 65ff.
[9] ibid.
[10] Künzel, Hartwig M.: Bieten begrünte Umkehrdächer einen dauerhaften Wärmeschutz? In: IBP-Mitteilung 271. Stuttgart, 1995
[11] Hauser, Gerd: Energieeinsparverordnung (EnEV) 2009/2012. Presentation at GRE general meeting. Munich, 2009

Moisture control

Hartwig M. Künzel

C 2.1

Moisture control in flat roofs is a key theme because waterproofing to the roof not only protects against rainwater, but also prevents the construction drying out towards the outside. In addition, the continual improvements to thermal insulation and the airtightness of buildings bring with them an increased risk of damage. This is due to the tendency for the air humidity in airtight buildings to be higher, but also to the fact that the greater temperature differences between the inner and outer surfaces of components increase the risk of condensation. As less heat escapes from the interior into the building envelope, less water can evaporate, which means that any unplanned moisture within the construction, e.g. condensation due to convection of the air or construction moisture (i.e. water generated by wet trades), represents a greater problem than was the case in the past.

It is necessary to analyse the climatic loads on building components before selecting suitable moisture control measures. An interior climate that deviates from the normal conditions in residential or office buildings will frequently have a significant effect on the moisture behaviour of a structure. Standard solutions, as can be found in standards, trade guidelines or product specifications, can quickly lead to the failure of a construction. The same is true for external climate conditions that deviate from the known standard climate. Whereas most designers are aware of the fact that a building in the tropics is exposed to thermal and moisture loads different to those of Central Europe, the climatic differences within one country or one region are often not adequately considered. Buildings within heavy shadows or buildings at high altitudes, whose surfaces experience only a minimal temperature rise even in the summer, can suffer from a reduced drying-out potential and hence a lower tolerance to moisture-related damage in flat roofs.

Principles

A number of fundamental terms and processes are explained below to help the reader understand the subject of moisture control.

Hygrothermics

The term hygrothermics describes the interaction between temperature and moisture. As the temperature has a decisive influence on the vapour pressure in damp building materials and the maximum possible moisture content of the air, it is – in addition to the relative humidity of the air – the governing factor with regard to vapour diffusion. Furthermore, the temperature has an effect on the capillarity of building materials because the viscosity of water increases significantly as the temperature drops, and below freezing point there is no more transport of liquids in larger capillaries. Conversely, moisture influences the temperature conditions and heat flows in a building component. Damp building materials exhibit a higher thermal conductivity than dry ones (see Fig. C 1.7, p. 52). This fact is critical for limiting the moisture content in insulating materials. In addition, there is a transfer of heat due to vapour diffusion with a change of phase, i.e. water evaporates upon absorbing energy on the warm side of the insulation, migrates by way of vapour diffusion to the cold side and condenses there again. In doing so, the energy required for the evaporation is regained, but in this case on the outside of the insulation, where it is quickly lost to the surroundings. In some circumstances, this phenomenon, which is also known as latent heat transfer, can reach a similar order of magnitude to that of heat transfer by conduction (see "Conduction", p. 50). Hydrothermics deals with the mutual influencing, or rather coupling, of the heat and moisture storage and the heat transfer processes in building components plus their description and analysis.

Air humidity and vapour pressure

Water vapour is a variable component in the composition of our atmosphere. The maximum possible amount of water vapour in the air is proportional to the saturation vapour pressure that becomes established over an open body of water. The saturation vapour pressure p_S of water increases exponentially with the temperature (Fig. C 2.2) and at 100 °C reaches the same value as the atmospheric pressure at sea level (1013 hPa). The water vapour partial pressure p_D describes the actual amount of water

C 2.1 Rainwater collecting on a roof surface
C 2.2 Relationship between water vapour saturation
       pressure and temperature
C 2.3 Relationship between water vapour concentration,
       temperature and relative humidity. The humidity of
       the air increases as the interior air cools down to the
       dew point. After that the air remains at 100 % relative
       humidity and the excess moisture condenses out
       (condensation).
C 2.4 Hygroscopicity curves for building materials plotted
       against relative humidity (sorption isotherms)
C 2.5 Schematic presentation of capillarity, the pheno-
       menon behind the transport of liquids in porous
       building materials
C 2.6 The climate factors to which a flat roof surface is
       subjected plus their effects on the transport of
       heat and moisture within a building component
       a during the day
       b at night

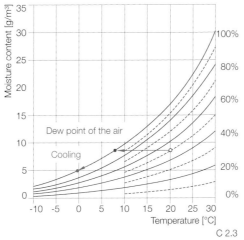

C 2.2

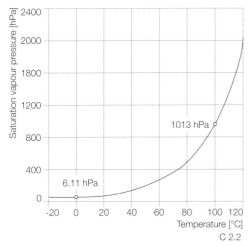

C 2.5

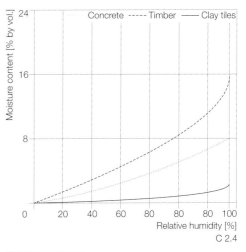

C 2.3

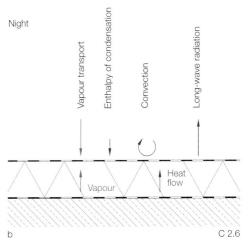

C 2.4

vapour in the air. The ratio of p_D to p_S is the relative humidity of the air:

$$\varphi = p_D / p_S$$

where:
φ relative humidity [-]
p_D water vapour partial pressure [Pa]
p_S water vapour saturation pressure [Pa]

In order to characterise the quantity of moisture in the air and the quantity of condensation that occurs upon cooling, it is better to specify the concentration of water vapour in the air instead of the partial pressure. As water vapour and the other air molecules behave almost like ideal gases at normal pressure, the concentration can be calculated directly from the partial vapour pressure with the help of the ideal gas equation:

$$c = p_D/(R_D \cdot T)$$

where:
c water vapour concentration [kg/m³]
R_D gas constant for water vapour,
    $R_D = 462$ J/kg·K
T absolute temperature [K]

The dew point of the interior air and the quantity of condensation per cubic metre of air upon further cooling is a consequence of the fact that the water vapour saturation pressure is dependent on the temperature (Fig. C 2.3).

Hygroscopicity
The hygroscopicity of building materials can exert a decisive influence on the humidity of the air in a room and hence the climate in that room. We distinguish between hygroscopic and non-hygroscopic building materials. A building material that is hygroscopic, and initially dry, absorbs water vapour from the air until it reaches its equilibrium moisture content for the respective ambient conditions; it does this by storing the water molecules in the pores of its microstructure, a process known as water vapour sorption. Storing a building material until it has reached its equilibrium moisture content (constant weight) in various humidity conditions enables a hygroscopicity graph to be drawn for that specific material (Fig. C 2.4); this is known as the sorption isotherm because it is normally determined for constant tempera-ture conditions. The upper limit of a sorption isotherm is found at a relative humidity of about 95 %. With a higher humidity or in the case of contact with water, building materials that exhibit capillarity also store the water in liquid form in their pores, and it is possible to achieve much higher moisture contents than those due purely to water vapour sorption. However, the majority of insulating materials exhibit neither hygro-scopicity nor capillarity, which means that an increased level of interstitial moisture occurs only in the event of condensation.

Moisture transport
The transport of moisture in flat roofs is nor-mally by way of water vapour diffusion. In forms of construction that are not absolutely airtight, the water vapour can also be transported with the air flow through a component (vapour con-vection). Once the moisture content has reached a sufficiently high level, capillary action begins in building materials that exhibit capillarity, which means that water is also transported in liquid form (Fig. C 2.5). Even if a much greater quantity of moisture can be transported by capillary action than by vapour diffusion, this still tends to play only a subsidiary role when assessing the moisture situation in a flat roof. The reasons for this are as follows:
• External waterproofing protects against water in liquid form
• Hydrophobic (i.e. water-repellent) insulating materials that do not exhibit any capillary action

The transport of moisture in liquid form gener-ally occurs only in those loadbearing elements whose moisture content is above a critical level. This is the case, for example, in new in situ concrete structures and aerated concrete affected by construction moisture, where capil-lary action can speed up the rate of drying.

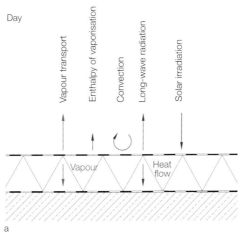

a

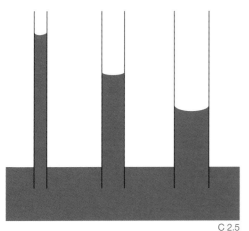

b

C 2.6

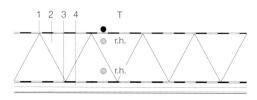

1 Flexible roof sheeting   Sensor positions:
2 Mineral wool insulation   T   Surface temperature
3 Vapour barrier   r.h.   Relative humidity
4 Timber decking   (1 top, 1bottom)

                                              C 2.7

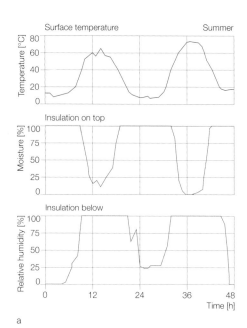

a

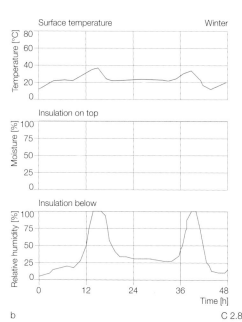

b   C 2.8

Non-steady-state hygrothermal processes in flat roofs

In the hydrothermal context, the roof to a building forms the interface between the interior and exterior climates. It protects the interior of the building against the natural weather conditions and attenuates the effects of the external temperature and moisture levels. In an unventilated flat roof, radiation, precipitation, wind and to a large extent the humidity of the external air, too, are already completely screened off by the roof waterproofing material, which means that it is only the temperature of the roof surface and the interior climate that are relevant for the conditions within the roof components. Whereas in the interior the temperature and moisture levels vary comparatively minimally over the course of a day or a year, the temperature of the roof surface can fluctuate by up to 70 K over the course of just 24 hours!

Solar irradiation represents the primary energy input during the day. During the warmer months of the year, primarily in the summer, the roof surface therefore heats up to a greater extent than the soffit of the roof construction. Part of this incoming energy is lost again into the atmosphere in the form of long-wave radiation emitted from the surface. There is also an exchange of energy at the roof surface by way of convection. And, with a damp roof surface, drying processes also take place, which remove further heat by way of evaporation. Within the construction itself, a temperature gradient from outside to inside is established, which gives rise to a vapour pressure gradient such that water vapour diffuses towards the inside. At night these processes are reversed; the long-wave radiation emissions from the surface and the convection continue, but, of course, there is no energy input from solar irradiation. On clear nights the energy loss due to long-wave emissions is so great that the temperature of the roof surface can often drop below that of the outside air. Once the surface reaches the dew point of the outside air, which is a regular occurrence with highly insulated buildings, condensation forms. The energy input due to condensation is relatively small. The enthalpy of condensation and enthalpy of vaporisation (Fig. C 2.6, p. 63) are indeed of equal magnitude, but in terms of energy it is the vaporisation (evaporation) that plays the more important part. This is because in addition to the condensation, precipitation, which is usually present in larger quantities, also evaporates. The cooling effect at the roof surface also reverses the direction of heat and water vapour flows within the construction (typical measurements: see Figs. C 2.7 and C 2.8).

Moisture always vaporises on the warmer side and then migrates by way of vapour diffusion to the colder side, where it condenses again. Vaporisation always requires a supply of heat. This heat escapes again when condensation forms on the cold side and so this process represents a form of heat transfer that takes place in addition to the conduction of heat through the layer of insulation (latent heat transfer). In insulating materials permeable to diffusion, studies have shown that this latent heat transfer can, when there is sufficient water present (approx. 2 % by vol.), lead to a three-fold increase in the heat flow compared with pure conduction in a dry insulating material. These not insignificant moisture-related heat losses or gains are, however, not dealt with in any current energy-saving standards or legislation. Nevertheless, such heat flows should generally be avoided by keeping the construction dry, even when using materials not susceptible to moisture.

Longer-term observations of the temperature and moisture conditions in flat roofs finished with a dark-coloured, radiation-absorbing waterproofing material show that in the summer the inward vapour diffusion flows are greater than the outward flows. This is why in summer moisture accumulates above the vapour barrier or the moisture in the room dries out when, instead of a vapour barrier, a vapour check with a lower diffusion resistance has been built into the roof. In the autumn, when the days are shorter, all the moisture migrates towards the outside again and collects beneath the roof covering. And in the winter, especially when there is snow on the roof, the vapour diffusion flow is not usually reversed, even with the sun shining; in other words, there is only an outward diffusion flow. This leads to the construction below the vapour barrier drying out completely and the moisture accumulating below the roof covering. Where instead of a vapour barrier a vapour check with a lower diffusion resistance has been used, the moisture distributed throughout the roof is joined by a small amount of moisture from the interior, which also migrates to the outside. In roofs with insulating materials less permeable to diffusion, e.g. rigid polystyrene foam or PU foam, the day-to-day fluctuations in the vapour diffusion flows are much less significant. Over the course of a year, however, there is still a transfer of moisture. Roofs with diffusion-permeable but highly hygroscopic insulating materials, e.g. cellulose or wood fibres, exhibit a similar behaviour to those with mineral fibre insulating materials. However, in these insulating materials, the moisture levels change at a slower rate and are less significant (see "Insulating materials", pp. 93–95).

Moisture loads

Moisture control is frequently seen exclusively as protection against precipitation (protection against rain) and protection against vapour diffusion from the interior (protection against condensation). However, the following moisture loads should not be ignored either:

- Condensation caused by infiltration of interior air in the winter
- Condensation caused by infiltration of exterior air due to overcooling of the roof covering

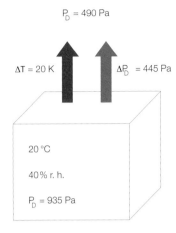

$P_D = 490$ Pa

$\Delta T = 20$ K $\Delta P_D = 445$ Pa

20 °C

40 % r. h.

$P_D = 935$ Pa

a

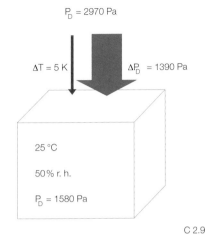

$P_D = 2970$ Pa

$\Delta T = 5$ K $\Delta P_D = 1390$ Pa

25 °C

50 % r. h.

$P_D = 1580$ Pa

b   C 2.9

C 2.7   Sensor positions for investigating relative humidity
C 2.8   Surface temperatures and relative humidities
        measured on a flat roof with a room temperature
        of approx. 20 °C during the measuring period
        a day in June
        b day in January
C 2.9   Comparison of the typical temperature and vapour
        pressure differences between interior and exterior
        air in Central Europe (a) and the tropics (b). Despite
        the much smaller air temperature difference in the
        latter, the vapour pressure difference in hot, humid
        climate zones is more than three times that in a
        Central European winter.

(ventilated forms of construction only)
• Construction and sorption moisture

On a flat roof, protection against the rain is provided by a watertight roof covering (waterproofing). The prerequisite here is absolute watertightness irrespective of the level of precipitation. However, the roof covering is generally simultaneously relatively vapour-tight, which means that only very small quantities of moisture can dry out through it to the outside. In order to protect the roof against interstitial condensation caused by moisture diffusing into the roof from the interior, an essentially vapour-tight vapour barrier is frequently installed on the warm side. We assume here that these two impermeable membranes on either side of the roof construction (waterproofing and vapour barrier) can protect against both rain and condensation.

Over recent years, the many cases of moisture damage in flat roofs, primarily on timber buildings, have made designers even more aware of the problems of using the moisture assessment method according to DIN 4108-3 [1]. It has been shown that moisture from the interior can infiltrate the roof construction by convection as well, and not only by way of vapour diffusion. In these circumstances, a vapour barrier could even be a disadvantage because in summer it hinders any moisture that may be present from drying-out. Practical experience shows that it is almost impossible to prevent a certain moisture infiltration due to air convection, especially in lightweight roofs (timber or metal sandwich forms) [2]. This moisture should therefore be properly considered in the moisture assessment of the roof construction, e.g. by considering leakage rates as proposed by the American moisture control standard ANSI/ASHRAE 160. The same applies to construction moisture, which is primarily a problem with concrete structures, but can also be troublesome in timber buildings. And there is also the possibility that with a highly insulated building, overcooling of the roof surface at night as a result of long-wave radiation emissions can lead to moisture from the outside air condensing on the roof. In the case of ventilated flat roof forms with little thermal mass above the air space, this phenom-

enon can lead to a net gain in moisture in winter. This fact must be properly considered during the planning, e.g. by analysing the moisture behaviour of the design with the help of hygrothermal simulation software. In order to prevent energy losses or damage, all potential moisture loads or wetting mechanisms must be properly considered when assessing the moisture situation of a roof. As both the non-steady-state moisture loads and also the drying-out potential are heavily dependent on the respective boundary conditions, moisture control design requires accurate knowledge of the anticipated interior and exterior climate conditions.

Moisture loads in other climate zones
An accurate analysis of the exterior climate and the resulting moisture loads is certainly necessary when designing a flat roof for a building in another climate zone. If the interior needs more cooling than heating, the direction of vapour diffusion is the opposite of that in Central Europe. The vapour pressure difference and hence the potential vapour diffusion flow are also reversed (Fig. C 2.9). Despite the smaller temperature difference between inside and outside, the vapour pressure gradient in the tropics can be more than three times the comparable value in Central Europe. A vapour barrier on the inside would be completely inappropriate in this situation. German standards (e.g. DIN 4108-3) or trade guidelines are therefore unsuitable for designing moisture control measures for flat roofs in other climate zones. Hygrothermal simulation software can help to assess the effects of the true climate conditions in other countries.

Protection against condensation as a result of water vapour diffusion

A temperature-induced vapour pressure gradient from inside to outside prevails in winter in Europe's cold and temperate climate zones. The consequence of this is that during the heating period the water vapour diffuses mostly from inside to outside, through the building component. If the amount of water vapour diffusing into the component is greater than that diffusing out of it, the difference remains within the

component in the form of either interstitial condensation or sorption moisture (see p. 63). In summer there is at best only a small temperature, and hence also vapour pressure, gradient across the component. However, if there is moisture within the component, a vapour pressure gradient towards the inside and also the outside generally ensues because the saturation vapour pressure prevailing in the component is higher than the vapour pressure outside or in the interior. This gradient enables any moisture to dry out. It is necessary to draw up a balance of the seasonal vapour diffusion flows in order to assess the long-term moisture conditions due to vapour diffusion.

Vapour diffusion flows and resistances
The water vapour flowing as a result of vapour diffusion through a building component layer of thickness d is proportional to the water vapour diffusion coefficient δ of a building material and the in situ partial pressure difference Δp:

$$g = - \delta \, (\Delta p / d)$$
$$= - (\delta_L / \mu) \cdot (\Delta p / d)$$
$$= - (\delta_L / s_d) \cdot \Delta p$$
$$= - \Delta p / Z$$

g   water vapour diffusion flow [kg/m²s]
δ   water vapour diffusion coefficient of building component layer [kg/m·s·Pa]
δ_L water vapour diffusion coefficient of a stationary layer of air [kg/m·s·Pa]
Δp water vapour partial pressure difference across building component layer [Pa]
d   thickness of building component layer [m]
μ   water vapour diffusion resistance factor (μ value) of building component layer [-]
s_d diffusion-equivalent air layer thickness (s_d value) of building component layer [m]
Z   water vapour diffusion resistance of building component layer [m²·s·Pa/kg]

Building component layer	Vapour permeability s_d [m]
Diffusion-permeable	$s_d \leq 0.5$
Diffusion-resistant	$0.5 < s_d < 1500$
Diffusion-tight	$s_d \geq 1500$

C 2.10

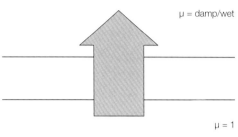

μ = damp/wet

μ = 1

a

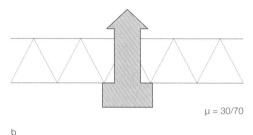

μ = 30/70

b

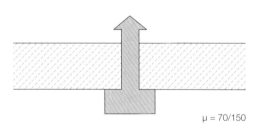

μ = 70/150

c

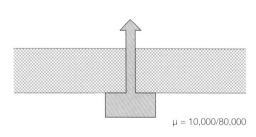

μ = 10,000/80,000

d

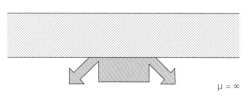

μ = ∞

e

C 2.11

The μ and s_d values are now acknowledged in Europe as the figures that characterise the vapour diffusion properties of a material or building component layer:

$$\mu = \frac{\delta_L}{\delta}$$

The s_d value specifies the thickness that a stationary layer of air must be in order to have the same water vapour diffusion resistance as the building layer component under consideration:

$$s_d = \mu \cdot d$$

The water vapour diffusion resistance Z is calculated as follows:

$$Z = \frac{s_d}{\delta_L}$$

The importance of vapour diffusion in a building component can be estimated by means of the vapour diffusion resistances of the materials used (Fig. C 2.11). Building materials with μ values < 10 can be regarded as highly vapour-permeable. However, for a more accurate specification we need to know their thickness. This is why s_d values, which are not dependent on thickness, are used in DIN 4108-3 to classify the vapour permeability of building component layers (Fig. 2.10):

- Building component layers with $s_d \leq 0.5$ m are regarded as permeable to diffusion because in practice they present no significant barrier to vapour diffusion. Such layers include, for example, the diffusion-permeable roofing felts used below the roof coverings on pitched roofs or the separating layer for draining rainwater laid above the thermal insulation of an inverted roof. As yet, no diffusion-permeable flexible sheeting has been produced for waterproofing flat roofs – despite the fact that a number of manufacturers have erroneously described their products as such.
- Building component layers with s_d values between 0.5 m and 1500 m are regarded as diffusion-resistant. And resistance to diffusion is essential if a building component layer is to be classed as a vapour check. As this range is rather wide, it is divided into two segments. Modern vapour checks are frequently characterised by a moderate resistance to diffusion ($0.5\text{ m} < s_d \leq 5\text{ m}$). The advantage of this is that the construction is protected against excessive condensation in winter without hindering the summertime drying-out potential excessively. Flexible sheeting for waterproofing flat roofs generally exhibits a much greater resistance to diffusion, which means that drying out towards the outside is unlikely with such materials.
- Building component layers with $s_d \geq 1500$ m are regarded as diffusion-tight. Only diffusion-tight layers may be used as vapour barriers. Examples of these are metal foil, glass and cellular glass. The use of a vapour barrier

would appear to be a reliable solution because it prevents diffusion of water vapour. However, it frequently does not prevent the infiltration of moisture in other ways, e.g. by air convection. In addition, a true vapour barrier makes it impossible for construction moisture to dry out.

Controlling vapour diffusion

In order to prevent interstitial condensation in building components, under European climate conditions it is advantageous when the diffusion permeability of the roof component layers increases from inside to outside. The situation is exactly the opposite in hot, humid climates. Flat roofs have a diffusion-resistant waterproofing on the outside and so in the majority of cases a certain amount of condensation cannot be avoided in winter. This means it is even more important to ensure that the construction can dry out again in summer. Protection against condensation therefore does not mean excluding condensation at all costs, but rather allowing only that amount of condensation that the construction can accommodate without damage and without impairing the thermal performance, and which can dry out again in summer.

Condensation protection to DIN 4108

The standard DIN 4108-3, which deals with climate-related moisture control, forms part of building legislation in Germany. In the introduction to this standard it points out that the effects of driving rain and condensation from the interior air must be limited under wintertime conditions. The idea behind this is to prevent moisture-related damage, e.g. mould growth, corrosion or an unacceptable reduction in the thermal performance. The requirements laid out in the standard refer exclusively to building components after they have released any construction moisture. However, it is emphasized that, as a building dries out, situations can occur that may need to be specially considered and call for additional measures necessary.

DIN 4108-3 states that a separate condensation analysis is unnecessary for flat roof constructions with a highly diffusion-resistant ($s_d \geq 100$ m) vapour barrier beneath the thermal insulation (Fig. C 2.12). However, for this exemption to apply, the materials below the vapour barrier may not account for more than 20 % of the total thermal resistance of the roof construction. Owing to numerous cases of damage caused by trapped moisture in lightweight forms of construction [3], however, this approach is only recommended in conjunction with an absolutely airtight solid form of construction. According to the standard, a separate condensation analysis is also unnecessary for roofs of aerated concrete without a diffusion-resistant layer on the soffit and without thermal insulation, also inverted roofs with a diffusion-permeable ballast (e.g. coarse gravel). All other types of flat roof construction will require a vapour diffusion calculation using the Glaser method or hygrothermal simulation software (see pp. 70–73).

Protection against condensation due to water vapour convection

Water vapour convection, i.e. the infiltration of moist interior air into a building component, has a similar effect to the vapour diffusion which has been known for a long time; but in contrast to vapour diffusion, the significance of water vapour convection was for a long time underestimated. The consequences were leaks and damage, especially in lightweight forms of construction that complied with the diffusion requirements of DIN 4108-3 but obviously not the airtightness requirements of DIN 4108-7.

Air flows through building components arise due to air pressure differences between interior and exterior. However, these only become a problem when they transport so much moisture with them that condensation occurs on the cold side of the component. Therefore, in Central Europe air flows are only a problem when they are from inside to outside and at the same time it is so cold outside that the temperature in the component adjacent to the air flow drops below the dew point of the interior air. Air that flows through the building component from the outside, and heats up as it does so, promotes drying out, i.e. an overpressure in the building in the winter is unfavourable from the moisture control viewpoint. In hot, humid climate zones, a similar situation can occur in the opposite direction when moist air from outside flows into an air-conditioned building. In that situation an overpressure within the building tends to be advantageous.

Air pressure differences between interior and exterior

The air pressure difference across a building component is the driving force behind an air flow through the component. There are various reasons for such pressure differences:
- Thermal currents
- Wind
- Total pressure differences due to mechanical ventilation systems, extractor fans and open fireplaces

Fans in wet rooms, extractor fans or open fireplaces generally lead to an underpressure in the building and therefore do not represent a problem in the climate of Central Europe. Centralised ventilation systems should be set to keep the pressure neutral, except for clean rooms of course, which must be treated separately. Wind can cause severe suction pressures on flat roofs, although these generally fluctuate considerably.

The most important component is thermal currents, which in winter act permanently on the building envelope. Hot air tends to rise in an enclosed space because it displaces the colder and hence heavier air entering that space. Towards the top of the building envelope this creates an overpressure which results in a flow from inside to outside. This overpressure is balanced by an underpressure at the bottom of the building which causes a flow from outside

to inside. If the leaks in the envelope are evenly distributed, the plane of neutral pressure is located in the centre of the building, with respect to the area of the envelope. But with an uneven distribution, the plane of neutral pressure is displaced in the direction of the bigger leaks (Fig. C 2.13).

So on the flat roof of a heated building an overpressure always prevails in winter. The magnitude of this pressure can be calculated as follows:

$$\Delta P = \rho \cdot \frac{T_e - T_i}{T_i} \cdot g \cdot \frac{h}{2}$$

where:
ΔP pressure difference between inside and outside [Pa]
ρ density of outside air $\rho = 1.3$ kg/m³
T_e temperature of exterior air [K]
T_i temperature of interior air [K]
g gravitational constant $g = 9.81$ m/s²
h height of coherent air space in building [m]

The overpressure increases in proportion to the height of the coherent air space and the temperature difference between inside and outside. Heated high-rise buildings tend to be more at risk of convection-related problems than low-rise ones. The pressure difference on the flat roof of a two-storey detached house is, for example, about 2.5 Pa in winter. Even with an airtight design to DIN 4108-7 (air change rate $n_{50} = 3$ h⁻¹), the outcome would be a flow through the building envelope amounting to about 150 l/m²h [4]. Only that part of the air flow that passes directly through the component has a negative effect, not that part that flows through joints and gaps, e.g. at windows and doors.

Airtight design of flat roofs

The airtight design of building components is an essential requirement for functioning moisture control. Protecting against air convection is not the same as protecting against vapour diffusion. A loadbearing structure of steel trapezoidal profile sheets is, for example, relatively vapourtight but never airtight at joints or where fixings penetrate the material. Such roofs must therefore be provided with an additional airtight layer, e.g. in the form of bonded plastic sheeting over the entire area.

As the majority of flat roofs require a vapour barrier, it would seem sensible to combine protection against convection and vapour diffusion in one layer attached to the warm side of the construction. Tear-resistant foils and sheets, whose overlapping joints can be readily and permanently glued or welded together, are ideal for this (Fig. C 2.14). Special attention must be paid to ensuring airtightness at all penetrations (Fig. C 2.15). Careful design and careful workmanship are also essential at the junctions between the airtight layer of the flat roof and the airtight layers of adjacent compo-

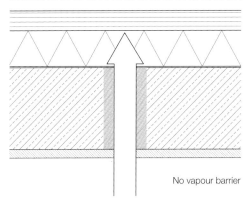

No vapour barrier

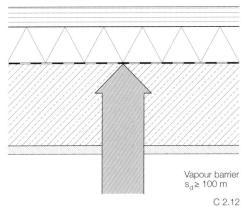

Vapour barrier
$s_d \geq 100$ m

C 2.12

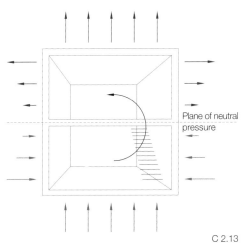

Plane of neutral pressure

C 2.13

C 2.10   Classification of vapour permeability of building component layers to DIN 4108-3
C 2.11   Orders of magnitude of water vapour diffusion resistance (μ value) for various flat roof materials compared with stationary air
     a   Air, mineral insulating materials
     b   Polystyrene
     c   Concrete
     d   Bitumen
     e   Metal, cellular glass
C 2.12   Protection against condensation for flat roofs to DIN 4108-3. Installing a vapour barrier with an sd value of min. 100 m means that it is not necessary to calculate the risk of interstitial condensation for the construction.
C 2.13   Pressure differences over the building envelope as a result of thermal currents for a building whose coherent air space extends over two storeys because of a staircase opening.

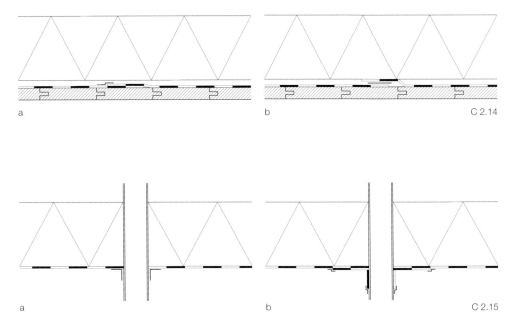

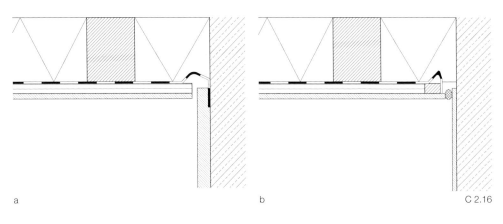

C 2.14

C 2.15

C 2.16

convection-related drying reserve is based on American research into practical lightweight forms of construction and has been worked out for Central European climate conditions [6]. In the meantime, the figure of 250 g/m² has become widely accepted, and is included in, for example, the timber flat roof publications of "Informations-dienst Holz" and in the draft version of the wood preservation standard DIN 68800-2 [7].

The recommendation for the moisture control design of lightweight and prefabricated flat roofs is therefore to ensure an adequate summertime drying-out potential. Providing the aforementioned drying reserve when using the Glaser vapour diffusion calculation to DIN 4108-3 is one way of doing this. A hygrothermal simulation, in which the moisture infiltration due to vapour convection is considered appropriately, can be used when a more accurate analysis is necessary [8].

Protecting against rain

On flat roofs, protection against the rain is not guaranteed by draining off the water, as is the case with pitched roofs. As a build-up of water on the roof surface is practically inevitable (see "Drainage", pp. 113–116), a watertight roof finish is vital. This is generally achieved by providing a layer of waterproof flexible sheets on the roof, which have to be either glued or welded together. One alternative is to use materials that are applied in liquid form, which are easier to use where there are complicated contours and high numbers of penetrations in particular. The design of the waterproof roof finish also depends on the thermal and mechanical loads due to the particular type of construction and use of the roof (see "Waterproofing materials according to the German Flat Roof Directive", pp. 86–87). Information on this can also be found in the draft version of DIN 18531-3.

The majority of waterproofing materials are not only watertight, but at the same time highly diffusion-resistant, i.e. no significant drying-out nor any moistening of the construction below can take place through these materials. Roof waterproofing materials with a somewhat lower vapour diffusion resistance, e.g. some synthetic materials, do permit a certain amount of drying out if there is no further roof covering. Depending on the colour of the waterproofing, some 300–400 g/m² of drying-out moisture per year can therefore pass through a roof covering with an s_d value of 20 m. If the material's s_d value is only 10 m, then up to 1000 g/m² would even be possible.

Such moderately diffusion-resistant materials are less suitable on permanently wet surfaces, e.g. beneath planting or other roof finishes that absorb water, because in such instances the moisture can also diffuse in the other direction, i.e. into the roof. And they are totally unsuitable as waterproofing to inverted roofs, as a number of cases of damage with water-filled blisters between the concrete slab and the waterproof-

nents, e.g. walls or ring beams. At these points simple gluing is usually insufficient. The roof-wall junction detail should include clamping strips, sealing compounds or compressible strips, or the airtight layer should be tucked behind render or plaster (Fig. C 2.16).

Considering air convection in airtight designs

Even state-of-the-art airtight flat roof designs are not completely free from faults. In practice, a small amount of moisture infiltration due to vapour convection can never be avoided completely. As we do not know exactly just how much moisture really does infiltrate building components due to vapour convection, providing a "drying reserve" has become established as a moisture control design principle for lightweight flat roofs. This means that in terms of vapour diffusion a roof should be constructed only as vapour-tight as necessary, but at the same time as vapour-permeable as possible. The aim of this is to allow any moisture to dry out which has infiltrated the building component not by vapour diffusion but by some other means. However, this approach only permits a

qualitative description of the moisture behaviour, i.e. it is not possible to specify the limits within which a certain type of construction will actually function.

The amount of moisture that infiltrates the construction through gaps in the vapour barrier, or rather airtight membrane, by means of convection is a multi-dimensional effect that cannot be handled directly by a one-dimensional calculation. As the exact form of leaks and flow paths is unknown, it would appear sensible to employ an approach that only models the condensation occurring, as a source of moisture within the construction, instead of the flow itself.

A first attempt to quantify moisture infiltration due to vapour convection was the inclusion of a vapour convection condensation quantity of 250 g/m² in the vapour diffusion calculation according to DIN 4108-3, which was proposed in 1999 for the moisture assessment of timber structures [5]. For that the building component "to pass" the moisture analysis, the vaporisation quantity calculated must exceed the diffusion-related quantity of condensation by at least 250 g/m² (the drying reserve). The size of the ing material have shown (Fig. C 2.17) [9]. On inverted roofs the water migrates through the waterproofing – from the film of moisture always present beneath the layer of insulation – by way of osmosis (Fig. C 2.18). The transport mechanism is comparable with the vapour diffusion in the partial pressure gradient, with the osmotic pressure responsible for the vapour transport being caused by salts in the concrete. The water subsequently collects on the saturated concrete slab and breaks down the bond between the concrete and the waterproofing, which leads to the formation of water-filled blisters. The use of exclusively highly diffusion-resistant or diffusion-tight membrane systems is therefore recommended for waterproofing inverted roofs.

Steady-state or non-steady-state moisture assessments

As with thermal performance calculations, steady-state assessment methods are preferred for hygrometric design because of their ease of use. They are characterised by the fact that they ignore storage effects and hence can only approximate the true building component behaviour when the duration of the boundary conditions is much greater than the dynamic reaction time of the component. For example, the thermal reaction time of a component, i.e. the time it takes for the temperature level in the component to adjust essentially to the ambient temperature level, is in the order of magnitude of hours to a few days, depending on the mass of the component. By contrast, typical cold periods in the winter – during which an almost constant temperature gradient from inside to outside prevails – are frequently several weeks long. So the duration of the boundary conditions is much longer than the time required for a steady-state temperature level to become established in the building component. The thermal mass effects of the external components therefore do not play a significant role in the mean heat transfer, and a steady-state assessment is adequate for determining the transmission heat losses. But the component reacts totally differently in the case of a summertime heatwave lasting only a few days. The heat that remains in a component following exposure to intense solar radiation is released into the interior after a certain delay, even if the outside temperature in the evening is much cooler. Steady-state methods cannot model this behaviour and are therefore not entirely suitable for analysing the summertime thermal performance.

Compared to thermal energy, moisture is far more sluggish in its behaviour. Whereas it generally takes only hours for a component to cool down fully, it can be weeks before the wintertime condensation has been fully released again. Construction moisture can even take several years to dry out! The applicability of steady-state methods for assessing moisture

must therefore be examined carefully for every application. However, realistic modelling of the moisture behaviour of forms of construction generally calls for the use of non-steady-state methods of calculation. This is particularly true for roofs because the cyclic heating effect of solar irradiation leads to severe fluctuations in temperature and moisture behaviour.

Moisture control design using steady-state vapour diffusion calculations

The steady-state Glaser method for calculating the vapour diffusion processes in components, which is described in DIN 4108-3, uses much simplified steady-state block boundary conditions. In doing so, the true climatic conditions are simplified to such an extent that two blocks with constant boundary conditions ensue, which model the effects of a cold winter and a moderately hot summer. At the same time, all thermic and hygrometric phenomena plus the moisture transport due to air convection and capillarity are neglected. According to normative Annex A of the standard, this method cannot be used on green roofs, nor for calculating the natural drying-out behaviour, e.g. the release of construction moisture.

Calculation principle

The Glaser method of calculation was introduced as a standard method in 1981 and designed in such a way that the results can be obtained graphically. Here, the component construction is first specified from inside to outside. The thermal conductivity, vapour diffusion resistance (μ value) and thickness has to be specified for each layer. Afterwards, the temperature gradient through the component cross-section is determined using the thermal resistances of the individual layers. The gradient of the saturation vapour pressure in the component can then be determined from this temperature gradient. Instead of drawing the temperature gradient in relation to the component thickness, it is entered via the s_d value of the component. The advantage of this is that the gradient of the water vapour partial pressure that has to be drawn subsequently is a straight line from its value on the room side to that on the outside. However, if the straight line representing the water vapour partial pressure intersects the saturation vapour pressure, the former must be corrected such that the straight line only touches the saturation vapour pressure line from below. This gives rise to a kink in the gradient which forms the starting point for two straight lines at different angles in the direction of the outside and the inside, and thus a difference between the incoming and outgoing vapour diffusion flows. This difference remains as interstitial condensation in the component, or rather dries out if condensation is already present and the boundary conditions permit this (Figs. C 2.19 to C 2.21).

C 2.17

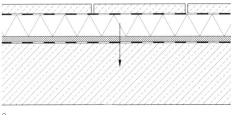

a

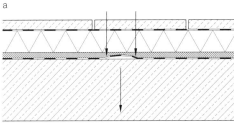

b

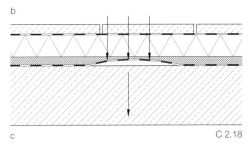

c   C 2.18

C 2.14   Sketches showing the principle of overlapping and bonding sheet materials in order to create an airtight membrane with the help of
    a   single-sided adhesive tape
    b   double-sided adhesive tape
C 2.15   Details of the penetration of an airtight membrane sealed with the help of
    a   self-adhesive tape
    b   a collar, with all joints again sealed with self-adhesive tape
C 2.16   Detail showing the junction between the airtight membrane in a roof and a wall, sealed
    a   by tucking the membrane behind the plaster
    b   with the help of a clamping strip plus a sealing compound or compressible preformed strip
C 2.17   Water-filled blisters between concrete slab and roof waterproofing on an inverted roof
C 2.18   Formation of water-filled blisters between concrete and waterproofing
    a   Moisture migrates from the film of water beneath the insulation to the underside of the waterproofing (osmosis)
    b   Saturation of the concrete surface
    c   Formation of blister

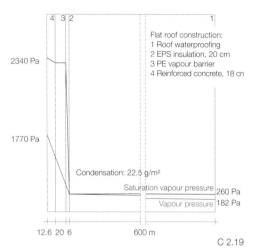

Flat roof construction:
1 Roof waterproofing
2 EPS insulation, 20 cm
3 PE vapour barrier
4 Reinforced concrete, 18 cn

2340 Pa

1770 Pa

Condensation: 22.5 g/m²

Saturation vapour pressure 260 Pa
Vapour pressure 182 Pa

12.6 20 6   600 m

C 2.19

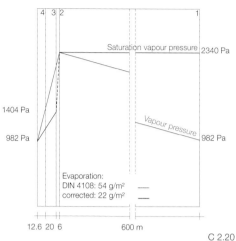

Saturation vapour pressure 2340 Pa

1404 Pa

982 Pa

Vapour pressure 982 Pa

Evaporation:
DIN 4108: 54 g/m² ——
corrected: 22 g/m² ——

12.6 20 6   600 m

C 2.20

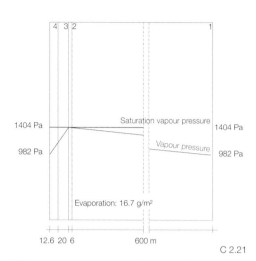

1404 Pa   Saturation vapour pressure 1404 Pa

982 Pa   Vapour pressure 982 Pa

Evaporation: 16.7 g/m²

12.6 20 6   600 m

C 2.21

Boundary conditions

The boundary conditions given in DIN 4108-3 for non-air-conditioned residential and office buildings are a 60-day condensing and a 90-day evaporation phase. During the condensing phase in the winter (exterior climate: -10° C and 80 % r.h.; interior climate: 20 °C and 50 % r.h.), the quantity of interstitial condensation accumulates in building components. This quantity of condensation must dry out again during the evaporation phase in the summer, with the same conditions prevailing indoors and outdoors (12 °C and 70 % r.h.). According to a special provision for roofs, 20 °C may be assumed for the external surface temperature because of the solar irradiation. But a correction is not necessary if during the graphical determination of the steady-state vapour diffusion flows local water vapour partial pressures occur that are greater than the saturation vapour pressure. This is actually impossible physically and owing to the steeper partial pressure gradient leads to a higher theoretical vaporisation quantity. A corresponding correction is therefore expected to be included in the draft of the new edition of DIN 4108-3.

Assessment criteria

Adequate protection against condensation for a given form of construction is deemed to be verified when the following assessment criteria are satisfied:
- The quantity of water drying out during the evaporation phase is at least equal to the quantity of condensation.
- The quantity of condensation is max. 1000 g/m², or max. 500 g/m² in the case of layers that do not absorb water by way of capillary action..
- In the case of timber the maximum increase in the moisture content is < 5 % by wt., in wood-based products < 3 % by wt.

The advantages of this condensation protection verification, which has been in use and included in the standard since 1981, are obvious. The calculation is very simple and if necessary can be carried out graphically, without any further aids. The criteria for assessing the result are so unambiguous that a plausibility check by a specialist is superfluous.

Limitations in the use of the Glaser method for flat roofs

Considering the surface temperature boundary condition of 20 °C, the condensation protection assessment for flat roofs according to DIN 4108-3 appears to be very optimistic when compared with the assessment for walls. This is especially relevant when construction moisture is present in the components over a longer period of time, which increases the occurrence of condensation in winter. The same is true for flat roofs with a light-coloured waterproofing material which reflects a large proportion of the solar irradiation. Studies have revealed that the temperature of white roof surfaces is on average no higher than the exterior air temperature [10]. So in this case the solar irradiation bonus when calculating the evaporation phase is entirely unjustified. The condensation protection assessment for flat roofs with light-coloured finishes should therefore be carried out exclusively with the standard boundary conditions, which also apply to external walls. In cases of doubt, and this also applies to roofs where construction moisture dries out only slowly, the moisture assessment should be carried out with the help of a hygrothermal simulation according to DIN EN 15026.

Moisture assessment using non-steady-state methods of calculation

Steady-state methods of calculation have been used in practice for many decades because they are easy to use. They are certainly suitable for assessing the seasonal vapour diffusion processes in building components. When using these methods, the materials used should be neither significantly hygroscopic nor exhibit significant capillarity, and their hygrothermal properties should not vary, i.e. no heat conduction dependent on moisture and no moisture-related change in the vapour permeability. If the aforementioned conditions are not satisfied or the daily changes to the boundary conditions are significant for the moisture behaviour of a particular type of construction, then the situation is different: the frequently pronounced temperature fluctuations at the surface of a flat roof and the consequences for vapour diffusion cannot be analysed using steady-state methods of calculation. The exponential increase in the water vapour saturation pressure as the temperature rises repeatedly results in errors in the determination of the mean value required for steady-state calculations. Furthermore, the moisture in roofs with diffusion-permeable insulating materials reacts very rapidly to temperature changes.

Hygrothermal simulation tools

Only by using appropriate numerical simulation tools is it possible to model and analyse the non-steady-state hygrothermal processes in flat roofs accurately. Non-steady-state methods for calculating heat and moisture transport in building components have been available for a number of years. And the growing number of specialist publications shows that these methods are being used more and more. Owing to the limitations of the steady-state vapour diffusion method according to Glaser, DIN 4108-3 refers to these non-steady-state hygrothermal calculation methods for assessing green roofs or components with construction moisture or rain infiltration. And since 2007 there has been a Euronorm (DIN EN 15026) covering these models and their application. This standard forms the basis for the non-steady-state moisture assessment of forms of construction by considering the temperature and moisture conditions that become established within a building component under natural climatic conditions. The aim of DIN EN 15026 is to replace the outdated Glaser method. It therefore refers exclusively to one-dimensional simulations because their results are now regarded as reliable and the corresponding numerical models are relatively easy to handle. In addition, for quality control purposes, the standard contains a validation example in normative Annex A for which an analytical solution is given. Information is provided regarding the range in which the numerical results must lie so that the simulation method satisfies the requirements of the standard.

To comply with the standard, a numerical model must include the following storage and transport phenomena:
- Thermal mass of the dry building material and the moisture it contains
- Heat transfer due to moisture-related conduction
- Heat transfer due to vapour diffusion (latent heat effects)
- Hygroscopicity due to water vapour sorption and capillary forces
- Moisture transport due to water vapour diffusion
- Transport of liquid water via surface diffusion and capillary action

Boundary conditions for the calculation
The following climate parameters must be included in the calculation:
- Interior and exterior temperatures
- Interior and exterior air humidities
- Short- and long-wave radiation (solar and thermal radiation)
- Precipitation and wind

The standard distinguishes between three different qualities for both the interior and exterior boundary conditions. For the exterior climate conditions these are:
- 1st choice: measurements from a period of at least 10 years.
- 2nd choice: the reference year that causes

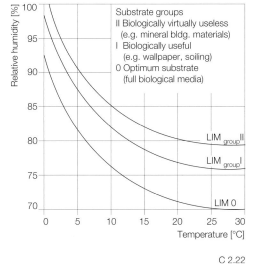

C 2.22

C 2.19 Water vapour partial pressure and saturation vapour pressure gradients for a flat roof during a 60-day condensing phase, calculated on the basis of temperature gradients
C 2.20 Water vapour partial pressure and saturation vapour pressure gradients for a flat roof during a 90-day evaporation phase. The uncorrected vapour pressure gradient intersects the saturation vapour pressure because of the higher surface temperature of 20 °C. This physically unfounded procedure leads to an increase in the theoretical drying-out of the construction.
C 2.21 Water vapour partial pressure and saturation vapour pressure gradients for a flat roof during a 90-day standard evaporation phase without the bonus of a higher surface temperature of 20 °C.
C 2.22 Substrate-related limits for the germination of mould spores in relation to temperature and air humidity. Besides the hygrothermal conditions, the quality of the substrate also plays a major role. The germination of spores, the prerequisite

for mould growth, does not take place beyond the respective limit.
C 2.23 Mean daily values for interior air temperature (a) and humidity (b) in dwellings and offices in relation to the mean daily temperature of the outside air according to DIN EN 15026 Annex C
 A Rooms with normal occupation
 B Rooms with high occupation
C 2.24 Distributions of temperature, relative humidity and moisture content and their respective fluctuations (light-coloured areas) in a concrete flat roof with an inadequate vapour barrier (s_d = 20 m) in the seventh year after its construction. The construction moisture that has escaped from the in situ concrete, through the vapour barrier and into the mineral wool insulation migrates in an annual cycle back and forth between the roof waterproofing and the vapour barrier according to the temperature conditions while the concrete slowly dries out towards the inside.

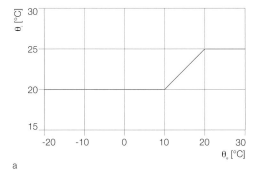

a

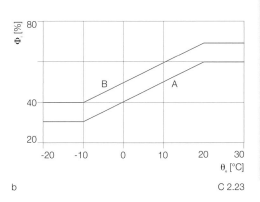

b   C 2.23

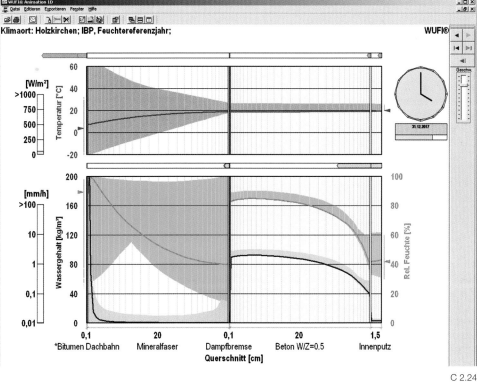

C 2.24

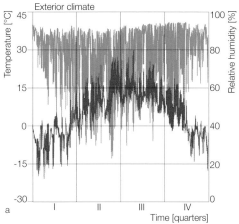

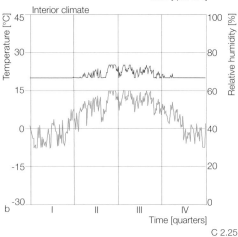

C 2.25

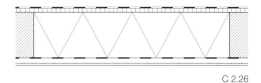

C 2.26

C 2.25 Diagram showing outside air temperature and humidity
a for the moisture reference years on the north side of the Alps
b for interior climate conditions to DIN EN 15026 with normal occupancy (see Fig. C 2.23)

C 2.26 Construction of a timber flat roof: 22 mm OSB, 240 mm mineral fibre insulation, 240 mm load-bearing joists, vapour barrier, soffit lining

C 2.27 Calculated progression of the total moisture content in the roof shown in Fig. C 2.26 in relation to the vapour diffusion properties of the vapour barrier used. The initial moisture content corresponds to the moisture content of the materials installed in the air-dry condition (equilibrium moisture content at 80 % r.h.). The moisture infiltration due to vapour convection based on the draft of the wood preservation standard DIN 68800-3 has been considered in the hygrothermal simulation.

C 2.28 Calculated progression of the moisture content of the timber in the OSB decking to the flat roof shown in Fig. C 2.26. The short-wave absorptance a forms the basis for the calculation.
a in relation to the vapour diffusion properties of the vapour barrier used
b with a smart vapour barrier (PA film) and a reflective (white) waterproofing material in relation to the exterior climate at the respective location

the most serious problems, probably once in 10 years; it can be a particular cold or hot year depending on the situation (wintertime or summertime condensation).
• 3rd choice: mean weather data record with which an extreme year is simulated by way of an annual temperature shift of ±2 K.

For the interior climate conditions these are:
• 1st choice: measurements for a similar building under the same climatic conditions or target values given by the air-conditioning plant.
• 2nd choice: results from the hygrothermal building simulation.
• 3rd choice: determination of the interior air conditions based on specified moisture production and air change rates.

If none of these choices is practicable, the interior climate conditions can be derived according to Annex C of DIN EN 15026 (alternatively with the help of the classification in DIN EN ISO 13788), although this is not to be recommended because these values are clearly excessive for the moisture.

Presentation of non-steady-state calculation results
The hourly changes in the temperature and moisture zones at the boundaries of the building component plus the heat and moisture flows via these zones are output as the results of the calculations. These results mean it is possible to determine both the long-term courses of temperature, relative humidity and the moisture content at various positions in the building component, also their local distribution at certain times. A film-like sequence of non-steady-state moisture and temperature profiles is an appropriate and descriptive way of presenting the results (Fig. C 2.24, p. 71).

Assessment of calculation results
The assessment of the results of hygrothermal simulation calculations focuses on the moisture contents and the temperature and humidity conditions in the individual building component layers. In a similar way to the evaluation according to the Glaser method, it relies on excluding the long-term accumulation of moisture in the construction. Following an analysis of the course of the total moisture content, the hygrothermal conditions in the individual material layers and at the surfaces and the interfaces must be examined. There is a risk of mould growth on the inner surfaces and in the region of air spaces in the warmer areas of a component if the corresponding temperature and moisture limits are exceeded (Fig. C 2.22, p. 71).
When assessing the moisture contents in the individual layers it is advisable to compare them with the maximum moisture contents of the respective building materials where these are available. For timber or wood-based products a limit of 20 % by wt. is usually assumed, and this should not be exceeded for more than a brief period if at all possible. But for the majority of mineral building materials such limits

are not generally available. In such cases other considerations will apply, e.g. not allowing the moisture content of materials susceptible to frost to exceed a certain value.

Example: timber flat roof
The above discussions show that the frequently inaccurate steady-state methods involve problems in assessing moisture control needs and can lead to damage to the building fabric. Timber flat roofs are particularly at risk and so the following section shows which analyses are possible with the help of hygrothermal simulation programs.
A non-steady-state calculation of the vapour diffusion processes is carried out and the influence of the vapour complexion on the moisture content of the construction is investigated as well on the basis of the stipulations in DIN 68800-2 using the example of a flat roof with 240 mm mineral fibre insulation and 22 mm OSB decking (Fig. C 2.26) [11]. The grey-coloured waterproofing to the roof has a diffusion resistance of 300 m and a short-wave radiation absorptance of 0.6.
Attached to the inside face is either a vapour barrier with an s_d value of 100 m or 2 m, or a so-called smart vapour barrier made from polyamide (PA film), which has a variable vapour diffusion resistance (see "Smart vapour barriers", p. 96). According to DIN EN 15026, the external climate to be used is the moisture reference year (meteorological data record on the basis of hourly measurements for a critical hygrometric year) for the northern side of the Alps; normal occupancy is assumed for the interior (Fig. C 2.23a, p. 71). The zone in which interstitial condensation as a result of vapour convection is likely to occur in this construction is confined to the area between the OSB decking and the mineral fibre insulation below it. The calculations were carried out with the help of hygrothermal simulation software for the roof construction to a two-storey house (h = 5 m) shown in Fig. C 2.26.
Assuming an air-dry condition for the construction, i.e. all the materials in the roof are initially in equilibrium with the mean external air humidity of 80 % r.h., the calculation begins in October and continues for a period of five years using the same climate data record. Fig. C 2.27 shows the resulting plots of the total moisture content in the roof for the constructions with the various vapour barriers. The constant increase in moisture in the roof construction with the highly diffusion-resistant vapour barrier (s_d = 100 m) shows that there is a problem here. In the more accurate evaluation it is advisable to consider the building component layer that involves the greatest moisture risk, in this case the OSB decking. Fig. C 2.28a shows the moisture contents output for the OSB for the three different construction variations. The fact that the moisture content of the OSB decking in the roof reaches the critical value of 20 % by wt. within five years in the roof with the highly diffusion-resistant vapour barrier, whereas the

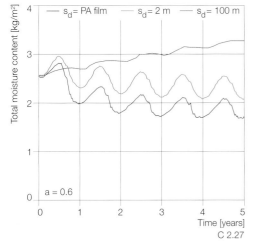

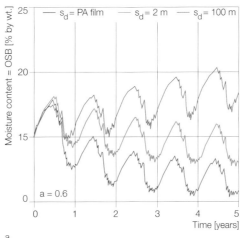

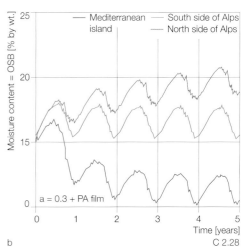

C 2.27

a

b

C 2.28

decking in the variations with the other vapour barriers dry out more and more, shows quite clearly the type of construction that should *not* be chosen. The construction with the "smart" PA film achieves the best results because its variable vapour diffusion resistance in the winter, when there is a risk of condensation, is greater than that in the summer, when the roof should be drying out.

However, the drying-out potential of the PA film also has its limits, as is shown by the results for the moisture content of the OSB decking in the same construction but in this case with a reflective (bright white) roof surface (Fig. C 2.28b). The high reflection of the solar irradiation means that the roof surface heats up only marginally in these climate conditions on the northern side of the Alps, which leads to an inadequate drying-out potential in summer. However, in a warmer climate, e.g. on the southern side of the Alps or around the Mediterranean, the roof would function without problems even with a highly reflective surface.

Practical moisture control advice

Hygrothermal simulation is today regarded as the most reliable – albeit the most involved – method of assessing moisture conditions. This method permits reliable forecasts of the known issues regarding condensation in winter and drying-out potential in summer, and also the influence of construction moisture. In contrast to the steady-state methods, the moisture assessment can be carried out for all possible internal and external climate conditions. However, this is only true when sufficiently accurate data is available for the materials being used. In addition, this approach does call for appropriate experience in the use of numerical simulation methods and sufficient expertise in applying the subsequent plausibility check.

Many standard forms of construction not susceptible to moisture can, however, be designed without the need for any calculations. For example, flat roofs of in situ concrete do not usually experience any problems with moisture infiltration from the interior; construction moisture is the largest moisture load in this case and the

thermal insulation above must be protected from this at all costs. The solution here is to install a highly vapour-tight vapour barrier. By contrast, in a timber flat roof the vapour barrier chosen should allow a certain drying out in the direction of the interior, and a smart vapour barrier is generally the most sensible solution. As with metal sandwich designs it is not the vapour diffusion but rather the vapour convection that represents the biggest problem, a continuous airtight membrane must be ensured. An accurate analysis of the moisture conditions in the roof at the planning stage is absolutely indispensable in the following situations:

- The internal boundary conditions are very different to the normal interior climate conditions of offices or dwellings.
- The location or the height above sea level results in a considerable deviation from the usual Central European climate conditions.
- The sequence of layers or building component interfaces does not correspond to the tried-and-tested standard details.
- New materials are to be used, the properties of which are very different to those of traditional building materials.
- The roof construction contains materials, the function and durability of which are impaired in the presence of marginally higher moisture levels.

A hygrothermal simulation to assess the moisture conditions will be necessary in these cases. If the results of the simulation are still not clearly favourable, either the choice of the roof construction should be reconsidered or more accurate observations of the roof should be carried out once it has been erected. These days, it is possible to monitor moisture levels in a roof over a longer period of time with the help of built-in sensors and to then react specifically to critical values as required.

References
[1] Mohrmann, Martin: Feuchteschäden beim Flach-dach. In: Holzbau – die neue quadriga. Berlin, 3/2007, pp. 13ff.
[2] Geißler, Achim; Hauser, Gerd: Abschätzung des Risikopotentials infolge konvektiven Feuchtetrans-ports. Final Report, AIF Research Project No. 12764, Kassel, 2002
[3] Oswald, Rainer: Fehlgeleitet. Unbelüftete Holz-dächer mit Dachabdichtungen. In: deutsche bauzeitung 7/2009, pp. 74ff.
[4] Zirkelbach, Daniel; et al.: Dampfkonvektion wird berechenbar – Instationäres Modell zur Berück-sichtigung von konvektivem Feuchteeintrag bei der Simulation von Leichtbaukonstruktionen. AIVC Conference. Berlin, 2009
[5] Künzel, Hartwig. M.: Dampfdiffusionsberechnung nach Glaser – quo vadis? In: IBP Mitteilung 355. Stuttgart, 1999
[6] Geshwiler, M.: Air Pressures in Wood Frame Walls. Thermal Performance of the Exterior Envelopes of Buildings. 7th Conference. Clearwater Beach, Florida, 1998
[7] Schmidt, Daniel; Winter, Stefan: Flachdächer in Holzbauweise. Pub. by Informationsdienst Holz. Bonn, 2008
[8] Künzel, Hartwig M.; Zirkelbach, Daniel: Trocknungs-reserven schaffen – Einfluss des Feuchteeintrags aus Dampfkonvektion. In: Holzbau – die neue quadriga, 1/2010, pp. 28ff.
[9] Finch, G.; Hubbs, B.; Bombino, R.: Osmosis and the Blistering of Polyurethane Waterproofing Mem-branes. 12th Canadian Conference on Building Science & Technology. Montreal, 2009
[10] Künzel, Hartwig M.; Sedlbauer, Klaus: Reflektierende Flachdächer – sommerlicher Wärmeschutz kontra Feuchteschutz. In: IBP Mitteilung 482. Stuttgart, 2007
[11] ibid. [2]

Fire

Ulrich Max

C 3.1

Fires in the building sense are all damaging fire events in or on buildings. The possible causes of fires range from spontaneous combustion to technical defects and arson (Fig. C 3.2). In Germany the building regulations of the federal states prescribe fire protection measures that are intended to limit the damage to persons and property caused by fires. These measures must be taken into account at the design stage and their scope depends on the particular use of the building. A fire will start when the following conditions for combustion occur simultaneously:
• The presence of combustible materials.
• The combustible materials must be preheated by one or more sources of ignition until their ignition temperature is reached.
• A supply of oxygen is available.

In practice, however, as combustible materials and oxygen are almost always present we should always assume that a fire could start at any time. But by limiting the possible sources of ignition, e.g. open fires or hot working processes (cutting, drilling, grinding and welding), it is possible to reduce the likelihood of a fire breaking out in a building, although this cannot be entirely eliminated.

The phases of a fire

Ignition occurs when a combustible material is preheated by a source of ignition to the extent that it or rather its products of decomposition (pyrolysis products) have reached their ignition temperature, at which point they react with the oxygen available and release heat while doing so. Once a material has been ignited, the combustion process continues autonomously. This is the ignition phase in which the fire is essentially limited to its point of origin (Fig. C 3.3). Normally a fire in this phase can be extinguished by the persons present at the scene without endangering themselves. But if left to burn uncontrolled, a fire will continue to spread until all the combustible materials have been consumed by the fire. The ignition phase is followed by the incipient or growth phase, the length of which depends on the nature of the materials.

The fire starts to spread in this phase, but only to a limited extent, and the heat flow (calorific value) is low. Fire detection systems (e.g. smoke detectors) and the automatic transmission of an alarm to the fire brigade are useful during this phase because a fire can then be extinguished before it engulfs a larger part of the room or the building. But if extinguishing the fire at an early stage is unsuccessful, it starts to spread further and the rate of spread increases as the temperature in the room rises. The suspended floors, or a flat roof, to the building on fire then take on a special significance.

The highest temperatures in the room, after those in the vicinity of the seat of the fire, are those near the soffit of the floor or roof above. This is because smoke and fumes rise and then spread in all directions beneath the soffit until they are stopped by deeper downstand beams or smoke curtains. The ignition of the combustible materials in the ceiling or floor or roof above takes place when the temperature there is sufficient to ignite the materials. The spread of smoke and fumes below the soffit preheat the combustible materials there such that they ignite over a wide area. This brief process is known as flashover, which occurs when the temperature in the vicinity of the soffit reaches 300–500 °C. In small rooms the fire then spreads to engulf the entire room. We speak of a fully developed or free-burning fire, i.e. combustion takes place over the entire floor area of the room. During this phase the temperatures in the room can reach 1000 °C and more. The free-burning phase is followed by the smouldering or decay phase, once most of the combustible materials in the room have been consumed by the fire.

In very large rooms (e.g. industrial sheds), flashover and the sudden spread of fire over the entire floor area of the room is less likely because the smoke and fumes cool as they spread across the soffit, and possibly partially escape through smoke and heat vents. The effects of the fire, although it is fully developed, remain confined to one area of the room.

C 3.1   Fire test on flexible sheeting to DIN 4102-7
C 3.2   Causes of fires in Germany according to information provided by the Institut für Sachverständigenwesen e.V.
C 3.3   Idealised diagram of temperature and heat flow (calorific value) in the various fire phases for a fire in a building
C 3.4   Overview of fire protection measures

Cause of fire	
Electricity	31 %
Human error	19 %
Other and unknown causes	16 %
Arson	15 %
Overheating	6 %
Open fire/naked flame	4 %
Spontaneous combustion	4 %
Work involving fire risk	2 %
Explosion	2 %
Lightning	1 %

C 3.2

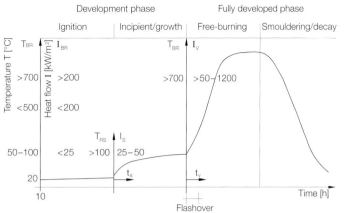

R   Smoke
V   Fully developed fire
S   Incipient phase
BR Room containing fire

C 3.3

Fire protection concepts

Fire protection measures are necessary because the goal is to guarantee adequate safety in the event of a fire. The total range of fire safety measures taken ensure that the objectives of the building regulations are achieved (Fig. C 3.4). The specification or stipulation of these measures and the assessment of whether or not they meet the objectives form the fire protection concept. In order to comply with the requirements of the building regulations, it is necessary to draw up a fire protection concept as part of the building approval procedure. The statutory requirements encompass protection of persons completely and protection of property only to the extent of protecting neighbouring areas and limiting the spread of fire beyond a defined area. Additional regulations imposed by insurers may be necessary for buildings of high value or those containing high-value goods, which may lead to higher fire protection requirements than those necessary for complying with the building regulations.

Technical fire protection measures
The preventive technical measures cover the choice of building materials and the use of constructional measures, which involve the fire resistance of the materials and dividing the planned building into fire compartments.

Structural fire protection
The task of structural fire protection is to limit the spread of fire by way of constructional measures. For example, in multi-storey buildings in particular, it is important to guarantee the stability of the loadbearing construction for the rooms above the storey in which the fire occurs. Structural fire protection also includes the planning of escape and rescue routes plus the provision of and access to adequate hardstandings for fire-fighting vehicles.

Systems and installations
Systems and installations can be provided to detect fires and warn others, and also to limit the effects of any fire:
· Automatic extinguishing systems can prevent or control the spread of a fire.

· Automatic fire detection systems can detect a fire at an early stage and hence enable a building to be evacuated quickly. The fire brigade can then start its fire-fighting measures at an early stage of the fire and there is a realistic chance that they will be successful.
· Automatic alarm systems warn people to leave the building.
· Smoke and heat vents represent a useful addition to manual fire-fighting by the fire brigade and reduce the temperature effects on the building components.

Organisational fire protection measures
Organisational measures are preventive measures designed to minimise the risk of the outbreak of a fire. Training and instruction schemes prepare the users of a building so that they will behave correctly if a fire does occur.

Fire-fighting measures
Besides automatic extinguishing systems, it is the fire brigade that provides active fire-fighting measures. According to German legislation, all municipal authorities must maintain a functioning public fire brigade service such that every building and structure falls within the jurisdiction of a fire brigade. In the case of larger special buildings, principally industrial premises, there is sometimes a company or works fire brigade

as well. The most important constructional measure besides accessibility is the provision of a guaranteed supply of extinguishing water. In larger buildings the deployment plans for fire-fighting measures are backed up by fire brigade plans.

Operational measures
The operational measures are in the first instance the fire-fighting measures by the municipal, possibly also the works, fire brigade. Deployment plans must be drawn up if fire-fighting measures are to be successful. In addition, access routes to the building must be kept clear. Further operational precautions can be employed to minimise the effects of a fire, and in this case it is the protection of persons that is given priority. Apart from that, a company should use every opportunity to reduce the number of potential sources of ignition and thus reduce the chance of a fire breaking out in the first place.

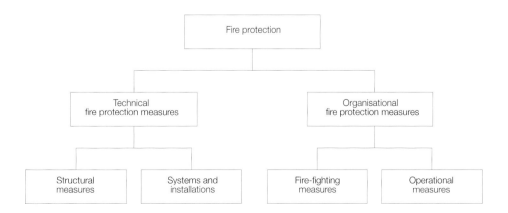

C 3.4

Assessing the structural fire protection aspects of materials and components

Standard fire tests, as laid down in DIN 4102-1 or equivalent European standards, are carried out to classify structural fire protection measures. The degree to which building components and building materials contribute to preventing the spread of fire within and beyond the area of the fire is also assessed. The technical evaluation of building materials with respect to their contribution to the spread of fire within the area of the fire during the individual fire phases is carried out with the help of test facilities with differing temperature conditions depending on the building materials class that is being aimed at. In order to prevent a fire spreading beyond the area of the fire, various fire protection cells are formed and defined by enclosing building components. The cell with the highest level of protection is the fire compartment, and containing the fire within this compartment should be given highest priority. All the building components forming the boundaries to this fire compartment (suspended floors, roofs and walls and their respective supports) must retain their stability for the duration of the fire resistance rating that has been specified.

Building materials classes

Demands are placed on the flammability of all the combustible materials likely to be used in buildings. In principle, we distinguish between incombustible (class A) and combustible (class B) materials. Defined fire tests establish whether combustible materials are not readily flammable (class B 1) or flammable (class B 2). The fire scenarios for not readily flammable materials are those of the ignition phase. The heat flow (calorific value) of the test corresponds approximately to a fire in a waste-paper basket, whereas flammable materials are tested with a heat flow corresponding to the flame of a cigarette lighter. Building materials that do not comply with all the requirements placed on not readily flammable or flammable materials are classed as highly flammable (class B 3) and may not be used in buildings. These designations for the building materials classes are used in the building regulations that must be observed when using building materials.

Fire tests to DIN 4102-1 are gradually being replaced by European tests and a classification according to DIN EN 13501 in the course of European harmonisation.

Fire resistance

In order to guarantee the room-enclosing function (integrity and insulation) plus the stability of the construction, the components concerned are classified by fire tests according to German (DIN 4102-2) or European (DIN EN 1363 to 1366) standards for different fire durations. The important thing here is that the components being tested in the furnace are exposed to a defined fire event (standard fire conditions) in which the temperature is controlled according

to the standard time–temperature curve. The length of time for which the building component fulfils the required functions (enclosure and/or stability) is its fire resistance rating. This allows it to be allocated to one of the classes specified for this type of component, for normal fire conditions a fire resistance of at least 30, 60 or 90 minutes. In the German building regulations the fire resistance ratings are specified by way of descriptive terms. The associated technical testing conditions depend on the requirements placed on the functions of the components specified in DIN EN 13501 and DIN 4102 (Fig. C 3.5). Requirements regarding the fire resistances of (flat) roofs and their supports are generally only laid down for special structures. What also has to be considered with roofs is that the fire could spread not only in the room(s) below the roof, but also from adjoining buildings. Larger roofs must therefore normally be finished with a so-called hard roof covering (i.e. resistant to sparks and radiant heat). The associated test is carried out according to DIN 4102-7 or the pre-standard DIN V ENV 1187. The German testing standard DIN 18234 specifies the design and construction requirements regarding the spread of fire via an extensive roof exposed to fire from below. These requirements are mainly relevant for the roofs to industrial sheds. This standard is unique to Germany and at present has no European equivalent.

Normative fire protection requirements for flat roofs

For the fire protection of roofs we distinguish between the requirements placed on the roof covering and the fire resistance of the loadbearing construction. Only in exceptional cases do special requirements apply to this space-enclosing element.

General requirements according to the Model Building Code (MBO)

With the exception of the provisions according to cl. 32 (6) of Germany's Model Building Code (MBO, *Musterbauordnung*), for the roofs to buildings adjacent at eaves level no requirements are placed on the fire resistance of the topmost floors, including a flat roof, provided there are no habitable rooms above these. Building services installations (e.g. ventilation plant rooms) are not classed as habitable rooms. The requirements given in Fig. C 3.6 therefore apply to superstructures in general, but not to roofs, especially flat roofs. In some regulations applying to special structures, the primary loadbearing structure does have to satisfy certain requirements (e.g. industrial buildings, places of assembly and retail premises). According to DIN 4102-7 and DIN V ENV 1187, hard roof coverings must resist a fire load from outside due to sparks and radiant heat for a sufficient length of time. Roof coverings that do not satisfy this requirement are permitted on buildings belonging to classes 1–3 only when those build-

ings comply with certain conditions (see MBO cl. 32 (2)). Roofs must therefore be designed and constructed in such a way that a fire cannot be spread to other parts of the building and neighbouring plots by way of sparks or heat radiation, for instance. The distance between roof openings, rooflights, etc, and a firewall must be at least 1.25 m (see MBO cl. 32 (5)). When dividing large buildings into fire compartments, care must be taken to ensure that there is sufficient distance between the fire walls and the external wall of the higher building. At a rising external wall, the roof must be fire-resistant and may not have any openings over a width equivalent to the difference in height between the two parts of the building, but 5 m at least. This requirement also applies when a fire wall is not specified below the rising external wall (see MBO cl. 32 (7)). In this case the fire resistance rating is that given in Fig. C 3.8 and only applies to a fire load from inside to outside. This can be provided in the form of a fire-resistant ceiling, for instance, although it must be ensured that the loadbearing components (beams and columns) supporting this ceiling have a suitable fire resistance in the event of a fire in the room concerned.

The above requirements must also be complied with in the case of special structures unless the relevant regulations specify other requirements.

Places of assembly

In places of assembly without automatic fire-extinguishing systems the loadbearing structure supporting the roofs above the assembly areas must be fire-retardant. Roof structures above grandstands and outdoor stages must be at least fire-retardant or made from incombustible materials (see cl. 3 (1) of *Versammlungsstättenverordnung*, Places of Assembly Act, VStättVO).

Ceilings and soffit linings above assembly areas (and hence also below flat roofs over assembly areas) must be made from incombustible materials. Linings made from incombustible or not readily flammable materials or enclosed, unventilated timber linings may be used in assembly rooms with a floor area not exceeding 1000 m^2 (see VStättVO cl. 3 (2)). Insulating materials must be made from incombustible materials (see VStättVO cl. 4(1)).

Retail premises

According to the *Muster-Verkaufsstättenverordnung* (MVkVO, Model Retail Premises Act), the requirements regarding the fire resistance of the roof structure are identical to those for places of assembly. Roofs above retail premises must be made from incombustible materials (including the insulation) apart from the roof covering itself and the vapour barrier.

Industrial buildings

Larger fire compartments are permitted in industrial buildings depending on the systems and installations and the fire-fighting measures, i.e. the distance between fire walls can be increased beyond that specified in the MBO (*Industriebaurichtlinie*, Industrial Buildings Directive, IndBauRL, section 6). Apart from that, there is the option of calculating the requirements for the loadbearing structure depending on the actual fire loads present. To do this, a fire load calculation is performed according to DIN 18230 (IndBauRL, section 7).

Without a fire load calculation, the loadbearing and bracing components plus the primary loadbearing structure to the roof (e.g. trusses) must be designed and constructed according to the fire resistance ratings given in Fig. C 3.8. They must be made from incombustible materials or employ a fire-retardant form of construction. Loadbearing structures made from timber or plastics are therefore only permitted when these are at least fire-retardant. Roofs with a surface area exceeding 2500 m² (with a construction consisting of, for example, roof covering, thermal insulation, vapour barrier, supporting structure) must be designed and constructed according to requirements given in IndBauRL section 5.11 in such a way that spread of fire via the roof is prevented. This requirement is deemed to be satisfied, for example, for roofs designed…

• according to DIN 18234-1 (including Beiblatt 1)
• with a loadbearing shell made from mineral building materials (e.g. concrete, aerated concrete)
• with roof coverings made from incombustible materials

Building authority designation	Loadbearing without enclosing function	Loadbearing with enclosing function	Non-loadbearing internal walls	Non-loadbearing external walls
Fire-retardant	R30 [F30]	REI30 [F30]	EI30 [F30]	E30[2] EI30[2] [W30]
Highly fire-retardant	R60 [F60]	REI60 [F60]	EI60 [F60]	E60[2] EI60[2] [W60]
Fire-resistant	R90 [F90]	REI90 [F90]	EI90 [F90]	E90[2] EI90[2] [W90]
Fire wall		REI-M90	EI-M90	
Enhanced fire wall [1]		REI-M180	EI-M180	

[1] Not a requirement of the building regulations, but rather the insurers
[2] From inside to outside

C 3.5

Building class	Requirement	Remarks
GKL 5	Fire-resistant	–
GKL 4	Highly fire-retardant	Height ≤ 13 m
GKL 2 and 3	Fire-retardant	Height ≤ 7 m

C 3.6

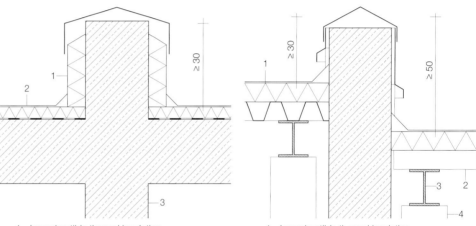

a
1 Incombustible thermal insulation
2 Incombustible roof covering, adhesive, thermal insulation
3 Fire wall

b
1 Incombustible thermal insulation
2 Steel trapezoidal profile sheeting
3 Steel beam
4 Steel column

C 3.7

No. of storeys in building	1		2			3		4	5
Fire resistance of loadbearing and bracing components	No require-ments	F30	F30	F60	F90	F60	F90	F90	F90
Safety category	Size of fire compartment [m²]								
K1	1800[1]	3000	800[2,3]	1600[2]	2400	1200[2,3]	1800	1500	1200
K2	2700[1]	4500	1200[2,3]	2400[2]	3600	1800[2]	2700	2300	1800
K3.1	3200[1]	5400	1400[2,3]	2900[2]	4300	2100[2]	3200	2700	2200
K3.2	3600[1]	6000	1600[2,3]	3200[2]	4800	2400[2]	3600	3000	2400
K3.3	4200[1]	7000	1800[2,3]	3600[2]	5500	2800[2]	4100	3500	2800
K3.4	4500[1]	7500	2000[2,3]	4000[2]	6000	3000[2]	4500	3800	3000
K4	10000[1]	10000	8500	8500	8500	6500	6500	5000	4000

[1] Width of industrial building ≤ 40 m and area of smoke and heat vents ≥ 5% (to DIN 18230-1)
[2] Area of smoke and heat vents ≥ 5% (to DIN 18230-1)
[3] For low-rise buildings MBO cl. 25 para. 1 permits a size of 1600 m² compared with MBO cl. 28 para. 1 No. 2

C 3.8

C 3.5 Building material requirements and classifications according to European and German testing methods
C 3.6 Requirements to be satisfied by loadbearing components (columns, suspended floors, also flat roofs if applicable) depending on the building class according to the Model Building Code (MBO)
C 3.7 Combustible roof finishes and other combustible components adjacent to a fire wall may not be built into or continue across the fire wall.

a Junction between solid flat roof and loadbearing fire wall
b Junction between flat roof of steel trapezoidal profile sheeting and non-loadbearing fire wall
C 3.8 Permissible sizes of fire compartments in industrial buildings without a fire load calculation depending on the fire resistance of the loadbearing and bracing components plus the number of storeys

Sound insulation

Phillip Leistner, Lutz Weber

The aim of the acoustic design of a building is to create suitable conditions for its users. Their health, well-being, and in workplaces their productivity too, are the critical aspects. And it is not only preventing annoyances, indeed noise pollution, that we must consider, but also the active design of agreeable acoustic conditions. Aspects such as airborne sound insulation to protect against external noises, impact sound and structure-borne sound insulation plus sound absorption within the interior are all vital factors here. As the individual criteria have an influence on each other, the sound insulation – or noise control – requirements must be considered in conjunction with the constructional, thermal performance, moisture control and architectural aspects in order to achieve the best design in terms of function and economics.

Principles

The physics of building and room acoustics mean that they affect each other to different degrees. For example, the resultant sound insulation between two rooms depends on both the acoustic properties of the separating components and the sound absorption within the rooms. A number of important terms are defined below (Fig. C 4.2).

Airborne sound insulation
The airborne sound insulation describes the resistance of a building component to vibrations of the air caused by external noise or the sounds from adjacent rooms within a building.

Sound reduction index
The sound insulating effect of components with respect to their excitation by airborne sound is described by their sound reduction index R. This is the logarithmic relationship between the incident and the transmitted sound power, P_1 and P_2 respectively, which is defined by the following equation:

$$R = 10 \lg \left(\frac{P_1}{P_2} \right) \ [dB]$$

A building component with a sound reduction index R of 30 dB, for example, reduces the sound power by a factor of 1000. However, the logarithmic relationship between the sensitivity of the human ear and the incident sound pressure means that the subjective decrease in the loudness level is only about a factor of eight. A 10 dB reduction in level in the mid-range of levels corresponds roughly to a halving of the perceived loudness level. If sound propagation takes place via flanking components in addition to the separating building component itself, then a dash suffix (') is added to the symbol R (the so-called apparent sound reduction index R').

Weighted sound reduction index
Generally, the rise in the sound reduction index rises rapidly as the frequency increases. The sound insulation is therefore normally measured in one-third octave bands in the frequency range from 50 to 5000 Hz. However, a frequency-based approach is not absolutely essential for many building acoustics applications, which means that the calculations can be approximated with single figures. In order to obtain a representative single figure, the weighted sound reduction index R_w is derived from the measured spectrum according to DIN EN ISO 717-1. For external building components, however, R_w should be treated with caution because it is based on the spectrum of customary internal noises in residential buildings, whereas external noise (e.g. traffic) mostly contains a much higher level of low-frequency noise components [1]. Exacerbating this problem is the fact that R_w is derived only for the frequencies from 100 to 3150 Hz, which means that important components of the range of human hearing – extending from about 20 to 20 000 Hz – remain unconsidered. This is particularly true for the frequencies between 50 and 100 Hz, which play a significant role in sound insulation against traffic noise. In order to take into account this particular acoustic situation for external building components, the spectrum adaptation term C_{tr} according to DIN EN ISO 717-1 has to be added to the weighted sound reduction index R_w. The C_{tr} value takes into account noise sources such

as urban traffic. The correction terms defined in VDI 2719 and VDI 4100 can be used as an alternative. However, this problem has not yet been included in current sound insulation requirements, which still refer to the weighted sound reduction index R_w for external components as well.

Sound insulation of roofs
In order to measure the sound insulation in the laboratory, the roof to be tested is erected between two rooms, one above the other, according to DIN EN ISO 140-3. In the upper room (source room), a loudspeaker generates a diffuse sound field. The resulting sound levels in the source and receiving rooms, L_1 and L_2 respectively, are then measured. The measurements allow us to calculate the sound reduction index of the roof:

$$R = L_1 - L_2 + 10 \lg\left(\frac{S \cdot T}{0.16 \, V}\right) \; [dB]$$

where:
S   area of building component [m²]
V   volume of receiving room [m³]
T   reverberation time in receiving room [s]

When measuring in the field, the roof surface is subjected to the sound from a loudspeaker positioned at an angle of 45° above the roof according to DIN EN ISO 140-5. The sound reduction indexes calculated according to DIN EN ISO 140-3 and 140-5 agree approximately, whereas those based on field measurements are generally about 1–2 dB higher because of the different type of noise excitation [2].
DIN 4109 contains minimum requirements for the airborne sound insulation of external components in order to protect building occupants against external noise. These requirements are included in building legislation and depend on the use of the room requiring protection as well as the level of external noise. Another factor that must be considered is the area of the external components in proportion to the floor area. A correction K must therefore be added to the defined requirements (Fig. C 4.3) depending on the ratio between the total external surface area S_{Atot} and the floor area S_G of the room under consideration:

$$K = \left[10 \lg\left(\frac{S_{Atot}}{S_G}\right) + 1\right] \; [dB]$$

As already mentioned, these requirements represent minimum values that according to DIN 4109 have been determined in order to protect persons in habitable rooms from unreasonable disturbances caused by sound propagation. For dwellings offering a higher standard of comfort, VDI 4100 therefore includes the proposal to increase the minimum requirements by 5 dB in its sound insulation category SSt III. In contrast to the minimum values, however, this proposal is not legally binding; it must be explicitly

agreed between the contracting parties. Nevertheless, in the court cases of recent years judges are increasingly taking the view that notwithstanding the minimum requirements of DIN 4109, the sound insulation achievable within the scope of the agreed form of construction, assuming proper design and workmanship according to the acknowledged codes of practice, must be provided [3].

Resultant sound reduction index
The resultant sound reduction index of all external components $R'_{w,res}$ expresses the fact that the external surfaces to rooms frequently comprise elements with different sound insulation properties, e.g. a flat roof with flat or raised rooflights, roof vents, etc. In this case the resultant sound reduction index of the total surface area is calculated as follows:

$$R'_{w,res} = -10 \lg\left(\frac{1}{S_{tot}} \sum_j S_j \, 10^{-R'_{w,j}/10 \, dB}\right) \; [dB]$$

with $S_{tot} = \sum_j S_j$

where:
S_j area of jth element
$R'_{w,j}$ weighted sound reduction index of jth element

As the sound insulation of rooflights etc. is often negligible, they frequently determine the resultant sound insulation of the entire roof surface (Fig. C 4.4). The same is true for ventilation equipment unless it is fitted with effective silencers. "Acoustic coordination" of the various roof elements is therefore necessary in order to guarantee effective sound insulation, i.e. the sound reduction indexes of the individual parts should not differ too greatly from each other.

Critical external noise level
The critical external noise level L_{mA}, which is used as a measure of the level of sound insulation required, describes the average noise level to which the external component under consideration is subjected during the day, i.e. from 6 a.m. to 10 p.m. The critical external noise level is normally determined by calculation. In Germany different sets of rules, e.g. the Directive for Noise Abatement on Roads (RLS 90) or the Provisional Method of Calculation for Environmental Noise on Railway Lines (VBUSch), apply depending on the source of the noise (e.g. road or rail traffic). Simplified methods for estimating the effective traffic noise are given in DIN 4109, section 5.5, and in DIN 18005-1. In the case of noise due to trade and industry, the valid recommended immissions values given by the zoning in the local development plan according to the Technical Rules for Protection Against Noise (*TA Lärm*) can often be used for simplicity.
Once the critical external noise level L_{mA} is known, provided it does not exceed a given internal level L_I on average over a certain period of time, then the sound reduction index required

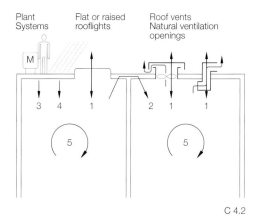

C 4.2

External noise level L_{mA} [dB(A)]	Sound reduction index required $R'_{w,res}$ [dB]	
	Dwellings	Workplaces
≤ 55	30	–
≤ 60	30	30
≤ 65	35	30
≤ 70	40	35
≤ 75	45	40
≤ 80	50	45
> 80	[1]	50

[1] Specified in relation to local circumstances

C 4.3

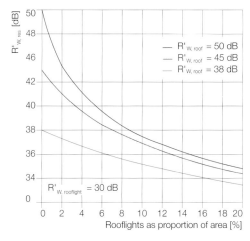

C 4.4

Source of noise	Correction term K [dB]
Railway lines with mainly passenger traffic	0
Other railway lines	3
Urban roads	6
Other roads	3
Airports	6

C 4.5

C 4.6   Standard tapping machine for measuring impact
         sound insulation
C 4.7   Sound transmission between two neighbouring
         rooms. There are three flanking paths F_f, F_d and
         D_f in addition to the direct transmission path D_d.
         It is mainly path F_f that is critical for insulation
         against flanking transmissions in roofs.
C 4.8   Schematic presentation of a rainfall noise test rig
C 4.9   Rainfall noise test rig
C 4.10   Reduction in sound transmission through the use
         of an elastic bearing plotted against the frequen-
         cy ratio f/f_R. In this case f_R designates the reso-
         nant frequency of the bearing. Positive values in-
         dicate a decrease in the sound transmission,
         negative values an increase.
         The curves shown here were calculated with a
         simplified model of a damped swing arm.

C 4.6

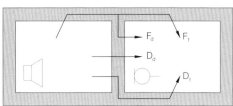

C 4.7

for the external components can be calculated according to VDI 2719 and VDI 4100 as follows:

$$R'_{w,res} = L_{mA} - L_i + 10 \lg \left(\frac{S_{ges} T}{0.16 V} \right) dB + K \; [dB]$$

S_{tot} total area of external components [m²]
K correction term which takes into account the frequency spectrum of the effective external noise (0 dB ≤ K ≤ 6 dB; Fig. C 4.5, p. 79)

This type of calculation must always be used when individual requirements regarding the internal noise levels in a building must be satisfied. The sound reduction index R used hitherto will be replaced by the standardised sound level difference D_{nT} in the next edition of DIN 4109, which is currently in preparation. As the D_{nT} value is related to the reverberation time instead of the equivalent absorption area, there is a systematic difference between R and D_{nT} which depends on the geometric relationships in the receiving room. The relationship between these two variables in the case of sound transmission through a roof into a square or rectangular room of height h is as follows:

$$D_{nT} = R + 10 \lg (h) \, dB - 5 \; [dB]$$

Using this equation we can establish that R and D_{nT} are equal when h ≅ 3.1 m, but there is a difference of 1.0 dB ($D_{nT} < R$) for the typical ceiling height of 2.5 m. As the figures for the sound insulation requirements of external building components according to the current status of the revision work will not change, this means that the requirements related to a ceiling height of 2.5 m will increase by 1 dB.

Insulation against flanking transmissions
It is not only the direct passage of sound from outside to inside that is important for sound insulation in roofs; in many instances flanking sound transmissions within the roof construction are also relevant. The transmission of sound along the roof results in an additional path for sound propagation between neighbouring rooms which can reduce the sound insulation between those rooms to a considera-

ble extent, especially with lightweight roofs and partitions with good sound insulation (Fig. C 4.7). In order to determine how the roof influences the apparent sound reduction indexes of internal walls, it is necessary to carry out a sound propagation calculation according to DIN EN 12354-1. This involves adding together the energy values of the various transmission paths:

$$R' = -10 \lg (10^{-R_{Dd}/10 \, dB} + \sum_{i,j} 10^{-R_{ij}/10 \, dB}) \; [dB]$$

This equation adds together the direct transmission through the partition (sound reduction index R_{Dd}) and twelve different flanking transmission paths (flanking sound reduction indexes R_{ij} between flanking components i and j) which in all cases are related to the area of the partition. The flanking sound reduction index R_{ij} is calculated from the sound reduction indexes R_i and R_j of the components involved and the sound reduction index of the junction K_{ij}.
The apparent sound reduction index R' is always smaller than the sound reduction index of each individual transmission path R_{ij}. The consequence of this is that the building component with the lowest sound reduction index determines the sound transmission in the end. If we assume that all transition paths R_{ij} have the same sound reduction index, R_i would lie 10 lg(13) dB ≅ 11.1 dB below the R_{ij} value. From this we can deduce that it is generally sufficient when the flanking sound reduction index of the roof is about 10 dB higher than the required apparent sound reduction index. Where this also applies to the other flanking paths, we can assume that, in total, the sound insulation is adequate.
Taking into account the aforementioned design rules, a detailed calculation according to DIN EN 12354-1 is generally unnecessary. In principle, there is a close relationship between the sound insulation for flanking and direct transmissions. Lightweight roofs with only a small amount of sound insulation for direct transmissions therefore frequently exhibit problems with respect to flanking transmissions as well. Designers must take great care when a lightweight flat roof is to be built on top of a heavyweight concrete or masonry building with a high degree of sound insulation. If the sound insulation against flanking transmissions in a roof is inadequate, a non-

continuous suspended ceiling, sometimes also improved details at the junctions between internal walls and roof, can help to improve the situation.

Impact sound insulation
A flat roof that is actively used will have to satisfy impact sound insulation requirements as well as those regarding airborne sound. A standard tapping machine (Fig. C 4.6) is activated on the top of the roof in accordance with DIN EN ISO 140-7 in order to determine the impact sound insulation. The resulting sound level L_i is measured in the receiving room below the roof and then used to calculate the normalised impact sound level L'_n of the roof as follows:

$$L'_n = L_i + 10 \lg \left(\frac{0.16 \, V}{10 \, T} \right) \; [dB]$$

where:
V   volume of room [m³]
T   reverberation time [s]

As with the airborne sound situation, a single figure for impact sound, the weighted normalised impact sound level $L'_{n,w}$ is derived from the frequency-dependent value L'_n. When assessing impact sound insulation requirements, it is vital to distinguish between minimum requirements laid down in building legislation and non-binding proposals for enhanced sound insulation. Although DIN 4109 contains no explicit requirements regarding the impact sound insulation of roofs, designers would be well advised to apply the values for terraces and loggias above habitable rooms, for which a minimum requirement of $L'_{n,w} \leq 53$ dB applies, or for enhanced sound insulation a proposal of $L'_{n,w} \leq 46$ dB.
Impact sound insulation to flat roofs is not included in the DIN 4109 catalogue of standard constructions. The values for suspended floors with a similar construction can be used as a guide, which can be found in DIN 4109 and also, for example, in a comprehensive research report published by Germany's national metrology institute, the Physikalisch-Technische Bundesanstalt [4].
Effective impact sound insulation in roof can be guaranteed by constructing it as a component

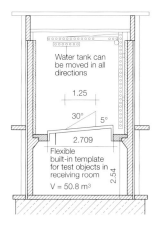

C 4.8

C 4.9

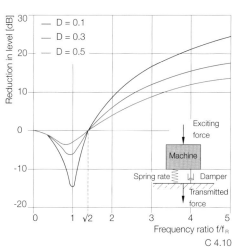

C 4.10

with two or more layers, similar to a suspended floor with a floating screed. In such an arrangement, a suitable separating layer made from a resilient material that attenuates impact sound should be laid between the top and bottom layers of the construction.

Noise of rainfall

Rainfall striking the surface of a roof can in some circumstances lead to a high noise level in the room(s) below which is regarded as very irritating. This is especially the case with lightweight roofs or flat or raised rooflights. The noise level of various lightweight roof constructions lies between about 50 and 70 dB(A), and with lightweight membrane designs the level can even be much higher. These levels are so high that they considerably impair communication and working conditions in the rooms affected. Wherever possible, lightweight roof designs should be subjected to a rainfall noise test. A standard method of measuring rainfall noise is still not available and so there are currently no requirements regarding corresponding sound insulation. However, in 1997 DIN EN ISO 140-18 introduced the first standardised method for measuring rainfall noise in the laboratory. The method uses a water tank with a perforated base to create the noise of rainfall; this is mounted 3.5 m above the roof element to be tested. The amount of artificial rain (40 l/m²h) with a defined droplet size falling on the roof corresponds to heavy natural rainfall and sets up vibrations in the test roof (Figs. C 4.8 and C 4.9) [5].
The variable for characterising rainfall noise in the laboratory is the sound level in the receiving room below, which should be more or less identical with that in a room in a dwelling or place of work with customary furnishings and fittings. Sound insulation and rainfall noise on building components are closely related, i.e. high rainfall noise usually means an inadequate level of sound insulation.
On a flat roof, rain strikes the surface at an angle of incidence of approx. 90° and so especially loud rainfall noise can be expected. The roof construction should therefore include a sufficient amount of insulation against airborne sound in order to avoid degrading the acoustics. With a

weighted sound reduction index of $R'_{w,res} \geq 45$ dB, we can assume that the noise of rain in the interior – apart from extreme weather conditions such as a cloudburst or hail – will not exceed a level of $L_i = 40$ dB(A), which means that the disturbing noise remains within reasonable levels. With an inhomogeneous roof surface (roof with rooflights, etc.), it should be remembered that the individual surfaces with the lowest level of insulation will determine the passage of sound for the roof as a whole.

Building services

Flat roofs are often used as mounting surfaces for building services such as ventilation and air-conditioning plant. Such installations generate vibrations that are transmitted into the roof as structure-borne sound and from there are radiated into the room(s) below. The ensuing noise can be disturbing and therefore strict sound insulation requirements apply. According to the minimum requirements of DIN 4109 (which form part of building legislation and are thus compulsory), in dwellings the maximum level may not exceed 30 dB(A), in classrooms and workplaces 35 dB(A); values 5 dB(A) higher in each case are permissible for ventilation systems without noticeable individual noises. The simplest and most effective method for avoiding noises from building services and plant is to mount all the components on suitable elastic bearings. Designed and installed correctly, such bearings can achieve effective isolation for structure-borne sound between installation and roof. Many manufacturers fit elastic bearings to their equipment as standard, usually integrated into the feet of the equipment, and these sometimes consist of steel springs or elastomeric elements.
Critical for the planning of an elastic bearing for isolating structure-borne sound is that the resonant frequency must be sufficiently far below the frequency of the installation noise because otherwise the acoustic effect may be severely impaired, possibly even reversed. The reason for this is that the isolating effect first takes effect well above the resonance and increases steeply with the frequency (Fig. C 4.10). On the other hand, in the vicinity of the resonance the sound transmissions are amplified, increasing

the noise level in this frequency range.
In the case of an elastic bearing over the whole area, the resonant frequency f_R is as follows:

$$f_R = \frac{1}{2\pi} \sqrt{s'\left(\frac{1}{m''_1} + \frac{1}{m''_2}\right)} \ [Hz]$$

where:
$s' = E/d$
m''_1, m''_2 masses of vibrating components
s dynamic stiffness [N/m³]
E elastic modulus [N/m²]
d thickness of elastomeric layer [m]

It is either necessary to increase the mass of the building component or decrease the stiffness of the bearing in order to lower the resonant frequency. In critical cases, a two-part elastic bearing may be useful for isolating structure-borne sound in some circumstances. This can involve the use of two mass-spring systems, one above the other. Information about the technical and acoustic properties of elastic bearings can be found in VDI 3727, VDI 2062 and VDI 3833. In practice, elastic bearings, assuming they have been designed and installed correctly, can on their own reduce the noise level by about 10–20 dB, depending on the frequency spectrum of the noise source.

Protection against sound immissions

In commercial or industrial buildings housing noisy plant or workplaces, it is also important to consider the noise emissions from inside to outside, radiated from the building envelope, as well as the sound insulation requirements for the interior. The noise immissions into the surroundings can lead to disturbances and noise pollution for others and so the recommended immissions values contained in *TA Lärm* may not be exceeded. We make a distinction between the daytime and night-time values for recommended immissions levels. The values in each case relate to the so-called rating level L_r, which refers to the sound level at the immissions point, averaged over a certain period of time, plus surcharges for pitch, information and impulse content and the times of day with a higher sensitivity to noise. The recommended immissions values depend on the use of the

C 4.11

Use of buildings at immissions point	Max. day rating level L_r [dB(A)]	
	Day[1]	Night[2]
Industrial zones	70	–
Commercial zones	65	50
Central, village and mixed zones	60	45
General residential zones and small settlements	55	40
Exclusively residential zones	50	35
Health spas, hospitals, nursing homes	45	35

Times valid exclusively for industrial noise
Time periods: [1] 6 a.m. to 10 p.m., [2] 10 p.m. to 6 a.m.

C 4.12

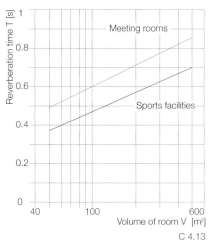

C 4.13

area affected and lie between 45 and 70 dB(A) during the day 35 dB(A), during the night 50 dB(A) (see Fig. C 4.12).

Diverse special rules, e.g. for brief noise peaks, infrequent events, masking of industrial noise by traffic noise, etc., apply in addition to the recommended values given in Fig. C 4.12. In some cases those may result in values more stringent than those given in the table because the recommended immissions values refer to the total of all noise events caused by the noise of trade and industry. If there are already operations in the vicinity of the immissions point which essentially exhaust the recommended value, new operations will have to comply with lower values.

The calculation of the sound level generated at a neighbouring immissions point due to the sound radiated from a building is carried out using DIN EN 12354-4 (Building acoustics – Estimation of acoustic performance of buildings from the performance of products – Part 4: Transmission of indoor sound to the outside), also DIN ISO 9613-2 (Acoustics – Attenuation of sound during propagation outdoors – Part 2: General method of calculation) and VDI 2720 (Noise control). Simplified methods for calculating the effective noise load can also be found in DIN 18005. In practice the sound propagation calculation is usually so complicated and time-consuming that it can only be carried out with the help of special software. Under certain conditions a simple theoretical estimate is often adequate for determining the noise load due to sound radiated from a flat roof:

- Unrestricted sound propagation between roof and immissions point (i.e. no screening elements in propagation path)
- Distance between immissions point and centre of roof surface is at least twice as large as the largest horizontal dimension of the roof (on a rectangular building the length of the diagonal)
- Distance between centre of roof and immissions point is no greater than approx. 200 m
- Uniform sound field within building (e.g. for adequately high single-storey sheds without internal walls)
- Approximately even frequency spectrum for plant noise

Under these conditions the sound radiated from the roof can be focused into a point sound source in the centre of the roof. In addition, there is no need to consider how meteorological conditions might affect the sound propagation, which means that the immissions level L_s can be calculated as follows:

$$L_S = L_i - R'_{w,res} +$$

$$[10 \lg (S_{tot}) - 20 \lg (s) - 17] \, dB(A) - A_{gr}$$

where $A_{gr} = \left[4.8 - \dfrac{2 \, h_m}{s} \left(17 + \dfrac{300}{s} \right) \right] dB(A) > 0$

and:
L_i internal level in building [dB(A)]
$R'_{w,re}$ resultant sound reduction index of roof surface [dB]
S_{tot} total area of roof [m²]
s distance between centre of roof and immissions point [m]
$h_{m'}$ mean height of sound ray above ground level [m]
A_{gr} ground attenuation [dB(A)]

As A_{gr} fluctuates between 0 and 4.8 dB, the ground attenuation can be ignored for an initial rough estimation of the immissions level, which means that the result of the calculation is on the safe side.

To compare the result with the recommended immissions values (Fig. C 4.12), a rating level must first be derived from the calculated immissions level. This is carried out in the simplest case for a uniform plant noise without pitch, information and impulse content by using an equation that relates to the day rating level over a period $T_r = 16$ h (6 a.m. to 10 p.m.):

$$L_r = 10 \lg \left(\frac{1}{T_r} \left[T_i \, 10^{(L_{s,i} + 6 \, dB(A))/10 \, dB(A)} + T_a \, 10^{L_{s,a}/10 \, dB(A)} \right] \right) dB(A)$$

where:
T_i duration of effect [h]
$L_{s,i}$ immissions level of plant noise in periods with a higher sensitivity to noise (6–7 a.m. and 8–10 p.m.) [dB(A)]

T_a duration of effect outside the aforementioned times [h]
$L_{s,a}$ immissions level of plant noise outside the aforementioned times [dB(A)]

The calculations are only suitable for a rough estimation of the resultant noise load. Where there is any doubt with respect to the applicability of the conditions or the result is ambiguous, a detailed sound propagation calculation must certainly be carried out according to the appropriate directive or regulations. If the calculation indicates that the valid recommended immissions values are exceeded, the resultant sound reduction index of the roof must be increased by an amount at least equal to that by which the values are exceeded. The measures necessary for this include increasing the mass, adding additional layers or providing a sound-absorbing soffit to the roof.

Room acoustics

Whereas disturbing noises infiltrating from outside can be handled by the sound insulation of the construction, the design of the room acoustics focuses on the critical sound sources in the room below a flat roof. Two variables, related to the use and nature of the room, are our prime concern here:

- The attenuation of noise
- The acoustic quality and intelligibility of speech, music and signals

These objectives lead to high demands in interiors intended for some form of communication, for example, because an appropriate noise level and also good intelligibility of speech are then essential. This can be achieved with effective room attenuation measures, especially on the underside of the roof. Measures for screening the sound in the room and directing or even scattering the sound may be useful as well in certain cases. The designer must therefore allow for the inclusion of sound-absorbing surfaces.

Building component	Material	Effective range of frequencies
Ceilings, also suspended (> 20 cm deep)	Fibreboards, soft foams	Broadband
	Perforated boards with fibre inlay	
	Microperforated elements	Medium and high frequencies
	Slit and strip-type absorbers	
Suspended acoustic panels, other screens in room	Fibreboards	Medium and high frequencies
	Soft foams	
	Microperforated boards or foils	
Plate resonator	Plain boards below air space, e.g. metal, plasterboard, wood-based product	Primarily low frequencies
Absorbent soffit	Absorber behind perforated or slit covering materials, e.g. trapezoidal profile sheeting	Medium and high frequencies
	Porous absorber with stretched fabric covering	

C 4.14

Requirements

Traditionally, the critical – and in many cases the only – variable for characterising the attenuation of an interior is the reverberation time T:

$$T = 0.16 \, V/A \, [s]$$

where:

A   equivalent absorption surface in receiving room ($\sum \alpha_n S_n$) [m^2]
S_n partial surface area of room [m^2]
α_n sound absorption coefficient of partial surface area of room
V   volume of room [m^3]

In smaller rooms there is a direct relationship between reverberation time and intelligibility of speech and a reduction in the noise level so that the reverberation time can be quantified appropriately and can be regarded as the primary requirement for acoustic design. The value should be considered in the customary hearing range of 63 Hz to 8 kHz, although this range should be extended in certain cases, e.g. production plants with low-frequency humming. The measurement and design of room acoustics plus its treatment with appropriately low-tuned sound absorbers requires considerable work for frequencies around 50 Hz. Many sound sources, e.g. transformers, engines and compressors, generate intensive disturbing noise in this range.

The recommendations of DIN 18041 apply to many rooms in which communication is important; and for rooms in schools and nurseries, the known guidelines should be consulted [6]. In dwellings personal responsibility applies and in noisy production plants reducing the reverberation time is often a welcome noise abatement measure. Halving the reverberation time results in an approx. 3 dB reduction in the mean noise level (see p. 79–80). Fig. C 4.13 shows the volume-related values for the reverberation time in meeting rooms and sports halls. A fixed reverberation time established by a specialist, which takes into account the specific usage, should always be adhered to for larger sports halls, production centres and special uses.

Sound-absorbing components

Many different products made from a diverse range of materials are available for designing sound-absorbent soffits beneath flat roofs (Fig. C. 4.14). A number important aspects must be considered when selecting a solution, e.g. fire protection, hygiene and mechanical loads (e.g. resistance to ball impact in sports facilities). Loose mineral fibre layers should therefore always be covered and a practical cleaning option for vertical panels suspended from the soffit must be considered. Regular renovation or repair work should not be ignored either; simply applying a new coat of paint, for example, is often not possible with many room acoustics solutions, which means that in such cases it may be necessary to replace an entire soffit. When positioning sound absorbers in a room, the principle of a uniform distribution over all wall and soffit surfaces is to be recommended. Where this is not possible, the soffit should be preferred. Panels or other measures suspended from the ceiling give architects further design freedoms.

C 4.11   Microperforated sound absorber as printed foil with additional low E coating as sunshade and anti-glare measure
C 4.12   Recommended immissions values for various building uses according to the German Technical Rules for Protection Against Noise (TA Lärm)
C 4.13   Reverberation times to be adhered to for separate rooms for meetings and sport
C 4.14   Examples of customary sound-absorbing forms of construction for the soffits to, or ceilings below, flat roofs and their range of effectiveness

References
[1] Weber, Lutz; Koch, Siegfried: Anwendung von Spektrum-Anpassungswerten. Teil 1: Luftschalldämmung. In: Bauphysik 21/1999, pp. 167ff.
[2] Weber, L., Schreier, H.; Brandstetter, K.-D.: Measurement of sound insulation in laboratory – comparison of different methods. In: Proceedings – International Conference on Acoustics. Rotterdam, 2009, pp. 701ff.
[3] Verdict of the Federal Court of Justice of 14 Jun 2007 on the sound insulation between semi-detached houses, ref. No. VII ZR 45/06
[4] Scholl, Werner; Bietz, Heinrich: Integration des Holz- und Skelettbaus in die neue DIN 4109. Final report. Stuttgart, 2005
[5] Weber, L.; Seidel, J.; Rotaru, D.; Zhou, X.: Messung von Regengeräuschen nach DIN EN ISO 140-18. In: Fortschritte der Akustik. Band 2. Berlin, 2006, pp. 465ff.
[6] Baden-Württemberg Environment Ministry (pub.): Lärmschutz für kleine Ohren. Leitfaden zur akustischen Gestaltung von Kindertagesstätten. Stuttgart, 2009

Part D Design principles

Fig. D Opera house, Oslo (N), 2008, Snøhetta

Materials

Wolfgang Zillig

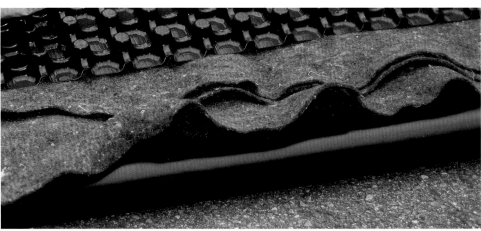

D 1.1

The right choice of materials is especially crucial for flat roofs if an impervious roof as well as the associated building physics requirements such as thermal performance, sound insulation and fire resistance are to be guaranteed over the long-term. The choice of materials is as diverse as the range of flat roof designs. But not all materials are equally suitable for every design, every application.

The descriptions of materials on the following pages essentially follow the constructional make-up from outside to inside; the loadbearing elements are described in Part B (pp. 23–47). Those roofing materials that are primarily used for pitched roofs but can also be used for flat roofs (up to 10° pitch) are granted special status. This mainly concerns sheet metal and glass; both are mentioned briefly here, but for more information the reader should consult the appropriate publications [1].

Generally, the valid fire protection stipulations require that the fire resistance of materials used for flat roofs comply with building materials class B 2 (to DIN 4102-1) or E-d 2 (to EN 13501-1) at least. The detailed fire protection requirements for flat roofs can be found in the chapter "Fire" (pp. 74–77).

Waterproofing

A waterproof layer protects the underlying construction and the building against precipitation. To guarantee the waterproofing function for the service life of the roof, the materials chosen must be able to withstand high temperature fluctuations (up to 80 K within 12 hours) and the UV radiation component of the solar irradiation (up to 1200 W/m²).

Waterproofing materials according to the German Flat Roof Directive

Harmonisation with Euronorms means that the normative requirements for waterproofing layers are divided among product, application and construction standards. DIN EN 13707 is the product standard for flexible bitumen sheeting, DIN EN 13956 the standard for plastic and elastomeric sheeting. In contrast to the national German standards valid up until now, there are hardly any limiting values, e.g. minimum thickness, included in these product standards. In order to achieve the current quality standard usual in Germany, the corresponding minimum requirements for waterproof sheeting for roofs have been incorporated into the application pre-standard DIN V 20000-201, which must be adhered to in Germany. This classifies various types of application:

- DE: flexible waterproof sheeting for single-ply proofing
- DO: flexible waterproof sheeting for the cap sheet of built-up roofing
- DU: flexible waterproof sheeting for the base sheet of built-up roofing
- DZ: flexible waterproof sheeting for the intermediate sheet(s) of built-up roofing

Among the aspects dealt with in DIN 18531 "Waterproofing of roofs – Sealings for non-utilized roofs" are the general properties and types of loads to which roof waterproofing materials are subjected. Application categories are specified depending on the quality of the flat roof waterproofing:

- K 1: standard design: roof waterproofing that satisfies normal requirements
- K 2: premium design: roof waterproofing that satisfies higher requirements needed for higher-value building uses or where access is difficult

The subdivision into loading classes is carried out based on the thermal and mechanical loads to which the waterproofing is subjected (Fig. D 1.2). The designer must allocate the respective roof construction to a loading class:

- I: high mechanical load
- II: moderate mechanical load
- A: high thermal load
- B: moderate thermal load

The properties classes for waterproof sheeting are in turn based on the loading classes (Fig. D 1.2):

- E 1: resistance to high thermal and high mechanical loads
- E 2: resistance to moderate thermal and high mechanical loads
- E 3: resistance to high thermal and moderate mechanical loads

D 1.1 Drainage element, protective fleece and water-proofing material
D 1.2 Loading and properties classes for roof water-proofing materials to DIN 18531
D 1.3 Building physics parameters for waterproofing materials to DIN EN ISO 10456
D 1.4 Classification of flexible bitumen sheeting

• E4: resistance to moderate thermal and moderate mechanical loads

Besides flexible sheeting made from bitumen and synthetic materials, the Flat Roof Directive also allows the use of liquid plastics for waterproofing roofs (Fig. D 1.8, p. 89). The draft of DIN 18531 includes these liquid waterproofing materials. Only those systems that have been awarded a National Technical Approval on the basis of ETAG 005 and are also approved by the German building authorities may be used. Fig. D 1.3 shows various building physics parameters for waterproof sheeting to DIN EN ISO 10456. Owing to their low insulating effect in comparison with thermal insulation and their minimal thickness, roof waterproofing materials can be neglected when calculating the thermal performance of a roof.

Bitumen flexible sheeting
Bitumen is black and is therefore frequently mistaken for tar; it is, however, a totally different product. The mistake is also partly due to the fact that up until the 1980s bitumen, tar, pitch and asphalt were grouped together under the heading of "bituminous materials". Today, of these substances, only bitumen is used for waterproofing flat roofs.

Manufacture
Bitumen is obtained through the distillation of petroleum and essentially consists of high-molecular hydrocarbons. We distinguish between straight-run bitumen (soft bitumen), vacuum asphaltic bitumen and hard paving-grade bitumen, as well as blown bitumen and polymer-modified bitumen depending on type of manufacture and application. Of these different types, blown bitumen and polymer-modified straight-run bitumen are used for waterproof sheeting (Fig. D 1.4).

Properties
Bitumen is a thermoplastic material; it becomes brittle upon cooling and passes smoothly through all the states from solid to viscous to runny upon being heated. As bitumen is almost insoluble in water, it is ideal for protecting building components against water. However, it is not permanently resistant to fuels such as petrol and diesel.
Bitumen flexible sheeting (often referred to simply as "roofing felt") does not consist of pure bitumen. It includes backings and inlays in order to improve its mechanical properties, e.g. tensile strength. Typical materials for this are polyester fleece (PV), glass cloth (G), glass fleece (V) and composite inlays where the main material is glass (KTG) or polyester (KTP). Metal foil inlays (e.g. aluminium, copper) are used as vapour barriers. In this case the inlay

is sandwiched between layers of the same type of bitumen.
Polymer-modified bitumen flexible sheeting is divided into elastomeric (PYE) and plastic (PYP) types depending on the elastic or plastic properties of the polymer added to the bitumen. Modification with polymers is necessary for improving the properties of the flexible bitumen sheeting, primarily reducing the sensitivity to temperature fluctuations. The plastic (e.g. good stability at high temperatures) or elastic (e.g. good flexibility at low temperatures) properties can be specifically controlled depending on the polymer used.
The UV resistance of bitumen flexible sheeting varies. Only sheeting of polymer-modified bitumen are suitable for the top layer.
PYP bitumen flexible sheeting exhibits better ultraviolet and infrared stability than the PYE

Requirements		Mechanical			
		High		Moderate	
		Loading class	Properties class	Loading class	Properties class
Thermal	High	I A	E1	II A	E3
	mod.	I B	E2	II B	E4

D 1.2

Material	Water vapour diffusion resistance factor μ [-]	Thermal conductivity [W/mK]	Density [kg/m³]
Polymer-modified bitumen	50,000	0.23	1100
PE-C [1]	30,000	0.33–0.50	920–980
PVC-P [1]	13,000–22,000	0.17	1390
EPDM	6000	0.25	1150
PIB [1]	260,000	0.20	930–1100
ECB	50,000–90,000		
FPO	150,000		
EVA	20,000–23,000		
IIR	200,000	0.24	1200

[1] Figures for flexible sheeting type PYE-PV 200 S 4 used as an example

D 1.3

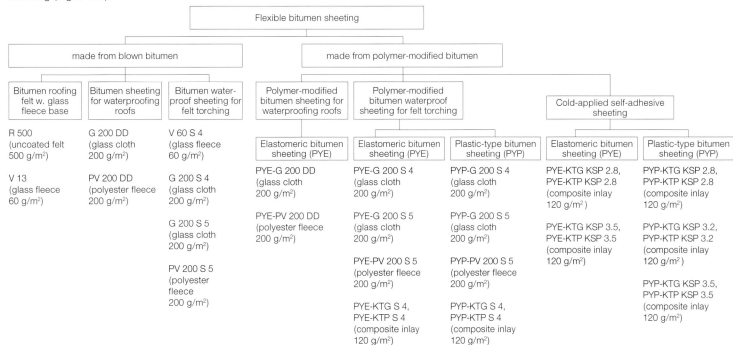

D 1.4

Code	Description
PYE	Elastomeric bitumen (bitumen modified with thermoplastic elastomers)
PYP	Plastic-type bitumen (bitumen modified with thermoplastics)
KSP	Cold-applied self-adhesive polymer-modified flexible bitumen sheeting with inlay
KSK	Cold-applied self-adhesive flexible bitumen sheeting with HDPE backing
V (figure)	Glass fleece (figure for V 60 = weight per unit area [g/m²], for V 13 = soluble matter content in 1/100th of content [g/m²])
PV (figure)	Polyester fleece (weight per unit area [g/m²])
G (figure)	Glass cloth (weight per unit area [g/m²])
R (figure)	Uncoated felt (weight per unit area [g/m²])
Vcu	Composite backing of glass fleece 60 g/m² to DIN 52141 with polyester-copper composite foil ≥ 0.03 mm
Cu01	Strip copper inlay 0.1 mm to DIN EN 1652
KTG	Composite inlay with predominantly glass component
KTP	Composite inlay with predominantly polyester component
S (figure)	Bitumen waterproof sheeting for felt torching (thickness of uncoated sheeting [mm])
DD	Between sheeting for waterproofing roofs
figure	Thickness of sheeting [mm]

D 1.5

D 1.5   Codes for classifying flexible bitumen sheeting
D 1.6   Applying flexible bitumen sheeting by means of felt torching
D 1.7   Applying flexible bitumen sheeting by means of pouring and rolling
D 1.8   Types of waterproofing with the respective welding or bonding methods and minimum joint widths
D 1.9   Requirements to be satisfied by roof waterproofing materials made from bitumen and polymer-modified bitumen according to the German Flat Roof Directive
D 1.10   Minimum nominal thicknesses for synthetic flexible sheeting to properties class E 1 in application categories K 1 and K 2
D 1.11   Joint in synthetic flexible sheeting produced by hot-air welding
D 1.12   Joint in synthetic flexible sheeting produced by solvent welding

types, which means that they it be used without protection on the surface. A factory-applied granular finish to the cap sheet is, however, recommended. Waterproof sheeting made from elastomeric bitumen must be covered with chippings, granules or a suitable coating in order to achieve a reasonable UV resistance.

Codes are used to indicate the different types of flexible sheeting (Fig. D 1.5).

Laying

The thermoplastic behaviour of bitumen is exploited during laying as well, in the form of a naked flame or hot gas (felt torching, see Fig. D 1.6) which melts the separate sheets together. As an alternative, the sheets can be bonded together (pouring and rolling, see Fig. D 1.7). Mopping is a method in which a sufficient amount of liquid bitumen (blown or polymer-modified) is spread out in front of the roll of sheeting material so that it forms a continuous bulge of bitumen just ahead of the roll of material as it is unrolled. However, this method is no longer used on flat roofs these days. There are also self-adhesive bituminous flexible sheets that can be applied cold. In this case a backing foil, which prevents the the roll of material adhering to itself during storage, is removed as laying proceeds. In all cases the method of laying must guarantee adequate protection against wind suction forces.

Bituminous waterproof sheeting is generally laid in several plies (built-up roofing) with overlapping joints and seams. Fine cracks form in the cap sheet due to the rolling and unrolling, and not just the weather. Water can seep through these cracks and so the waterproofing effect is only achieved by melting the plies together over their full area. A minimum of two plies is therefore the norm. The full bond ensures that any cracks are closed off. Fig. D 1.9 lists the requirements to be met by bituminous waterproof sheeting depending on the application and loading categories.

Bitumen flexible sheeting is supplied in rolls 1 m wide and up to 20 m long. These rolls must be stored vertically and protected against UV radiation, heat and moisture.

Plastic and elastomeric flexible sheeting

Unlike flexible sheeting based on bitumen, synthetic waterproofing materials achieve their moisture barrier effect not by melting two or more plies together, but by the water-repellent effect of the sheeting cross-section itself. Careful design and workmanship at joints and seams are therefore especially important.

The application standard DIN V 20000-201 lists groups of materials suitable for waterproofing flat roofs.

Laying

Synthetic flexible sheeting is generally laid loose, i.e. to protect against wind uplift it requires either mechanical fasteners or ballast. At most, only dabs or strips of adhesive are applied because otherwise the material's positive properties and flexibility cannot be retained and exploited. However, laying loose is an advantage for repairs and maintenance: it is relatively easy to remove synthetic flexible sheeting and so it should be preferred to bitumen flexible sheeting where this aspect is relevant.

Great care must be taken at seams and joints when laying synthetic flexible sheeting. Various jointing methods can be used at the joints between synthetic flexible sheeting or between the sheeting and other materials:

- Hot-air welding (the most common method, see Fig. D 1.11)
- High-frequency or hot-wedge welding (generally only in the factory, not on the building site)
- Solvent welding (Fig. D 1.12)
- Sealing tapes or strips

Flexible sheeting complying with properties class E 1 can be laid in a single layer on all roofs irrespective of loading class. Application categories K 1 and K 2 require different minimum thicknesses of waterproof sheeting which must be complied with depending on the materials group (Fig. D 1.10). The nominal thickness is defined as the thickness of the waterproof sheeting without any facings and/or self-adhesive coatings.

Synthetic flexible sheeting is supplied in rolls up to 2 m wide and max. 25 m long.

D 1.6

D 1.7

Ethylene copolymer bitumen (ECB)
Waterproof sheeting made from ethylene copolymer bitumen (ECB) is manufactured from granulate in an extrusion process with a central glass fleece inlay. The granulate, the raw material, consists of a mixture of ethylene and bitumen, which leads to a combination of the properties of the two substances. The ethylene component increases the mechanical strength and toughness, also the chemical resistance, whereas the bitumen component functions as a plasticiser, leads to good durability at low temperatures and guarantees compatibility with bitumen. The glass fleece ensures dimensional stability.
ECB sheeting is resistant to aqueous solutions of acids and bases. It can still be welded even after many years, thus facilitating repairs to damaged areas.

Thermoplastic polyolefin (TPO) and flexible polyolefin (FPO)
Waterproof sheeting made from thermoplastic or flexible polyolefin consist of polyolefins based on polyethylene or polypropylene with integral reinforcement, usually a polyester fabric. FPO flexible sheeting is generally compatible with bitumen and polystyrene, but not with PVC. A slip layer (e.g. 0.2 mm PE film) should be included when this material is laid directly on concrete, or flexible sheeting with an integral fleece underlay should be specified.

Polyvinyl chloride (PVC)
Polyvinyl chloride (PVC) is produced through the polymerisation of the monomer vinyl chloride. Only through the addition of stabilisers can PVC be made resistant to UV radiation and the effects of the weather. As pure PVC is hard and relatively brittle, plasticised polyvinyl chloride (PVC-P) is added in order to achieve flexibility. Waterproof sheeting made from PVC is available unreinforced or with an inlay of glass or polyester fleece, for instance.
The range of possible modifications and the associated material properties is very large and so is it difficult to describe the general features of PVC flexible sheeting; the properties extend right up to bitumen-compatible sheeting. As emissions of plasticisers contained in PVC are possible under certain circumstances, contact with certain materials should be avoided, e.g. by using a separating membrane. Admissible and inadmissible material combinations are listed in the data sheets of the individual manufacturers.
PVC flexible sheeting is not resistant to organic solvents (e.g. petrol, toluene), greases or oils. Compared with other synthetic waterproofing materials, it has a low resistance to vapour diffusion.
One disadvantage is that in a fire PVC gives off CFC and other gases such as chlorine and dioxin. But a positive aspect is that flexible sheeting made from PVC can still be readily repaired even many years after it has been laid. Used PVC sheeting recovered from a

Type	Base material	Welding/bonding method	min. joint width
Flexible bitumen sheeting with inlay and facing plies made from	Blown bitumen	Rolling and pouring Felt torching Mopping Cold application	80 mm
	Bitumen modified with thermoplastic elastomers (PYE)		
	Bitumen modified with thermoplastics (PYP)		
Synthetic flexible sheeting [1] made from	Chlorinated polyethylene (PE-C)	Solvent welding Hot-air and hot-wedge welding	30 mm 20 mm
	Ethylene copolymer bitumen (ECB)	Hot-air and hot-wedge welding	30 mm
	Ethylene vinyl acetate copolymer (EVA)	Solvent welding Hot-air and hot-wedge welding	30 mm 20 mm
	Flexible polyolefins (FPO)	Hot-air and hot-wedge welding	20 mm
	Polyisobutylene (PIB)	Solvent welding	30 mm
	Polyvinyl chloride (PVC-P)	Solvent welding Hot-air and hot-wedge welding	30 mm 20 mm
Elastomeric sheeting made from	Ethylene propylene diene rubber (EPDM)	Contact adhesive Sealing strips/tapes Hot vulcanising	50 mm 40 mm
	Isobutylene isoprene rubber (IIR)		
	Thermoplastic elastomers (TPE)	Hot-air and hot-wedge welding	20 mm
Liquid waterproofing made from	Flexible unsaturated polyester resins (UP)	–	–
	Flexible polyurethane resins (PUR)		
	Flexible reactive methyl methacrylates (PMMA)		

[1] Joints in synthetic flexible sheeting on site can be achieved with sealing strips/tapes and in a factory by high-frequency welding as well.

D 1.8

Roof waterproofing application category	Roof waterproofing loading class	Type of roof waterproofing (No. of layers and properties class [1])		
K 1	I A, I B, II A, II B	2 plies	Cap sheet: DO/E 1	
			Base sheet: DU/E 2	
	II A, II B	2 plies	Cap sheet: DO/E 1	
			Base sheet: DU/E 4 [2]	
	I A, I B, II A, II B	1 ply	DE/E 1 [3]	
K 2	I A, I B, II A, II B	2 plies	Cap sheet: DO/E 1	
			Base sheet: DU/E 1	

[1] Properties class according to product data sheet for flexible bitumen sheeting, tab. 1
[2] V 60 S 4 at least
[3] Only permissible for sheeting according to product data sheet for flexible bitumen sheeting, tab. 5, line 10

D 1.9

Materials group	Minimum nominal thickness [mm] for application category...	
	K 1	K 2
Ethylene copolymer bitumen (ECB)	2.0	2.3
Flexible polyolefin (FPO)	1.2	1.5
Polyvinyl chloride (PVC-P), plasticised, not bitumen-compatible, homogeneous	1.5	1.8
Polyvinyl chloride (PVC-P), plasticised, not bitumen-compatible, with inlay, reinforcement or facing	1.2	1.5
Polyvinyl chloride (PVC-P), plasticised, bitumen-compatible	1.2	1.5
Ethylene vinyl acetate copolymer (EVA)	1.2	1.5
Ethylene propylene diene rubber (EPDM) with reinforcement	1.3	1.6
Ethylene propylene diene rubber (EPDM) with reinforcement and polymer-modified bitumen ply on one side (PBS)	1.3	1.6
Ethylene propylene diene rubber (EPDM), homogeneous	1.1	1.3
Polyisobutylene (PIB)	1.5	1.5 [1]
Chlorinated polyethylene (PE-C)	1.2	1.5
Thermoplastic elastomers (TPE)	1.2	1.5
Isobutylene isoprene rubber (butyl rubber, IIR)	1.2	1.5

[1] Additional conditions: see trade guidelines for waterproofing – Flat Roof Directive, Oct 2008 edition, tab. 5

D 1.10

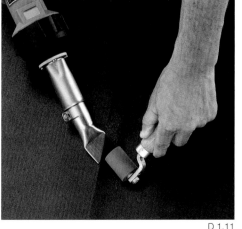

D 1.11

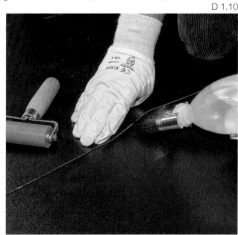

D 1.12

D 1.13

D 1.14

building can be recycled via the "Roofcollect" system operated by the European Single Ply Waterproofing Association (ESWA).

Ethylene vinyl acetate copolymer (EVA) and vinyl acetate ethylene copolymer (VAE)
Flexible sheeting made from ethylene vinyl acetate copolymer (EVA) and vinyl acetate ethylene copolymer (VAE) are based on ethylene vinyl acetate and PVC, with the PVC content in EVA sheeting being much higher. Chalk is the main filler added to the synthetic materials. Chalking can occur and that can lead to welding problems during later repairs. EVA and VAE products exhibit a low diffusion resistance compared with other types of synthetic flexible sheeting. After use, these materials can also be recycled via the "Roofcollect" system.

Ethylene propylene diene rubber (EPDM)
Flexible sheeting made from ethylene propylene diene rubber (EPDM) belongs to the group of rubber waterproofing materials. It are available with and without an inlay, also with an integral glass or polyester fleece underlay. EPDM flexible sheeting is frequently factory-prefabricated and supplied to site in large sheets already cut to size. We divide these products into three basic types:
• Single-ply, calendered rubber sheeting: this type of sheeting can be joined by way of hot vulcanising, but this process is generally only employed in the factory.
• Three-ply rubber sheeting: only the central ply is fully vulcanised, the outer plies consist of thermoplastic, non-vulcanised rubber; solvent welding with citric acid is used for joining the plies together.
• Rubber sheeting with a layer of hot-melt adhesive: a polymer-modified base sheet enables the welding to be carried out with hot air; joining to other layers is achieved by bonding the bitumen base to the rubber surface.

EPDM is UV-resistant and can be used without any protection to its surface. In addition, it is resistant to many chemicals.
Welding EPDM sheeting is comparatively problematic. Ageing of the material exacerbates this

problem, which is why flexible sheeting made from EPDM is regarded as less conducive to repairs.

Polyisobutylene (PIB)
Polyisobutylene (PIB) is manufactured through the polymerisation of the monomer isobutene (2-methylpropene). The first sheeting made from polyisobutylene appeared on the market in the 1960s. However, not until a synthetic fleece underlay was included did these products become really suitable for roof applications. PIB flexible sheeting is available with factory-fitted self-adhesive or solvent-weldable sealing edges. Waterproof sheeting made from PIB is compatible with bitumen products but is not resistant to organic solvents (e.g. petrol, toluene and petroleum) nor substances containing solvents (e.g. lacquers and paints), nor greases or oils.

Chlorinated polyethylene (PE-C)
Waterproof sheeting made from chlorinated polyethylene (PE-C) is soft without the addition of plasticisers, is compatible with bitumen and generally has a high resistance to chemicals. Self-adhesive PE-C sheeting is also available. Used materials can be recycled via the "Roofcollect" system.

Thermoplastic elastomers (TPE)
Thermoplastic elastomers (TPE) are based on EPDM and PP. As their name suggests, they can be shaped upon the application of heat and exhibit partially elastic properties. Seams and joints are connected by thermal welding. Flexible sheeting made from TPE exhibits good weathering resistance.

Isobutylene isoprene rubber (IIR)
Isobutylene isoprene rubber (IIR), also known as butyl rubber, belongs to the elastomers group. The material standard DIN 7864-1 "Sheets of elastomers for waterproofing; terms of delivery" does not define the material composition any further. It merely states that waterproof sheeting made from rubber combined with additives and processing agents is drawn out into sheets and vulcanised to form an elastomer. DIN 20000-201 requires, in addition, that the butyl

elastomer content must be min. 25 %. The further components are other polymers, flame retardant, stabilisers, fillers, processing agents and/or pigments. IIR exhibits good weathering resistance and is resistant to acids and alkalis, but not oils or greases.

Liquid waterproofing
The Flat Roof Directive lists synthetic liquid waterproofing materials made from flexible unsaturated polyester resins (UP), flexible polyurethane resins (PUR) and flexible reactive polymethyl methacrylate (PMMA). It is therefore identical with the draft of DIN 18531-2.
A waterproofing material applied as a liquid is advantageous when there are numerous details, junctions and penetrations to be waterproofed in a confined space (Fig. D 1.13). A layer of fleece is always essential because without it the material would only be classed as a coating and would therefore not comply with the requirements. The fleece increases the tearing strength, limits the expansion and helps to guarantee the necessary layer thickness. The main disadvantage of this type of roof waterproofing is the higher cost of the materials in comparison to waterproofing in the form of flexible sheeting. This is compensated for to some extent by the improved workability, especially when there are many details and penetrations.
Liquid waterproofing can be used as a single-ply waterproofing product according to application type DE and properties class E 1 provided the following requirements are satisfied:
• Pre-treatment of substrate
• Surface temperature min. 3 K above dew point temperature
• Suitable substrate
• Reinforcing fleece 110 g/m^2
• Fleece overlapping at joints
• Min. thickness 1.8 mm for UP and PUR, 2.1 mm for PMMA

Laying
The liquid waterproofing material is spread over the roof (Fig. D 1.14). If the liquid plastic consists of several components, these are mixed on site prior to applying them to the surface.
Three coats are usually applied: the fleece inlay is laid in the first coat of liquid plastic and then two further coats of liquid plastic are spread over this. At details, pieces of fleece are cut to size as requiered, soaked in the liquid plastic and then fitted into place. Any air bubbles can be removed with a brush or roller.
Curing takes place either through physical drying or by a chemical reaction. This process depends on the temperature, and in some cases the curing can be influenced somewhat by employing an additive that accelerates or retards the reaction. Such waterproofing materials cannot be applied at temperatures below about 5 °C.

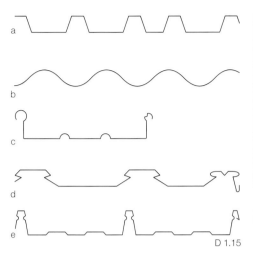

D 1.15

D 1.16

D 1.13 Laying the fleece for a liquid waterproofing
        material
D 1.14 Applying the liquid waterproofing material
D 1.15 Common profiles for metal roofing
        a Trapezoidal
        b Corrugated
        c Standing seam (aluminium, galvanised steel)
        d Ribbed profile (galvanised steel)
        e Standing seam (copper)
D 1.16 Flexible sheeting with integral photovoltaic
        elements

Waterproofing layers with integral photovoltaic elements

Flexible waterproof sheeting with integral photo-voltaic elements, like that shown in Fig. D 1.16, has been available for about 10 years. This represents a fairly straightforward way of using a roof surface to generate electricity; there is no need to mount the photovoltaic modules above the roof surface on complicated frame-works – the traditional solution. The weight per unit area of the flexible sheeting including the photovoltaic elements is very low, which means there is hardly any extra load to be carried by the structure. However, the comparatively shallow pitch of a flat roof places the photovoltaic elements at a less favourable angle to the sun and reduces their degree of efficiency. Ponding of water must be avoided and a self-cleaning effect guaranteed, which calls for a minimum pitch of 3°.

Both synthetic and bitumen flexible sheeting can be used as a backing material for the photovoltaic elements, which are laminated to the sheeting and wired up in the factory. The photovoltaic elements used frequently consist of high-performance thin-film solar cells based on amorphous silicon (a-Si). The manufacturers of such flexible sheeting with integral photovoltaic elements generally offer a warranty of 80 % of nominal output after 20 years.

Sheet metal roofing

Metal can be used as a flat roof covering as an alternative to flexible sheeting. The difference between a metal roof covering and the types of flexible sheeting described above is that because of the joints in a metal roof covering it can only be classed as rainproof, not water-proof. In Germany there is therefore a separate set of regulations and guidelines for metal roof coverings. Steel, stainless steel, aluminium and titanium-zinc are the metals most frequently used in conjunction with flat roofs. In addition, copper and lead are sometimes used as parapet cappings. Metal roof coverings are available in the form of coils or profiled sheets (Fig. D 1.15; see "Metal roofs", p. 108).

Metals can corrode in atmospheres with a high humidity or through direct contact with water. Corrosion is the chemical or electrochemical reaction of a metallic material with its environment, which leads to a measurable change in the material. There are different forms of corrosion: atmospheric corrosion, colloquially known as rusting, and galvanic corrosion. The former involves the oxidation of iron or steel in the presence of water and oxygen. Galvanic corrosion occurs when different metals are in contact in the presence of an electrolyte (e.g. water), which results in the degradation of the less noble metal. The electrochemical series must therefore be considered when combining different metals, a ranking that extends from the non-noble metals magnesium and aluminium right up to the noble metals silver and gold. One of the effects of this is that, for example, in pipes for roof drainage a metal higher in the series should be used further downstream behind a less noble metal in the direction of flow.

As sheet metals can be regarded as practically diffusion-tight, special building physics requirements must be taken into account (see "Protection against condensation as a result of water vapour diffusion", pp. 65–66).

Glass

Glass is used as a roofing material primarily on pitched roofs. Flat roofs of glass are therefore regarded as special forms of construction (see "Glass", pp. 44–47, and "Glass roofs", pp. 109–110).

Glass is an amorphous solid with inorganic constituents. The raw materials used for its production are quartz sand, soda, limestone and dolomite. During production these are heated to such a high temperature that they become runny. The amorphous condition is achieved by cooling the melt so rapidly that there is no time for a crystalline structure to form.

Protection against falling fragments of glass from damaged panes is absolutely paramount when using glass in roofs. Laminated safety glass is comprised of at least two panes of glass with an elastic interlayer between each pane which holds the fragments of glass together upon breakage. Polyvinyl butyral (PVB) in sheet form is the material normally used for the inter-layer.

Multi-pane insulating glass is employed in order to meet thermal performance requirements. The individual panes are interconnected via a hermetic edge seal, which creates an enclosed cavity between the panes. Inorganic substances (normally noble metal or metal oxide coatings) can be applied to certain pane surfaces in order to improve the thermal performance of the insulating glass. These coatings reduce the emissivity in the range of infra-red radiation and hence cut down the heat flow through the glazing. Another method for improving the thermal performance is to fill the cavity with a noble gas such as argon or krypton. The inner panes of all insulating glass units in flat roofs must be made from laminated safety glass.

Protecting the surface

Where the waterproofing material itself is not durable enough, then additional materials will be required to protect the waterproofing material against mechanical loads and the effects of the weather. We distinguish between light-weight and heavyweight protective materials. Lightweight surface protection in the form of chippings is used for bituminous flexible sheeting and essentially provides protection against UV radiation. Heavyweight surface protection can be in the form of loose gravel, stone or concrete flags, or planting. The high mass of such finishes reduces the effects of temperature fluctuations and therefore increases the service life of the waterproofing material below.

Chippings

Of the bituminous flexible sheeting types, only PYP products do not require any additional protection because this type of flexible sheeting exhibits adequate resistance to UV radiation. But even with this type, a lightweight covering of chippings is often used merely to improve the appearance. The chippings consist of fine fragments of slate with particle sizes from 0.6 to 1.2 mm which are available in various colours (e.g. blue-green, green, brown, red-brown). The chippings are worked into the roof surface with a roller while the so-called chippings com-

D 1.17 Example of extensive planting
D 1.18 Example of intensive planting, raised roof finish
D 1.19 Plants for extensive planting (selection)

Botanical name	Common name	Height [cm]	Bloom colour	Bloom time	Leaf colour	Particular features
Allium schoenoprasum	chives	20	violet	V–VI		Loose cushion, grey foliage
Anthemis tinctoria	yellow chamomile	40	yellow	VI–VIII		
Campanula rotundifolia	harebell	10–20	light blue	VI–VIII		Tolerates some shade
Dianthus carthusianorum	Carthusian pink	30	pink	VI–IX		Loose cushion
Dianthus deltoides in varieties	maiden pink	15	pink, white, red	VI–VI	green-brown	Lawn-like
Festuca ovina	sheep's fescue	0–30		V–VII	grey-green	
Hieracium aurantiacum	orange hawkweed	20	red	VI–VIII		Wool-like hair, cont. flowering, robust
Petrorhagia saxifraga	tunic flower	20	pink-white	VI–VIII		Prolific seeding, without becoming burdensome
Potentilla verna	creeping potentilla	10	yellow	III–V		Green blankets
Prunella grandiflora	large self-heal	20	violet	VI–VIII		Self-propagating, good blanket formation
Sempervivum species and varieties	houseleek	10	pink, red	VI–VII		Many colours
Thymus serpyllum	wild thyme	5	pink	VII–IX		Forms a flat cushion
Origanum vulgare	wild marjptram	30	pink-lilac	VII–X		Honey flora
Sedum album "Coral Carpet"	white stonecrop	5	white	VI–VIII	green, red	Leaves red in winter, moderately vigorous
Sedum album "Murale"	white stonecrop	3–5	pink	VI–VII	brownish	
Sedum cauticolum "Ruby Glow"	–	20–25	red	VIII	red-brown	Valuable autumn flower
Sedum flor "Weihenstephaner Gold"	–	15	golden yellow	VII–VIII	green	Deciduous, attractive blossoms
Sedum hybridum	immergrunchen	10	yellow	VI–VIII	green	Winter greenery, tolerates shade
Sedum lydium	stonecrop	3–5	white-pink	VI–VII	light grey-green	Green moss carpet
Sedum reflexum	prickmadam	15	yellow	VI–VII	green	
Sedum sexangulare	six-sided stonecrop	5	yellow	VI–VIII	yellowish green	Very persistent
Sedum spurium "Album Superbum"	Caucasian stonecrop	10	white	VII–VIII	green	Vigorous, tolerates shade
Sedum spurium "Fuldaglut"	Fulda glow	10	dark red	VII–VIII	red	
Sedum telephium "Autumn Charm"	stonecrop	40	rust red	IX–X	light green	

D 1.19

pound is still hot. A complete covering of chippings achieves the desired protection for the waterproofing underneath.

Loose gravel

A heavyweight material, normally a min. 50 mm layer of loose gravel with 16/32 mm grading, can be used to provide protection against the effects of the weather (see "Roof finishes", pp. 98–99). Such a heavy protective layer also simultaneously acts as ballast against wind uplift, although in this case, as prescribed in DIN 1055-4, corresponding structural verification is required (see "Wind pressure and wind suction", p. 28). A roof with a loose gravel covering is regarded as a hard roof covering according to DIN 4102-4 and therefore provides protection against sparks and radiant heat.

Stone and concrete flags

Stone or concrete flags can be laid on protective sheeting, in a bed of chippings or gravel, or raised above the waterproofing on proprietary supports. Any frost-resistant material can be used for the flags. To ensure protection against wind uplift, flags should not be smaller than 400 × 400 × 40 mm. A layer of flags is, like loose gravel, considered to be a hard roof covering meeting the requirements of DIN 4102-4.

Rooftop planting

Besides the ecological and urban planning benefits, planting on flat roofs has a constructional advantage because it protects the waterproofing material against extreme temperature fluctuations and the effects of the weather. The actual construction of a green roof is discussed in detail later (see "Green roofs", pp. 104–106).

Planting

The actual plants used must be coordinated with the substrate (growing medium) and the entire sequence of layers in the green roof construction. We distinguish between intensive and extensive planting depending on the plant species used.

Extensive planting

This type of planting employs species that essentially require little care and little water, and tolerate with respect to high temperatures and dry periods. Sedum varieties and mosses are generally used with thin layers of 4 cm and

more (Figs. D 1.17 and 1.19). Thicker layers, up to about 10 cm, can support herbs as well, whereas taller flowering plants and grasses will require a layer up to 15 cm deep, the point at which intensive planting begins.

Intensive planting

This type of planting includes shrubs, bushes, lawns and small trees (Fig. D 1.18). The choice of plants can therefore be similar to a normal garden at ground level. However, care should be taken to choose species that are less sensitive to dry periods, wind and frost.
Semi-extensive (or semi-intensive!) planting is a cost-saving special form of intensive planting. Only those species – bushes, grasses and small shrubs – that require little care are planted.

Substrate

The substrate provides the nutrients for the plants on a flat roof. Commercial substrates are mostly used because normal topsoil is often too heavy, too loamy. Furthermore, the latter often contain great numbers of seeds from other, unwanted plants. Commercial substrates consist primarily of a mineral loose fill (often recycled materials such as clay brick fragments) to which a certain amount of organic material has been added. Slabs made from mineral wool or modified foams, enriched with organic substances, are often used for the substrate, likewise pre-cultivated vegetation mats, which enable particularly thin layers to be achieved (see "Extensive planting", p. 106).

According to Germany's Fertilisers Act (*Düngemittelverordnung*), substrates for rooftop planting must be properly labelled. The raw materials and constituents plus the content of nutrients and pollutants must be specified when the concentrations are above the labelling threshold of the Act. Furthermore, compliance with the materials stipulations of the Fertilisers Directive (*Düngemittelrichtlinie*) is essential, which contains a list of the raw materials permissible (and the materials must be chosen from this list) [2]. Loose materials are available in 50 l sacks, 1 m³ big-bags, directly from a silo or as loose bulk fill. These are distributed over the roof manually with spades and wheelbarrows. Alternatively, the substrate can be delivered in a special vehicle and pumped via a hose onto the roof where it is then distributed.

Slab-type substrate materials are available in sizes up to 1 × 1 m.

Filter membrane
The filter membrane generally consists of a thin rotproof fleece. This is not a root barrier and so it can be penetrated by the roots of plants which can then reach the water in the drainage layer. These fleeces can be made from various materials, e.g. polyamide, polyacrylnitrile, polyester, polyethylene, polypropylene or mineral fibres. Filter membrane materials are supplied in rolls up to 3 m wide and max. 100 m long.

Drainage layer
Besides its primary function of draining excess precipitation, the drainage layer must also store water so that the plants have some water during dry periods. The drainage layer can consist of loose materials such as expanded shale or clay, or plastic elements.
In the case of loose materials, the choice of particle size depends on the intended thickness of the drainage layer:
· Drainage layers 4 – 10 cm: particle sizes between 2/8 and 2/12 mm
· Drainage layers 10 – 20 cm: particle sizes between 4/8 and 8/16 mm
· Drainage layers > 20 cm: particle sizes between 4/8 and 16/32 mm

The proportion of the smallest particle sizes ($\leq$ 0.063 mm) should not exceed 5 % by wt. in order to prevent a build-up of sludge impairing the drainage function.
Drainage elements made from rigid plastics or foams include special chambers in their top surface for retaining water. Continuous channels on the underside ensure that excess precipitation can drain away. Water-resistant elements made from EPS that do not absorb any water can in some instances be included in the ther-

mal performance calculation for the roof. The respective National Technical Approval contains the thermal resistance figure that should be used in such a situation.

Protective membrane
This material protects the waterproofing or root barrier against mechanical damage. In the case of extensive planting, a geotextile represents the best choice here. Such materials are mainly used in conjunction with retention works in civil engineering and are available in the form of fabrics, fleeces and composites. They are used for protecting against erosion, as a filter membrane, as reinforcement, for drainage or as a separating membrane. When these textiles are used as a protective membrane on a flat green roof, they must satisfy the following minimum requirements: weight per unit area 300 g/m², geotextile robustness class GRK 2. The geotextile robustness class characterises how well the material withstands mechanical loads such as the tipping of loose materials and building site operations [3]. A fleece or a sheet of rubber granulate is used as the protective membrane below intensive planting.

Root barrier
An additional root barrier must be provided if the waterproofing itself is not resistant to root penetration. Rolls of sheet material, like those for the waterproofing, are used, but in this case their resistance to root penetration must be verified. In Germany this verification is according to the so-called FLL test method (developed by the German Landscape Research, Development & Construction Society) or DIN EN 13948. This standard is based on the FLL method, but includes additional requirements and so it is regarded as a higher standard. Verification of resistance to root penetration does not necessarily mean that the material is resistant to the infiltration and penetration of plants with vigorous rhizome growth, e.g. bamboo and reeds. The trade association Fachvereinigung Bauwerksbegrünung e.V. (FBB) publishes a list of plants that require additional measures [4], also a list of waterproofing materials that have been classed as root barriers according to the FLL test method [5].

Insulating materials

The thickness of thermal insulation required is derived from the building physics requirements (see "Recommended thicknesses for layers of insulation", pp. 59 – 60). Compliance with the stipulations of Germany's Energy Conservation Act (*Energieeinsparverordnung*, EnEV) often results in insulation thicknesses of 20 cm or more these days. The combustibility, the type of application and the compressive strength of the insulating material are relevant as well as its thermal conductivity (Figs. D 1.20 and 1.21, p. 94).
Pre-standard DIN V 4108-10 defines the types of application. The following types are relevant for flat roofs:
· DAD: external insulation to roofs and suspended floors, protected from the weather, insulation below finishes
· DAA: external insulation to roofs and suspended floors, protected from the weather, insulation below waterproofing
· DUK: external insulation to roofs, exposed to the weather (inverted roof)

In addition, DIN V 4108-10 specifies the following compressive strength classes:
· dm: moderate compressive strength (e.g. roof for maintenance traffic only)
· dh: high compressive strength (e.g. roof for foot traffic, terraces)
· ds: very high compressive strength (e.g. industrial floors, parking decks)
· dx: extremely high compressive strength (e.g. heavily loaded industrial floors, parking decks)

Only in the case of ventilated roofs can any insulating material be used. On other types of flat roof construction the insulating materials must comply with compressive strength class dm as a minimum, and in the case of higher loads class dh or even dx may be necessary.

Laying
Insulating materials can be laid loose or bonded to the substrate over their full area or by way of just dabs or strips of adhesive. More than one layer is permissible in order to achieve the desired thickness.
Insulating materials are normally supplied in board form; popular sizes are 1 × 0.5 m and 1 × 1 m, in thicknesses up to 30 cm. Cellular glass, on the other hand, is only available in sizes up to 0.6 × 0.6 m and thicknesses up to 18 cm.

Foamed plastics
Foamed plastics offer the chance to produce boards with rebated or interlocking edges. However, rebates must be such that movements cannot have an effect over a large area and hence lead to the creation of thermal bridges.

Rigid polystyrene foam
Polystyrene is produced through the polymerisation of the monomer styrene. We distinguish

Insulating material	Type of application	Compressive strength class	Applications	Building materials class (DIN 4102-1)/ combustibility class
Mineral wool	DAA	–		A1–B1/up to A1
Expanded polystyrene	DAA	dh	Terrace, rooftop planting	B1–B2/A1
Extruded polystyrene	DAA, DUK	dh ds dx	Terrace, rooftop planting Inverted roof, car parking deck	B1/up to B
Rigid polyurethane foam	DAA	dh ds	Terrace, rooftop planting Car parking deck	B1–B2/up to B
Cellular glass	DAA	ds dx	Car parking deck Truck parking deck	A1/A1
Wood fibres	DAA	dh ds		B2/up to D

D 1.20

Insulating material	Density [kg/m³]	Thermal conductivity category (WLS) [W/mK]	Water vapour diffusion resistance factor µ [-]	Standard
Mineral wool	12–250	0.035–0.050	1/2	DIN EN 13162
Expanded polystyrene	15–30	0.032–0.040	20/100	DIN EN 13163
Extruded polystyrene	25–45	0.030–0.040	80/250	DIN EN 13164
Rigid polyurethane foam	≥ 30	0.025–0.035	30/100	DIN EN 13165
Cellular glass	100–150	0.040–0.060	practically vapour-tight	DIN EN 13167
Wood fibres	45–450	0.043–0.072	1/5	DIN EN 13171

D 1.21

between expanded polystyrene (EPS) and extruded polystyrene (XPS) depending on the method of manufacture. Owing to its low temperature range in service (75–85 °C), rigid polystyrene foam cannot be bonded with hot bitumen.

Expanded polystyrene (EPS)
The first step in the production of insulating boards made from expanded polystyrene (Fig. D 1.22a) is to expand the granulate raw material to 20 to 50 times its initial volume with the help of steam. The ensuing plastic beads are subsequently pressed into boards, blocks or other shapes. Other steps in the production process enable various material properties (e.g. density) to be modified as required.
EPS insulating boards are standardised in classes according to DIN EN 13163 and the relevant class must be included with the CE marking. They are essentially water-resistant, but long-term water absorption up to well over 5% by vol. is possible. EPS insulating boards are resistant to mould, decay and rotting, also alkalis and salts, but not UV radiation or solvents. EPS board products initially exhibit some shrinkage due to the emission of the blowing agent and therefore adequate storage time prior to being installed in a building is vital.

Extruded polystyrene (XPS)
In the production of extruded polystyrene (Fig. D 1.22b) the raw material in the form of a granulate is foamed up in an extruder through the addition of a blowing agent (normally CO_2) at a temperature of approx. 200 °C and then formed into a continuous ribbon of foam. The resulting foam material is characterised by its closed-cell structure, which makes XPS less sensitive to moisture and gives it a higher compressive strength than expanded polystyrene. DIN EN 13164 describes the standardised classes. XPS is very popular for inverted roofs because its closed pores mean that the boards do not absorb any water. The long-term water absorp-

tion can reach only about 0.3% by vol. at most. Insulating boards made from XPS are resistant to bitumen, salts and diluted acids and alkalis, and exhibit good resistance to microbes. However, they are not resistant to UV radiation, tar-based products, paint thinners or other solvents.

Rigid polyurethane foam (PUR)
The raw materials for polyurethane (Fig. D 1.22c) are polyol and polyisocyanate. Polyol can be obtained from petroleum, but also from renewable raw materials such as maize, sugar beet or potatoes. Rigid PUR foam is produced through a chemical reaction after mixing the components and introducing a blowing agent (usually pentane or CO_2). The insulating boards are manufactured using the lamination method: the PUR output from the mixing head is spread over a conveyor belt where it foams up and adheres to facing materials top and bottom [6], normally a mineral or glass fleece, or aluminium or composite foil.
Insulating boards made from polyurethane complying with DIN EN 13165 are water-and resistant thanks to their closed-cell structure and exhibit only a low long-term water absorption of max. 3% by vol. They are characterised by their high compressive strength and good resistance to microbes, petroleum products, solvents and fuels. However, they are not resistant to UV radiation.

Mineral wool insulating materials
Mineral wool (MW) insulating materials consist of inorganic fibres. Glass or rock wool are the materials used. Their production involves melting the raw materials at a high temperature and forming fibres. The fibres achieve their stability through the curing of the binder. The material properties can be influenced by binder content, fibre orientation and the degree of compaction of the fibres. According to DIN EN 13162 mineral wool insulating materials are available as boards, batts and blown or caulking material.

The fibres used for producing mineral wool are water-resistant and exhibit only a low degree of water absorption. Nevertheless, the binder can be affected by high moisture levels, which leads to a decrease (even the complete loss) of the material's compressive strength. The fibres are resistant to microbes, but the insulating material itself can provide a growing medium for fungal mycelium. The high levels of dust during installation of mineral wool which were common in the past have now been considerably reduced by modifying the binder and also the fibre geometry and properties. Modern mineral wool products are no longer considered to be potentially carcinogenic. This is not necessarily the case, however, when removing old mineral wool insulating materials, and in such cases special protective measures will be necessary.

Glass wool
Glass wool (Fig. D 1.22d) consists of quartz sand, limestone and approx. 60–70% scrap glass (cullet) to which 3–9% binder (frequently phenol formaldehyde resin) and 1% water repellent based on silicone or petroleum is added. The properties can vary considerably depending on density and product thickness.

Rock wool
Rock wool (Fig. D 1.22e) can be made from various rock types (e.g. diabase, basalt and dolomite) that are mixed with suitable additives and binders. The content of binder and water repellent is somewhat lower than with glass wool. The properties can vary considerably depending on density and product thickness.

Cellular glass
The raw materials for producing cellular glass (CG; Fig. D 1.22f) are quartz sand, dolomite, calcium carbonate and sodium carbonate. These are melted to form glass at a temperature of about 1400 °C and subsequently processed into a glass powder. The foaming process is carried out by mixing in a blowing agent, normally carbon, which is why cellular glass has a dark colour. The insulating boards are cut from the cellular glass blocks.
Insulating boards made from cellular glass to DIN EN 13167 are practically impermeable to water and vapour-tight. They exhibit a high compressive strength, but are brittle. Owing to the inorganic raw materials, cellular glass has a high thermal shock resistance. The boards are resistant to insects, rodents and microbes. One disadvantage is the strong smell given off during cutting.
As a rule, cellular glass boards are bonded to the substrate with a hot bitumen compound over their full area. Joints between individual boards are also filled with this compound. This type of bonding is necessary only for cellular glass and it prevents precipitation seeping underneath the insulation. Where steel trapezoidal profile sheets have been used for the loadbearing structure, the cellular glass boards can be bonded to the top flanges with a cold

adhesive. Full bonding at the joints between boards is necessary with this form of construction, too.

The first ply of built-up roofing, consisting of polymer-modified bitumen flexible sheeting, can be applied directly to the cellular glass boards using the pouring and rolling method. If a welding method is to be used, a coat of hot bitumen should be applied to the insulation first. The waterproofing can also consist of synthetic flexible sheeting or a liquid plastic in conjunction with a base sheet.

Tapered cellular glass boards are available for creating falls, also boards with a facing laminated to the upper surface, which allows the waterproofing to be welded directly to the cellular glass boards, rendering a layer of hot bitumen unnecessary. Hot bitumen is also unnecessary when using cold adhesives.

Wood fibres

The production of wood fibre insulating boards (WF) makes use of low-strength or scrap wood from the sawmill industry (mostly spruce, fir and pine) which is broken down mechanically and mixed with water. This mixture is dried to a residual moisture content of only 2 % and then cut into boards. The strength of wood fibre insulating boards is due to the "felting" (interlocking) of the fibres and also the adhesive properties of lignin, a constituent of wood. Additives such as aluminium sulphate, paraffin or glue are sometimes added during production in order to improve the bonding effect within the boards or to achieve a hydrophobic surface. Wood fibre insulating boards to DIN EN 13171 exhibit a hygroscopic material behaviour, i.e. they can store moisture, and are comparatively diffusion-permeable. Owing to their poor biological resistance, protection against moisture is especially important in flat roofs (see "Protection against condensation as a result of water vapour diffusion", pp. 65–66).

Layers for roof falls

A flat roof should always be designed and constructed with a minimum fall of 2 % in order to drain precipitation reliably. This fall can be achieved, for example, with the loadbearing construction. However, it is this is not possible or undesirable for architectural reasons, then the fall must be created within the roof finishes. A layer of normal-weight concrete, a screed laid to falls or a layer of loose bitumen-bonded perlite directly above the loadbearing structure represent possible solutions. Another option is to use tapered insulating boards. However, it must be ensured that the thermal performance required is still achieved at the thinnest point of the insulation. The advantage of creating the fall within the insulation is the substantial reduction in the load on the roof compared with, for example, concrete.

Insulating boards made from rigid polystyrene foam, mineral wool and cellular glass can all be supplied in tapered form for creating falls. The manufacturers of these products will require information regarding the positions of drainage outlets, the direction of fall and the slope of the roof for the production. Layout drawings showing the position of every individual insulating board can then be prepared (see "Falls in the insulation", p. 101) [7].

Protective and separating membranes, vapour barriers and bond enhancers

Protective and separating membranes, vapour barriers and bond enhancers all have to perform diverse functions in a flat roof. These functions range from the mechanical separation of layers to mechanical protection for waterproofing materials and building physics tasks.

Protective membranes

Protective membranes protect the waterproofing against mechanical damage. They are highly resistant to perforation and are laid over the entire area of the waterproofing.

Germany's Flat Roof Directive, for instance, lists the following materials:
- Synthetic fleece (min. 300 g/m²)
- Sheets of semi-rigid PVC (min. 1 mm thick)
- Sheets of PVC-P (min. 1.3 mm thick)
- Rubber granulate mats or boards (min. 6 mm thick)
- Plastic granulate (min. 4 mm deep)
- Drainage mats and boards for subsequent rooftop planting
- Boards made from XPS

Fleeces can be supplied in rolls up to 2 m wide and max. 50 m long, rubber granulate mats in widths up to 1.25 m approx. 8 m long.

Separating membranes

A separating membrane isolates building components or layers over their whole area or just parts thereof. Such a separation may be neces-

a

b

c

d

e

f   D 1.22

D 1.20 Types of application, compressive strength classes, applications and building materials classes for various insulating materials used in flat roofs
D 1.21 Building physics parameters for selected insulating materials
D 1.22 Insulating materials (selection)
a Expanded polystyrene (EPS)
b Extruded polystyrene (XPS)
c Rigid polyurethane foam (PUR)
d Glass wool
e Rock wool
f Cellular glass (CG)

Vapour barrier/ check	Water vapour diffusion resist- ance factor [-]	Material thickness [mm]	s_d value [m]
Foil/film			
Aluminium foil	practically vapour-tight	≥ 0.05	> 1500
PE film	100	0.25	100
PVC film	20,000	0.25	30
PA film	not constant	–	3.9/0.2[1]
Coatings			
High-build bit. (KMB), 1-part	2000	4	8
High-build bit. (KMB), 2-part	4000	4	16
Reaction resins	20,000		

[1] Values are valid for 40 % and 80 % relative humidity, for variability see Fig. D 1.24

D 1.23

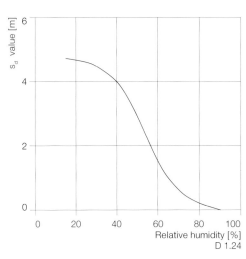

D 1.24

D 1.23 Hygrometric parameters for various vapour barriers/checks
D 1.24 Relationship between diffusion-equivalent air layer thickness (sd value) of smart vapour barrier based on polyamide (PA) and the relative humidity of the air
D 1.25 The function of a smart vapour barrier compared with a standard vapour barrier: wet wood was placed in bags made from these materials but the wood in the PA film (right) has dried out much more rapidly.

sary because of the different thermal expansion characteristics of the two materials or to prevent the transfer of movements and stresses between neighbouring layers. Separating membranes may also be necessary because of the chemical incompatibility of certain materials. Separating membranes are laid loose or with a partial bonding.
Examples of materials suitable for separating layers:
· Bitumen flexible sheeting with perforated glass fleece inlay
· Bitumen flexible sheeting
· Bitumen flexible waterproof sheeting
· Uncoated glass fleece
· Synthetic fleece
· Plastic film
· Materials laminated to the waterproofing material
· Foam mats

Plastic films and synthetic fleeces can be supplied in rolls up to 8 m wide and max. 250 m long.

Vapour barriers, vapour checks

Vapour barriers or vapour checks are necessary in order to prevent or reduce respectively the infiltration of water into the roof construction by way of diffusion (see "Protection against condensation as a result of water vapour diffusion", pp. 65–66). The degree of moisture control must be achieved through the choice of type of material and/or its thickness.
Vapour barriers and vapour checks must be installed airtight. Sheeting (see pp. 86–90), foil or coatings can be used. The foil and film materials most commonly used are aluminium, polyethylene (PE), polyvinyl chloride (PVC) and polyamide (PA). Vapour barriers and vapour checks applied as a liquid include polymer-modified high-build bitumen coatings (KMB) and reaction resins. The water vapour diffusion resistance factors of these materials are given in Fig. D 1.23.
Plastic films can be supplied in rolls in widths up to 8 m and lengths up to 250 m. Liquid coating systems are usually supplied in 1, 5, 10, 25 or 30 kg containers.

Smart vapour barriers
A smart vapour barrier behaves like a diffusion-retardant vapour check in the winter. Good conditions for the drying-out of the building component are, however, ensured, i.e. in summer and also at other times of the year when, due to the weather conditions, inverse diffusion takes place; at these times the smart vapour barrier becomes more diffusion-permeable long before condensation occurs and in this way promotes drying. It is the variable sd value of the smart vapour barrier that accounts for this behaviour.
The relative humidity behind a vapour barrier fluctuates considerably with the weather conditions (see "Steady-state or non-steady-state moisture assessments?", pp. 69–73). When the moisture migrates outwards, the area of construction around the vapour barrier becomes very dry. In this situation the vapour barrier has to ensure that no moisture enters from the inside which would augment this vapour flow. It therefore must be as impervious as possible. When the moisture migrates inwards (inverse diffusion), the relative humidity at the vapour barrier increases. In the extreme case condensation forms. At this point a vapour barrier with a high permeability is advantageous because then the incoming moisture can be released into the interior air, allowing the building component to dry out (Fig. D 1.25). Fig. D 1.24 shows the relationship between the relative humidity and the sd value of a smart vapour barrier based on polyamide (PA). The pronounced dependence between the vapour permeability and the ambient moisture conditions can be attributed to the storage of water molecules between the long chains of the polymer molecules. At normal interior air temperatures the sd value of the polyamide sheet varies between approx. 4.5 m in the dry state and 0.1 m when in contact with water (e.g. condensation on the sheet or contact with a wet building material).

Bond enhancers

Bond enhancers are required to guarantee the mechanical contact between layers or components when this cannot be guaranteed by direct application. For example, a bituminous

primer is required to bind dust and seal pores when a bituminous vapour barrier is fully or partially bonded to reinforced or aerated concrete. The primer functions as a bond enhancer. A bituminous primer is also necessary when a bituminous vapour barrier is to be applied to plastic-coated steel trapezoidal profile sheeting [8]. Bituminous primers are supplied in small containers up to 30 kg or large ones up to 1000 kg.

Concretes with high water impermeability

Concrete is frequently used in the loadbearing structure to a flat roof. The mechanical properties of concretes in general will not be discussed here, but rather those of concretes with a high resistance to water infiltration [9]. Where this type of concrete is used in a flat roof, additional waterproofing to the roof is unnecessary (see "Impermeable concrete slab as waterproofing layer", p. 104). Concrete is a mixture of aggregates of various sizes that are bonded together with the help of a mineral binder, the hydrated cement. This forms as a result of the chemical reaction between cement and water and, if present, certain additives/admixtures. The volume of the hydrated cement is smaller than the volume of the raw materials and this leads to micropores in the concrete. By reducing the number of pores and the interlinking between those pores, the concrete gains a high resistance to the infiltration of water. This can be checked by way of the water penetration depth according to DIN EN 12390-8. Impervious concretes can be produced with a water/cement ratio (weight-based ratio of water to cement) of ≤ 0.60 (for component thicknesses ≤ 40 cm). The cement content must be at least 280 kg/m^3, the minimum compressive strength grade C 25/30.
Moisture transport in cracks by way of capillary action can occur as well as moisture transport within the microstructure of the concrete itself. It is essential to limit the cracks widths in impermeable concrete if imperviousness is to be guaranteed (see "Impermeable concrete slabs", pp. 35–36). Cracks form in the hard-

ened concrete when the tensile stresses in the component – due to mechanical loads and/or deformations unrelated to the loads – exceed the tensile strength of the concrete. The deformations can be due to moisture-related swelling and shrinkage or different temperatures between the construction and usage phases, but also as the result of uneven temperature distributions due to diverse dissipation of the heat of hydration during the curing process. These effects can be reduced by using cements with a low heat of hydration and by covering the concrete with an insulating blanket during the curing process.

Rooflights

The following plastics, as well as glass, are popular for the light-permeable parts of rooflights.

Polymethyl methacrylate (PMMA)
PMMA, often referred to as Perspex, is a high-quality plastic with a high light transmittance, which is why opaque panels are recommended for preventing glare. PMMA has good weathering properties, which means that there is no appreciable change to its optical behaviour over the long-term, plus good sound insulation properties (sound reduction index R_w = approx. 23 dB). In the form of solid panels formed by extrusion, the material satisfies the requirements of building materials class B 2 (flammable). PMMA easily splinters during mechanical machining and is sensitive to impact.

Polycarbonate (PC)
Polycarbonate exhibits a high impact strength and therefore provides good security against hail and the stones thrown by vandals. The light transmittance is similar to that of PMMA; once again, opaque panels are recommended to prevent glare. PC panels are low in weight and therefore the sound reduction index R_w (approx. 20 dB) is slightly lower than that of PMMA. Polycarbonate is a class B 1 material (not readily flammable). Approved forms of construction made from double- and triple-wall PC panels are generally classed as a soft roof covering and can be used as smoke and heat vents.

Glycolised polyethylene terephthalate (PETG)
PETG is an impact-resistant plastic with a high optical quality, is resistant to UV radiation and easy to work. In building applications it is used for roofing or clear glazing.

Glass fibre-reinforced polyester resin (GFRP)
Glass fibre-reinforced polyester resin has a high strength and therefore offers security against hail and the stones thrown by vandals. It can be produced with a high light transmittance, but is never transparent. The glass fibres embedded in the material give it a good light-scattering property. Opaque panels are recommended for rooflights. GFRP is a class B 2

D 1.25

material (flammable) but does not melt or form droplets during a fire. Multi-layer constructions exhibit good thermal insulation behaviour, with U-values as low as 1.2 W/m²K. This material must be processed or coated in order to prevent yellowing and the exposure of glass fibres, which then collect dirt.

Incompatibilities between materials

Compatibility is an important issue that must be addressed when choosing building materials. A complete list of all known incompatibilities would exceed the scope of this book, but a number of typical problems have been mentioned in the descriptions of waterproofing materials (see pp. 86–91) and insulating materials (see pp. 93–95). Where various materials are to be combined in a roof construction (e.g. in a roof refurbishment project), the information given in the product data sheets must be read carefully. In cases of doubt, the designer should ask the manufacturer to confirm the compatibility of a certain combination of materials. One way of overcoming an incompatibility between two materials is to fit a suitable separating membrane between them. Such a separating membrane can be factory-applied to insulating materials, for example, in the form of a laminated facing material.
Another risk in the sense of material incompatibility can ensue when combining products of different manufacturers, even though the material designations may be identical or similar. This applies to sundry items such as adhesives, cleaning agents, welding solvents, etc. For warranty reasons, the information given in the manufacturer's installation instructions and product data sheets must always be adhered to. Material incompatibility as a result of galvanic corrosion can also occur in metal roof coverings and/or metal gutters and pipes. Advice regarding various combinations of metals has been given in the description of metal roof coverings (see p. 91).

References
[1] Schunck, Eberhard et al.: Roof Construction Manual. Munich/Basel, 2003;
Schittich, Christian et al.: Glass Construction Manual. 2nd ed. Munich/Basel, 2007
[2] Kennzeichnungspflicht für Dachsubstrate. In: Dach + Grün 1/2009, p. 6
[3] Köhler, Martin: Geotextilrobustheitsklassen. Eine praxisnahe Beschreibung der Robustheit von Vliesstoffen und Geweben gegenüber Einbaubeanspruchungen. In: tis Tiefbau Ingenieurbau Straßenbau. 11/2007, p. 52
[4] Fachvereinigung Bauwerksbegrünung e.V. (FBB): Pflanzenarten mit starkem Rhizomwachstum, wie Bambus und Schilf. Saarbrücken, 2005. http://www.fbb.de/PDFs/Prospekt%20Dachbegr%C3%BCnung%20Downloads/SRW.pdf; position as of 26 Feb 2010
[5] Fachvereinigung Bauwerksbegrünung e.V. (FBB): Wurzelfeste Bahnen und Beschichtungen. Saarbrücken, 2009. http://www.fbb.de/PDFs/Prospekt%20Dachbegr%C3%BCnung%20Downloads/Wurzelfeste-Bahnen.pdf; position as of 26 Feb 2010
[6] Riegler, Rosina: Fachgerechte Ausführung und Sanierung von Flachdächern und Gründächern. Merching, 2009, chap. 3.3, pp. 6f.
[7] ibid. chap. 2.2, p. 8
[8] ibid. chap. 2.2, p. 3
[9] Verein Deutscher Zementwerke e.V. (VDZ) (pub.); Freimann, Thomas: Zement Merkblatt Hochbau H 10: Wasserundurchlässige Betonbauwerke. Düsseldorf, 2006

Flat roof construction

Christian Bludau, Eberhard Schunck

D 2.1

Flat roofs are forms of roof construction with only a shallow pitch, possibly even none at all, and a covering of an impervious layer of material over their entire area. The borderline between a shallow-pitch and a pitched roof lies at an inclination of approx. 18% (10°). Pitches below 2% (approx. 1°) are regarded as forms of construction on which water run-off is not automatically guaranteed, and pitches up to 5% (approx. 3°) are likely to experience ponding due to deflection of the roof structure and/or unevenness in the roof finishes.

General construction

Flat roofs consist of at least a loadbearing layer, an insulating layer and a weather-resistant waterproof finish, which, depending on the form of construction and usage requirements, are arranged in different sequences and supplemented by other layers as necessary.
Fig. D 2.2 shows the flat roof types according to their construction principle. They can be divided into two types in terms of their building physics: the ventilated and unventilated flat roof. And the unventilated flat roof can be further subdivided into flat roofs with the waterproofing above the insulation (warm roof or warm deck) and the inverted roof (upside-down roof), a special variation of this form of construction in which the insulation is on top of the waterproofing. These three forms of construction can be finished with different materials although it is important to take into account the respective requirements that each type of roof has to satisfy.

Standards and regulations

There are many standards and regulations that deal with flat roofs, but most of these cover only certain parts of the flat roof construction. In Germany the so-called Flat Roof Directive (*Fachregel für Abdichtungen – Flachdachrichtlinie*) produced and published by the Zentralverband des Deutschen Dachdeckerhandwerks (ZVDH, German Roofing Contractors Association) provides a good summary of the most important regulations concerning the waterproofing of flat roofs.

Important publications dealing with just the waterproofing are DIN 18195 "Waterproofing of buildings" and DIN 18531 "Waterproofing of roofs – Sealings for non-utilised roofs". Another critical part of the flat roof is the thermal insulation. Here, the design rules for calculating the thickness of insulating material, the minimum thermal performance according to DIN 4108-2 and the requirements of the Energy Conservation Act (*Energieeinsparverordnung*, EnEV) must be adhered to (see "Energy-related thermal performance according to Energy Conservation Act", p. 59).
The design, construction and maintenance of planting on roofs is covered by Germany's Rooftop Planting Directive (*Dachbegrünungsrichtlinie*) published by the Forschungsgesellschaft Landschaftsentwicklung Landschaftsbau e.V. (FLL, German Landscape Research, Development & Construction Society).

Layers in the roof construction

The different forms of construction each consist of several layers with defined functions. Their arrangement varies depending on the type of flat roof.

Roof finishes
The covering to a flat roof can perform the following functions (with respect to the surface):
- Protection against mechanical actions
- Protection against UV radiation
- Protection against wind suction forces
- Protection against floating of the insulation (inverted roofs)
- Fire protection

It is the climate factors in particular, e.g. severe temperature and moisture fluctuations, UV radiation, that place great strains on the waterproofing materials. An additional protective layer above the waterproofing can therefore prolong the service life of the material underneath.
Heavyweight protection, e.g. loose gravel, on top of flexible sheeting laid loose not only protects against UV radiation, but also against wind uplift. Heavy materials are also advantageous

D 2.1 Laying rock wool insulation on a flat roof
D 2.2 Flat roof categories
D 2.3 Loose gravel finish on a flat roof with a low perimeter kerb and roof-mounted ventilation units
D 2.4 Laying rolls of waterproofing material with a standard overlap
    a 2-ply
    b 3-ply
D 2.5 An alternative method for laying rolls of waterproofing material
D 2.6 Flat roof with waterproofing in the form of plastic flexible sheeting which requires no protection to the surface

for fire protection. And when installing additional rooftop structures, e.g. solar panels or safety harness anchorages, loose gravel or rooftop planting may provide enough weight and thus obviate the need for fixings that penetrate the waterproofing material.

Furthermore, the roof covering may allow the roof to be used by the building's occupants.

Loose gravel

The simplest method of providing ballast, and a very popular solution, is to finish the roof with a layer of gravel (Fig. D 2.3). If at the same time the gravel is to provide protection against wind uplift, then its weight must be such that it counteracts the wind suction forces acting on the roof as calculated according to DIN 1055-4 (see "Wind pressure and wind suction", p. 28). Washed rounded gravel, 16/32 mm grading, is generally preferred for a flat roof finish.

The depth of the layer of gravel in the service condition must be at least 50 mm [1]. A protective membrane should be laid below the gravel to protect single-ply waterproofing. On an inverted roof, a filter fleece should be laid on top of the insulation to prevent fine particles being washed into the joints between the insulation boards.

Stone and concrete flags

Stone and concrete flags represent another way of securing the waterproofing material against wind uplift. Concrete paving slabs (min.

size 40 × 40 × 4 cm) or precast concrete blocks are laid in gravel or sand directly on top of the waterproofing or raised above this on proprietary mountings. It is also possible to use in situ or precast concrete slabs (max. size 2.5 × 2.5 m). However, in this case a protective membrane and two slip layers must be laid on top of the waterproofing in order to reduce the effects of changes in length due to thermal actions or other shear forces. Flags can be used to form an accessible surface.

Waterproofing

The most important layer in a flat roof is always the waterproofing. The task of this is to protect the building against precipitation and other environmental effects. When choosing a waterproofing material, its resistance to UV radiation is one of the most important factors to consider. A UV-resistant waterproofing material is absolutely essential on roofs where no further finishes, e.g. gravel or planting, are planned. Most types of synthetic flexible sheeting satisfy this requirement (Fig. D 2.6), but other waterproofing materials will require an additional protective layer, or may not be used as the topmost ply (cap sheet) in built-up roofing (see "Waterproofing", pp. 86–91).

All customary waterproofing materials are suitable for use on flat roofs, provided they comply with the requirements of the relevant standards, e.g. the Flat Roof Directive, DIN 18531-2 for inaccessible roofs, DIN 18195 accessible roofs

(or the corresponding ÖNORM standards in Austria and SIA standards in Switzerland). Flexible sheeting based on bitumen, elastomeric or synthetic materials and also liquid waterproofing products can all be used in most instances. Roof waterproofing comprising bitumen flexible sheeting generally consists of at least two plies (built-up roofing, Fig. D 2.4). With two plies the rolls of material overlap by half, with three plies by one-third. The minimum lap between rolls in the same layer should be 8 cm and with the method of laying shown in Fig. D 2.5 each third ply should overlap the first one by min. 4 cm. Ballast is not absolutely essential for the waterproofing material; instead, it can be bonded to the substrate, in which case the substrate must ensure good adhesion or, if not, it must be suitably prepared. The form of bonding used must be matched to the waterproofing material (e.g. synthetic flexible sheeting, bitumen flexible sheeting, etc.). Fig. D 2.8 (p. 100) specifies the bonding area or number of strips of adhesive. Bonding is normally only used on buildings up to 25 m high. Above that, additional ballast is required in order to provide adequate protection against wind suction loads of DIN 1055-4. Attaching the waterproofing with mechanical fasteners is another option popular for roof decking or steel trapezoidal profile sheeting. Individual fasteners in a line (Fig. D 2.7) or continuous fixings (metal profiles or strips, Fig. D 2.9) represent the normal fixing solutions. Both types of fixing can be used along the edges

D 2.2

D 2.3

D 2.6

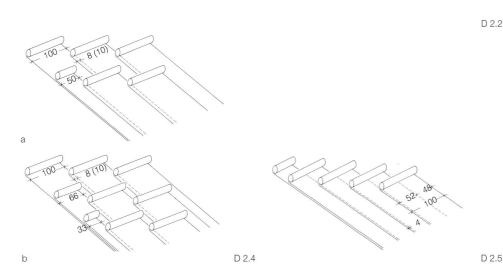

a

b

D 2.4

D 2.5

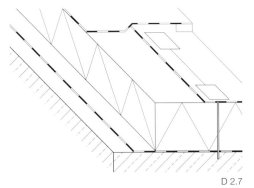

D 2.7

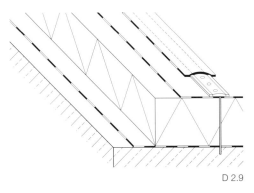

D 2.9

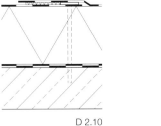

D 2.10

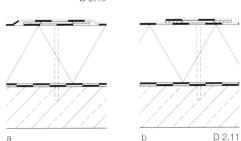

a　　　　　　　　b　　　　　　D 2.11

D 2.12

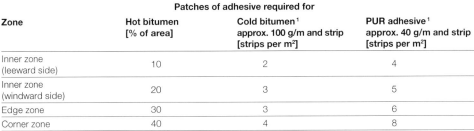

	Patches of adhesive required for		
Zone	Hot bitumen [% of area]	Cold bitumen[1] approx. 100 g/m and strip [strips per m²]	PUR adhesive[1] approx. 40 g/m and strip [strips per m²]
Inner zone (leeward side)	10	2	4
Inner zone (windward side)	20	3	5
Edge zone	30	3	6
Corner zone	40	4	8

[1] Always follow the manufacturer's instructions when using cold-applied adhesives.
  The following information (provided by the manufacturer) is particularly important for cold bitumen and PUR adhesives:
  • "Use by" date
  • Application and ambient conditions
  • Application instructions, e.g. quantities, distribution, preparation of substrate

D 2.8

(Figs. D 2.10 and 2.12) or in the middle (Fig. D 2.11) of rolls.

Individual fasteners in a line are normally specified when fixing along the edges of rolls, where the rolls overlap. There must be a minimum distance of 10 mm between the washer or disc of the fastener and the edge of the roll. The overlap is made up of 50–60 mm welded or bonded material, 10 mm safety margin, approx. 40 mm for the width of the fixing and 10 mm edge clearance. Manufacturers' recommendations must also be followed. A suitably stable substrate is essential to achieve the clamping effect.

When fixing in the middle of a roll, the individual or linear fixings are positioned at the recommended spacing and covered with an additional strip of waterproofing material. The number of fixings required is determined according to DIN 1055-4. Here again, manufacturers' instructions must be followed. Only approved self-drilling screws, wood screws or anchor systems with appropriate washers or discs to spread the load may be used for attaching waterproofing materials. These days, rustproof metal fixings and ageing- and temperature-resistant plastics are used. To achieve concealed fixings for plastic or elastomeric waterproofing materials, individual fasteners with strips or discs made from the waterproofing material itself, for example, are bonded or welded to the underside of the waterproofing layer and anchored in the supporting construction. The number of fixings depends on the wind loads expected. With linear fixings, at least two per m² should be provided irrespective of the number calculated. When using other forms of fixing at larger spacings, it may be necessary to consider the fixings and the forces they transfer to the loadbearing structure as point loads and analyse these separately. The minimum joint widths for waterproofing materials are given in Fig. D 1.8 (p. 89). Manufacturers' recommendations regarding minimum and maximum spacings must also be followed in order to achieve optimum seam strengths. Where secret nailing is used to fix bitumen flexible sheeting, a spacing between the rows of nails of 45 cm near corners and edges and 90 cm elsewhere has proved to satisfy the requirements [2]. The layer of insulation and the vapour barrier can be attached with the fixings for the waterproofing. But if the fixings are not adequate to hold the insulation in place as well, it will be necessary to secure the insulation additionally with adhesive or other mechanical fasteners. Roofs without any ballast are always subjected to higher temperatures than those with a covering of gravel etc., and the short-wave absorptance of the surface is very relevant here. In sunshine, dark-coloured roofs heat up more than light-coloured ones. Whereas a white roof reaches a temperature of 40°C on a summer's day, a black one can reach 80°C on the same day! Light-coloured roof coverings are therefore used in hot countries to minimise the heat flow into the roof construction (see "Thermal actions on flat roofs", pp. 53–54).

Waterproofing during construction
Temporary waterproofing is necessary for forms of construction that remain exposed to the weather for a longer period. Timber structures or uncovered thermal insulation in particular can be damaged by precipitation or high levels of UV radiation.

Separating membrane
A slip, separating or protective membrane, possibly a levelling layer, may be necessary in some circumstances. The function of these layers is to compensate for forces that build up due to, for example, the different thermal expansion characteristics of different materials, or to protect the waterproofing against being penetrated by fixings from below, or to create a suitable substrate for the waterproofing material. In addition, a separating membrane can isolate incompatible materials and help to reduce the drumming noise of rain. There are also vapour pressure equalisation membranes available that are intended to even out the high vapour pressure differences that can prevail between the inner and outer faces of the waterproofing. On an inverted roof the separating membrane can be used to prevent fine particles in the gravel topping being lost in the joints between the boards of insulating material. Diffusion-permeable water run-off separating membranes can be used here to drain most of the water away before it reaches the layer of insulation.

Decking to support waterproofing

The waterproofing must lie on a flat, even surface if additional stresses and strains are to be avoided. A sagging material must be avoided at all costs because then any loads will lead to damage. To avoid this problem, an additional upper decking of, for instance, timber, wood-based products or other board-type materials, must be installed in ventilated roofs or those with a soft insulating material.

Air space

The air space (or cavity) in a ventilated roof extends over the entire area of the roof and must be linked to the outside air via inlets and outlets. It should be possible for air to flow unhindered throughout the air space. Ensuring an adequate depth for the air space is another important requirement [3].

The natural movement of the air caused by the wind enables an exchange of air with the surroundings, allows moisture to be removed from the construction. However, ventilation only functions when the cross-sectional area of the air space and the sizes of inlets and outlets are adequate. In contrast to a pitched roof, the low inclination of a flat roof means that there is very little circulation of the air due to thermal currents. Wind-induced pressure differences are therefore necessary to ensure an adequate air change rate. As it is difficult to ensure continuous ventilation, and the anticipated flow of air through the space is questionable, ventilated flat roofs should be used in special cases only.

In the 1981 edition of DIN 4108-3 there was a requirement that roofs with a pitch < 10° should have a ventilation cross-section at two opposite eaves amounting to 2 ‰ of the total plan area of the roof. The unrestricted air space above the thermal insulation had to be at least 5 cm deep and the sd value (water vapour diffusion resistance, see "Vapour diffusion flows and resistances", pp. 65–66) of the construction above the air space had to be min. 10 m. However, the current edition of DIN 4108-3 no longer includes these requirements. Roofs are exempted from verification of the ventilation

when the pitch is < 5° and they are fitted with a diffusion-resistant layer with an s_d value of min. 100 m. All that is required is that the proportion of the thermal resistance below this layer may not exceed 20 % of the total thermal resistance of the roof. According to this standard all other forms of ventilated construction must be verified.

Thermal insulation

Various materials can be used for the thermal insulation in flat roofs (see "Insulating materials", pp. 93–95). Depending on the requirements of the particular form of construction, the insulating material may need to satisfy a certain compressive strength requirement as well as particular thermal properties (see "Thermal resistance of building components", p. 58).

More than one layer of insulating material is often required these days in order to satisfy the latest thermal performance requirements, which call for very thick layers of insulation. The layers of boards are bonded together with a suitable adhesive or bitumen, although here the manufacturer's advice regarding incompatibility issues must be taken into account. It is not essential to stagger the joints between different layers, but this is certainly an advantage. Germany's Flat Roof Directive calls for the boards to be pressed together at the joints, which can lead to a problem in the case of very stiff boards, e.g. expanded polystyrene (EPS), because with large overlaps between the boards the result is a plate effect (possibly over the entire area of the building) that can lead to considerable stresses within this plate. If relieved abruptly such stresses can damage the waterproofing. The Flat Roof Directive therefore prescribes boards with a maximum side length of 1.25 m [4]. In order to avoid stresses, 10 cm is prescribed as the maximum offset between the joints of adjacent layers of EPS boards that are bonded together in such a way that they can transfer loads across the joints. Extruded polystyrene (XPS) boards used as insulation on inverted roofs must have profiled edges (e.g. shiplap) if they are laid loose and currently may not be installed in more than one layer (Fig. D 2.13).

Falls in the insulation

If the fall of the roof cannot be created in the loadbearing structure, tapered insulation or a screed laid to falls represent the alternatives. For example, the majority of insulating material manufacturers can provide tapered insulating boards to suit the complete geometry of a roof, and supply custom layout drawings for the particular structure. A basic layer of insulation is frequently laid over the entire roof area first and the fall required built up on this in stages, a method that requires fewer custom tapered elements. Complex roof geometries can be dealt with in this way (Fig. D 2.14). The layout drawings of the manufacturer must, however, be followed. The drawings should also specify whether incompatibilities between the insulating material and adhesives or other substances are to be expected.

Vapour barrier/check

Measures to prevent the infiltration of moisture into the insulating material are necessary between insulation and supporting construction or insulation and interior. This is normally achieved with the help of a vapour barrier or vapour check, but there are also forms of construction in which the vapour barrier function is provided by materials such as solid timber boards, sheet metal or other diffusion-resistant or diffusion-tight materials. An airtight installation is especially important here because otherwise there is a risk of large quantities of moisture entering the construction by way of convection (see "Airtight design of flat roofs", pp. 67–68). When using a type of insulation that is diffusion-tight, e.g. cellular glass or a closed-cell plastic, it may be possible to omit the vapour barrier between the loadbearing construction and the insulation in some cases. Here again, it is important to guarantee the airtightness of the entire insulating layer.

D 2.13

D 2.7   Waterproofing material fixed with a line of individual fasteners

D 2.8   Patches of adhesive required for fixing flexible sheeting without ballast on buildings up to 25 m high

D 2.9   Flexible sheeting fixed with strips

D 2.10   Mechanical fasteners fitted in the overlap between rolls of flexible sheeting

D 2.11   Mechanical fasteners fitted in the middle of rolls of flexible sheeting
       a Below the sheeting
       b Above the sheeting

D 2.12   Inserting individual mechanical fasteners along the edge of a roll of flexible sheeting

D 2.13   XPS insulation laid loose with a diffusion-permeable filter membrane

D 2.14   Example of the layout of tapered insulation boards to create a fall within the insulation, including valley and outlet

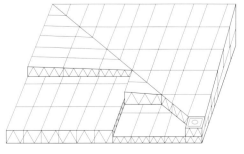

D 2.14

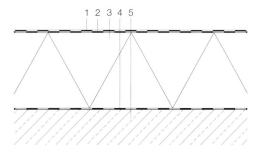

1 Waterproofing laid to falls
2 Separating membrane if required
3 Thermal insulation (possibly tapered to create falls)
4 Vapour barrier/check
5 Loadbearing structure

D 2.15

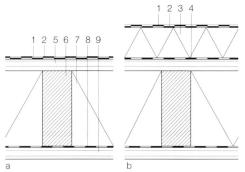

a   b

1 Waterproofing laid to falls
2 Separating membrane if required
3 Additional insulation above loadbearing structure
4 Temporary waterproofing for construction phase
5 Decking to brace structure
6 Loadbearing structure
7 Insulation between loadbearing members
8 Vapour barrier/check
9 Soffit lining/services installation level

D 2.16

D 2.15   Roof construction: waterproofing above insulation
D 2.16   Roof construction: waterproofing above insulation
         a Insulation between loadbearing members
         b Insulation between loadbearing members plus
            additional insulation above
D 2.17   Cut-away view of flat roof showing vapour barrier,
         insulation and waterproofing
D 2.18   Construction of ventilated flat roof in concrete
D 2.19   Construction of ventilated flat roof in timber
D 2.20   Roof construction: waterproofing below insulation
         (inverted roof)

D 2.17

Forms of flat roof construction

Different requirements apply to the construction
and the materials used depending on the position
of the thermal insulation and the waterproofing.
On the next pages we shall therefore distinguish
between the following forms of flat roof construction, categorised according to the position of
the waterproofing layer:
• Waterproofing above insulation
• Waterproofing below insulation
• Waterproofing between layers of insulation

This results in various requirements with which
the construction, insulation and waterproofing
have to comply.

Waterproofing above insulation
The form of flat roof construction employed most
frequently is that in which the waterproofing is
positioned above the thermal insulation (Figs.
D 2.15 and 2.17). From top to bottom, the
sequence of layers is: waterproofing, insulation,
loadbearing structure. The construction below
the waterproofing can vary to some extent, e.g.
with or without vapour barrier, or additional
insulation between loadbearing timber joists, but
the form of construction is similar for all variants.
One variation favoured with timber structures is
to place the insulation between the joists, i.e.
loadbearing structure and thermal insulation
are in the same plane. The voids between the
joists are filled completely with insulation. If this
depth of insulation is still inadequate, further insulation is installed above the joists (Fig. D 2.16).

Ventilated flat roof
The ventilated flat roof, also called cold roof or
cold deck, is another form of construction in
which the waterproofing is above the insulation.
This is a double-skin form of construction with
an air space between insulation and waterproofing which ventilates the roof and dissipates
any moisture. But ventilation can also *introduce*
moisture, e.g. the infiltration of drifting snow.
Even more significant, however, is the introduction of moisture due to condensation in the air
space along the underside of the decking supporting the waterproofing when the temperature
of the waterproofing drops below the dew point
of the outside air at night (due to long-wave
radiation emissions) or during an abrupt change
in the weather, e.g. a warm spell in winter.
The cost of providing the ventilated air space is
very high, especially with a concrete roof because
a second loadbearing deck is required to carry
the waterproofing. Such forms of construction
are generally unsuitable for accessible roofs
because of the high loads involved.
On concrete roofs there is the option of creating
the falls necessary by raising the decking for
the waterproofing clear of the concrete (Fig.
D 2.18). Most timber structures create the falls
necessary in the loadbearing structure itself
(Fig. D 2.19).
The ventilated roof was the preferred form of
construction for pitched roofs prior to the intro-

duction of diffusion-permeable secondary
covering materials. But today ventilated forms
of construction are not so common among
pitched roofs because it is difficult to guarantee
adequate ventilation and this form of construction
is costly. Ventilated designs represent the
exception for flat roofs except where special
conditions prevail.

Particular features of the construction
Where the waterproofing is above the insulation,
it does not usually have a stiff underlay, indeed
might even be laid on the soft thermal insulation
itself. Movements in the waterproofing material,
caused by thermal stresses, for example, can
lead to higher mechanical stresses at junctions
and edges in particular. The thermal insulation
material selected should therefore have a compressive strength that can handle the maximum
loads.

Waterproofing below insulation
The inverted or upside-down roof is the name
given to the reversed sequence of waterproofing
and thermal insulation, i.e. waterproofing below
the insulation (Fig. D 2.20). Such an arrangement
protects the waterproofing against severe
temperature fluctuations and mechanical loads.
But on the other hand, the insulation is now
subjected to higher moisture loads because
precipitation seeps underneath. The insulating
materials used in inverted roofs must therefore
be not only diffusion-resistant, but resistant to
water and frost as well. Currently, only rigid
boards made from extruded polystyrene (XPS)
are permitted in inverted roofs. In contrast to
the other types of roof described here, an
inverted roof cannot be built without a roof finish
because the insulation must be protected
against floating, wind suction and other environmental influences.
The fact that the insulation is on the outside in
an inverted roof means that this type of roof
exhibits certain peculiarities that must be taken
into account, the main one being the fact that
precipitation can seep below the insulation.

Water run-off
The drainage to an inverted roof generally takes
place at two levels. Most of the precipitation
drains away in or on the roof finishes above the
insulation. But some precipitation seeps slowly
through the joints between the insulating
boards down to the waterproofing where it then
drains away. It must be remembered here that
cold precipitation or meltwater on the waterproofing leads to heat losses, which must be
taken into account in the thermal performance
calculation. The roof drainage must therefore
be designed in such a way that a build-up of
water around the insulation is ruled out. A brief
build-up during heavy rainfall is harmless, however [5]. The prescribed minimum fall of 2 %
should always be adhered to.

Ballast

Ballast on an inverted roof is absolutely essential and should consist of min. 50 mm deep gravel in the service condition. Such a layer secures the insulation against wind suction calculated according to DIN 1055-4 and against floating in the case of excessive precipitation collecting underneath. At the same time, it protects the insulation against the direct effects of UV radiation, hail, sparks and excessive mechanical actions. Inverted roof designs are suitable for special uses such as rooftop planting, foot and vehicular traffic or helicopter landing pads. On a green roof the planting can simultaneously function as the ballast, provided the weight is adequate; otherwise, "grasscrete" paving blocks can be laid, for example. The insulation protects the waterproofing underneath against mechanical actions (e.g. gardening work). Other forms of ballast that can be considered are gravel, stone and concrete flags, paving stones for foot or vehicular traffic, etc. It must be ensured that the water can drain away.

A separating membrane, to prevent fine particles in the ballast from being washed into the joints between the insulating boards, must always be provided between ballast and insulation. Furthermore, sand and gravel mixes should not be used as the topmost layer because any segregation that occurs will lead to a layer with a high proportion of fine particles that absorbs water and possibly increases the moisture content and the weight. In addition, there is the possibility that fine particles could find their way down between the insulating boards and prevent proper drainage. Instead of the traditional separating membrane, a diffusion-permeable material can be installed that acts as a water run-off layer so that most of the precipitation is drained away before it even reaches the insulation. The effect of this is to decrease the heat losses caused by water seeping below the insulation to such an extent that they no longer need to be considered when calculating the U-value (see "Inverted roofs", pp. 57–58).

Loadbearing structure

According to the Flat Roof Directive, inverted roofs should be built on a thermally sluggish supporting structure (e.g. in situ concrete). This requirement appears in DIN 4108-2 and applies to forms of construction with a weight per unit area exceeding 250 kg/m². This is intended to prevent the occurrence of condensation in the event of an abrupt change of weather (e.g. thunderstorm) because cold water flowing beneath the insulation can cause severe cooling of the interior soffit. On lightweight loadbearing structures, the thermal resistance below the waterproofing must be at least 0.15 m²K/W (DIN 4108-2 stipulation) in order to prevent condensation forming on the soffit in unfavourable conditions (see "Inverted roofs", pp. 57–58).

Waterproofing between layers of insulation

Another option for flat roofs is to lay the waterproofing between two layers of insulation (Fig.

D 2.21, p. 104). This represents a combination of flat roof with waterproofing above insulation and inverted roof. This design minimises the heat losses due to precipitation draining away below the insulation. The insulating effect of the lower layer of insulation remains fully effective even if there are large quantities of rainwater temporarily seeping away below the upper layer of insulation. When the thermal resistance of the lower layer of insulation accounts for more than 50 % of the total thermal resistance, a precipitation-related surcharge (DIN 4108-2 stipulation) is not added to the U-value of the roof.

Particular features of the construction

Where the waterproofing is positioned between two layers of insulation, the requirements for inverted roofs still apply to the insulation above the waterproofing. Here again, a heavyweight roof finish is essential. The conditions below the waterproofing are the same as those in a flat roof with the waterproofing above the insulation. As with the inverted roof, the waterproofing is protected against severe temperature fluctuations and mechanical actions. But in contrast to the inverted roof, the waterproofing is installed on top of the lower layer of thermal insulation, not on the stiff loadbearing structure. Thermal movements can therefore lead to higher mechanical strains, particularly at edges and junctions.

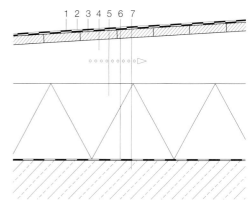

1 Waterproofing
2 Separating membrane if required
3 Decking laid to falls to support waterproofing
4 Ventilated air space
5 Thermal insulation
6 Vapour barrier/check
7 Loadbearing structure

D 2.18

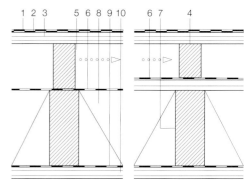

1 Waterproofing laid to falls
2 Separating membrane if required
3 Decking for supporting waterproofing
4 Spacers (firrings) to create falls
5 Ventilated air space
6 Secondary waterproofing layer
7 Loadbearing structure
8 Thermal insulation
9 Vapour barrier/check
10 Soffit lining/services installation level if required

D 2.19

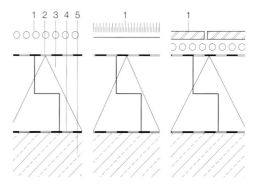

1 Ballast (water- and diffusion-permeable)
2 Separating membrane (also prevents loss of ballast)
3 Thermal insulation (Rigid foam boards with shiplap joints)
4 Waterproofing
5 Loadbearing structure

D 2.20

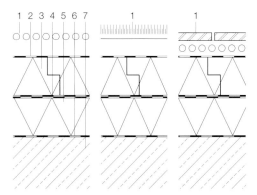

1　Ballast (water- and diffusion-permeable)
2　Separating membrane (diffusion-permeable)
3　Thermal insulation (XPS) with shiplap joints
4　Waterproofing
5　Rigid foam insulation
6　Vapour barrier
7　Loadbearing structure (preferably
　 reinforced concrete)　　　　　　　　D 2.21

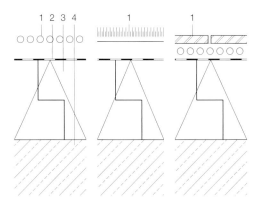

1　Ballast (water- and diffusion-permeable)
2　Separating membrane (diffusion-permeable)
3　Thermal insulation (XPS) with shiplap joints
4　Impermeable concrete slab
　　　　　　　　　　　　　　　　　　D 2.22

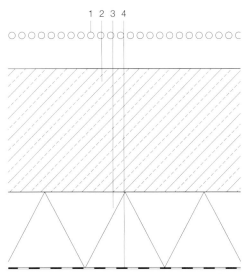

1　Protective finish
2　Impermeable concrete slab
3　Mineral fibre insulation
4　Smart vapour barrier
　　　　　　　　　　　　　　　　　　D 2.23

Impermeable concrete slab as waterproofing layer

One way of providing waterproofing is to construct a concrete slab with a high resistance to water penetration; the loadbearing structure thus functions as the waterproofing layer also. A separate waterproofing material is unnecessary because the concrete is watertight and diffusion-resistant. However, it must be ensured that all joints are properly sealed and no cracks can form which would then allow water to infiltrate the structure (see "Impermeable concrete slabs", pp. 35–36).

The insulation can be attached to the inside or the outside. If it is placed on the outside, the construction should be treated like an inverted roof (Fig. D 2.22). Insulation on the inside of the concrete slab presumes a particular type of insulating system because although the concrete is impervious, it is not diffusion-tight (Fig. D 2.23). In such a situation moisture could collect between the internal insulation and the concrete slab over the long-term.

Impermeable concrete slabs can be built for any length of span encountered in normal structures and can even be constructed in prestressed concrete, for example, to carry heavier loads [6].

Particular features of the construction

The concrete slab must be designed and constructed according to the requirements for concrete with a high resistance to water penetration as outlined in DIN 1045-2 (see "Concretes with high water impermeability", pp. 96–97). Where the insulation is on the inside, the roof should be finished with a layer of gravel or planting in order to protect the slab against excessive temperature variations over the course of a day and hence reduce thermal movements. Mineral wool insulation in conjunction with a smart vapour barrier represents a good solution for internal insulation. A normal vapour barrier is not advisable because its reduced drying-out potential can lead to moisture problems. In some cases it may be necessary to check the internal insulation with respect to its hygrothermal behaviour (see "Hygrothermal simulation tools", pp. 70–73). The variation with insulation on the outside has proved itself superior in practice because internal insulation can quickly lead to moisture-related damage in the event of incorrect design or poor workmanship.

The edges of the roof can be finished with kerbs, cornices, overhangs or parapets. A kerb, which must be at least 15 cm high, can be built to surround the whole of the roof or just parts thereof. This functions like a bund wall, retaining precipitation so that it cannot simply drain over the edges of the roof. Kerbs can also enclose layers of gravel or planting and also help to stiffen the roof structure.

If an in situ concrete parapet is built instead of a kerb, this should be at least 20 cm wide in order to simplify the concreting operations. Parapets should include joints that extend down as far as the kicker. Additional longitudinal reinforcement will be required below these joints to prevent cracking. Where the roof slab overhangs,

then this should be on all sides of the roof. Large concrete slabs include joints to divide them into sections. Roof overhangs, cantilevering slabs and parapets also need joints to compensate for the higher thermal loads to which they are exposed. Joints within the roof surface must be bordered by kerbs that are at least 5 cm higher than the kerbs around the perimeter of the roof to ensure that any water on the roof surface stays clear of these joints. Kerbs adjacent to joints are also necessary for cantilevering slabs. Additional dowels bars to control cracking are necessary across such contraction joints. Joint filling materials must allow sufficient movement of the concrete slab at the joints. All joints must be reliably sealed and liquid synthetic materials are ideal here. The waterproofing is laid in a loop over the joint so that it can be accommodate any movement. Penetrations, openings for rooflights, access hatches, etc. should also be framed by reinforced concrete kerbs. Roof outlets for roof drainage, vents, pipe penetrations, electric cables and conduits must be integrated into the formwork accurately and cast in to the concrete [7].

An impermeable concrete slab should be supported on a ring beam via some form of sliding bearing because the thermal loads on the slab can cause considerable changes in length.

Green roofs

A green roof represents a quality increase for any living or working environment, especially when the roof is accessible or least visible. In addition, it certainly improves the visual appearance of the roof. Intensive planting implies that the roof will be used as an extension to the indoor living areas and also for sport or recreation. Where extensive planting is planned, according to the Flat Roof Directive and DIN 18531, the roof is not generally accessible.

A green roof performs a number of functions. It protects the waterproofing against the effects of the weather and high temperature fluctuations, so that, compared with unprotected waterproofing materials, it can last up to twice as long before repairs or refurbishment are necessary [8]. Rooftop planting can function as ballast to secure rooftop structures (e.g. photovoltaic panels, air-conditioning plant, etc.), to guarantee their ability to withstand wind loads. In this case the rooftop structures do not require fixings that penetrate the waterproofing. Plants reduce the radiation reflected from the roof surface, which is an important advantage for photovoltaic panels because they are then subjected to less infrared radiation on the underside than is the case on a normal roof. That in turn results in a lower operating temperature and hence a higher degree of efficiency for the system.

As they create new areas of greenery, green roofs can be classed as measures that compensate partially for the intervention in the natural environment according to the provisions of the Federal Building Code (*Baugesetzbuch*, BauGB).

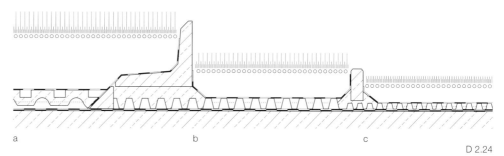

a   b   c

D 2.24

Standards and regulations

The Rooftop Planting Directive (*Dachbegrünungs-srichtlinie*) published by the Forschungsgesell-schaft Landschaftsentwicklung Landschafts-bau e.V. (FLL, German Landscape Research, Development & Construction Society) is regarded as the standard work on green roofs. It is also used outside of Germany [9].
A green roof requires special care and mainte-nance. The Rooftop Planting Directive as well as information published by the trade associa-tion Fachvereinigung Bauwerksbegrünung e.V. (FBB), part of the German Roofing Contractors Association (ZVDH), and the Bundesverband Garten-, Landschafts- und Sportplatzbau e.V. (Gardens, Landscaping & Sports Facilities Association) are very useful here.
Planting on industrial roofs is the subject of Technical Rule AGI B11.
Furthermore, DIN 18035-4 "Sports grounds – Part 4: Sports turf areas" and DIN 18915–18917 plus 18919 "Vegetation technology in land-scaping" also include information relevant to green roofs.

Construction

All flat and shallow-pitch roofs are suitable as a base for a green roof. The Rooftop Planting Directive classifies green roofs according to their fall. A special drainage layer is required for designs with a fall < 2 %. Green roofs should therefore be designed with a fall of at least 2 %. In practice, however, the supporting structures for rooftop gardens and green roofs over underground parking are often built with-out any fall.
As the fall increases (> 5 %), the water drains away faster. This can be compensated for by a form of construction that retains more water, or by using plants that need less water. On pitches > 15° construction measures will be required to prevent slippage, e.g. perimeter kerbs or proprietary anti-slip systems [10].
Flat roofs with the waterproofing above the insulation, inverted roofs and impermeable concrete slabs are all suitable for use as green roofs. The effects on the building physics func-tions must be checked in each case. With an impermeable concrete roof, it is essential to

ensure that roots cannot infiltrate the joints in the structure.
The loadbearing structure must be able to carry the increased load of a wet green roof. But when calculating the load resisting wind suction, only the load of the dry roof may be used.

Layers

The substrate (growing medium) forms the base for the layer of vegetation on a green roof. Below this there is a filter membrane that prevents fine particles from the substrate layer being washed into the drainage layer underneath. The drainage layer carries away excess precipitation, but can also retain water which is then available to the plants during periods of dry weather. The depth of the drainage layer lies between 1 and 20 cm.
A protective membrane installed between the drainage layer and the waterproofing, or root barrier, protects the waterproofing material against mechanical damage. However, the pro-tective membrane itself does not prevent roots from penetrating as far as the waterproofing. Only a waterproofing material resistant to root penetration or an additional root barrier can ensure that the waterproofing is not damaged by the roots of plants. On inverted roofs the root barrier is laid directly on the waterproofing and not on the insulation [11].
A detailed description of the individual layers

in a green roof can be found in the chapter "Materials" ("Rooftop planting", pp. 92–93)

Planting

There are three different types of rooftop plant-ing, which differ in terms of usage, construction and choice of plants (Fig. D 2.24).

Extensive planting

Extensive planting is essentially natural forms of vegetation that mainly look after themselves, growing and spreading without the need for any specific gardening work. Such a roof is not accessible, except for maintenance; it is purely an ecological protective covering to the roof. The depth of the green finish, a single-, multi-or thin-layer design, can lie between approx. 4 and 20 cm.
In a single-layer design (Fig. D 2.27a), the vege-tation also provides the filter and drainage functions. This is guaranteed by choosing a substrate that consists of a suitable mixture of mineral and organic materials with a range of particle sizes. The roof must also be designed with a minimum fall of 2 %. Sedum and moss varieties, but also grasses, can be used for the vegetation.
Another possibility is to use pre-cultivated veg-etation mats which are about 2 cm thick and in the saturated state weigh 20–50 kg/m². These are laid on a 5–8 cm deep substrate and merge

D 2.25

D 2.26

with this over time. Such mats can also be used to protect against erosion in areas of the roof subjected to high wind suction loads, or on roofs where other forms of planting are not possible.

A multi-layer design consists of the vegetation layer, an additional drainage layer (Fig. D 2.25, p. 105, and Fig. D 2.27b) and a filter membrane that prevents fine particles being washed out of the substrate and clogging the drainage layer. This form of construction is more costly than the single-layer variation but does result in a better water-retention capacity and enables the use of other plants. Multi-layer designs weigh between 60 and 150 kg/m² in the saturated state. Generally speaking, green roofs with extensive planting require less care and maintenance and are less costly to build than roofs with intensive planting.

Intensive planting
The planting, design and usage options for the more elaborate intensive planting option can be compared with open areas at ground level. This type of planting requires a regular supply of water and nutrients, also constant care and maintenance. The depth of the roof finishes starts at 15 cm and depending on requirements can reach 40 cm or even more, and weights of 150 to more than 500 kg/m². A roof with intensive planting involves a high capital outlay and high maintenance costs, but does offer a high return on the investment. The potential for use can even match that of a well-tended garden.

Semi-intensive planting
Also called semi-extensive planting, this is a less expensive, more easy-care, special form of intensive planting that makes use of a planned layout of ground-cover planting consisting of grasses, bushes and shrubs (Fig. D 2.26, p. 105). The depth of finishes required ranges from 12 to 25 cm, and weights between 150 and 200 kg/m² are usual. The plant species chosen do not place any particular demands on the sequence of layers, nor the supply of water and nutrients. Less care and maintenance are required. The costs lie between those of extensive planting and intensive planting [12].

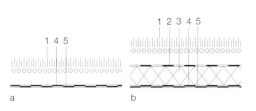

1 Vegetation
2 Filter membrane
3 Drainage layer
4 Root-resistant waterproofing
5 Separating membrane

D 2.27

Irrigation and drainage
At least one water supply point is required for irrigating the vegetation on a green roof. Automatic irrigation systems with sprinklers, drip hoses, etc. are normally used. On roofs with intensive planting it can be advantageous to carry out the irrigation by way of a build-up of water in the drainage layer.

Excess water, especially on roofs with intensive planting, must be drained away to one or more outlets via a drainage layer. Roofs with extensive planting do not necessarily have to include a drainage layer, in which case the substrate itself performs this function, but it then has to be able to store all the precipitation falling on the roof. This can briefly lead to flooding of the substrate during heavy rainfall, which then has to drain away on the surface.

All drainage installations must be permanently accessible for inspections and maintenance (Fig. D 2.28).

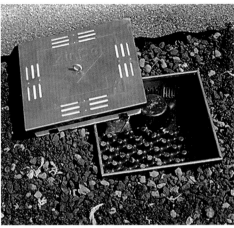

D 2.28

Roofs for foot and vehicular traffic

Flat roofs can be used in many ways, e.g. relaxation, recreation, sports, parking areas, helicopter landing pads, etc. But even if the roof is not intended to be used at all, it must be accessible for inspection, maintenance and installation purposes.

Roof surfaces intended for foot and vehicular traffic are subjected to much higher imposed loads than other flat roof surfaces. The loads must be transferred through the roof finishes to the loadbearing structure, which calls for an appropriate form of construction (Figs. D 2.30 and D 2.31).

DIN 1055-3 and DIN 1072 deal with the types of loads and their frequency for design purposes (Fig. D 2.29). These increased dead, imposed, braking and snow load assumptions must be taken into account during design and construction by providing thermal insulation with the necessary compressive strength according to DIN V 4108-10, and roof waterproofing suitable for such loading cases according to DIN 18195. Two slip layers are generally laid on the waterproofing so that the shear forces generated by braking vehicles, for example, cannot be transferred to the waterproofing material.

Loading due to	Load	Additional loads	
Individual persons accessing the roof for care or maintenance purposes (usage category H)	Point load: 1 kN at most unfavourable point		to DIN 1055-3
Loading class 1: Foot and cycle traffic on rooftop terraces and accessible roof surfaces (usage category Z)	Distributed load: 4.0 kN/m²		
Loading class 2: Motor vehicles up to 2.5 t	Total load: 25 kN Equivalent distributed load: 3.5–5.0 kN Single wheel load: 10 kN/m²	Horizontal loads due to braking, steering and accelerating	to DIN 1072
Loading class 3: Goods vehicles up to 16 t	Total load: 160 kN Equivalent distributed load: 8.9 kN/m² Max. single wheel load: 50 kN/m²	Horizontal loads due to braking, steering and accelerating	
Heavy goods vehicles up to 60 t	Total load: 600 kN Equivalent distributed load: 33.3 kN/m² Max. single wheel load: 100 kN/m²	Horizontal loads due to braking, steering and accelerating	

D 2.29

D 2.27 Schematic drawing of extensive planting
    a Single-layer approach
    b Multi-layer approach
D 2.28 Roof outlet designed for easy inspection
D 2.29 Types of loading and design loads on roofs
D 2.30 Roof construction for an accessible flat roof
D 2.31 Inverted roof with…
    a flags on proprietary supports
    b paving bricks
    c in situ concrete slab
D 2.32 Load transfer through base course and sub-base
D 2.33 Shear-resistant perimeter details

D 2.30

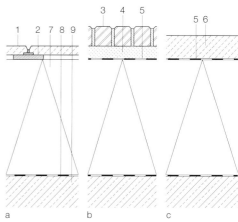

1 Concrete flags
2 Proprietary support
3 Paving bricks (with sand joints)
4 Frost-resistant base course
5 Separating membrane
6 In situ concrete slab
7 Thermal insulation (XPS) with shiplap joints
8 Waterproofing
9 Loadbearing structure

D 2.31

According to the FFL's "Recommendations for the design and construction of trafficked areas on structures", a build-up of water is not permitted beneath footways and carriageways. Water collecting beneath the finishes can freeze in winter and cause severe damage. A fall of min. 2 % is therefore required for surfaces for foot traffic belonging to loading class 1, and min. 2.5 % for surfaces for vehicular traffic belonging to classes 2 and 3. Irrespective of this, the minimum fall of the surface finish must also take into account the material used; this ranges from 1 % (for permeable finishes) to 3 % (stone paving). Finishes raised above the roof surface on special mountings which have permanently open joints do not necessarily require any falls.

All construction variations must be surrounded by a shear-resistant form of construction (Fig. D 2.33) that can withstand the forces and thus ensure that neither wearing course nor base course start to slide as a result of shear or other loads.

Forms of construction that satisfy this more demanding specification weigh at least 650 kg/m^2 and can have depths of 30 cm or even more (Fig. D 2.32).

Wearing course

A frost-resistant, non-slip wearing course made from, for example, flags, paving stones, "grasscrete" paving blocks, asphalt or concrete can be used as the wearing course for footways and carriageways, laid on a suitable base course. Finishes for sports facilities are also possible, as specified in DIN 18035. The prescribed minimum thickness for the material selected must be adhered to depending on the design loads and the form of construction. For example, timber can be used for a rooftop terrace, where the low loads involved do not necessarily require an elaborate base course.

Base course and sub-base

The task of the sub-base is to guarantee the load-carrying capacity of the overlying wearing course. This is where the loads are distributed evenly to the layers of the construction below (Fig. D 2.32). In practice these are sand and gravel mixes with 0/22, 0/32 or 0/45 grading specifications and minimum depths (which

depend on the largest aggregate size) of, normally, 10–15 cm.

A base course is laid on top of the sub-base to compensate for dimensional inaccuracies and to transfer the actions applied to the wearing course. The base course must exhibit an adequate compressive strength and must be permeable. Sands and chippings with grading specifications 0/4, 0/5 or 0/8 are typically used.

Drainage layer

The drainage layer collects the water infiltrating the layers above and drains it towards one or more outlets. Furthermore, it can compensate for the pressure of any ice that might form. Either loose mineral materials or preformed drainage elements are used for the drainage layer. Depending on the loading classes and the largest aggregate size of the loose material, the minimum depth of the drainage layer will be 10–15 cm. Plastic drainage elements must guarantee that water can drain in both the vertical and horizontal directions. They allow a stable and permanently functioning drainage layer to be formed in one operation.

Special features of inverted roofs

As an inverted roof differs from other forms of construction because the thermal insulation is laid on top of the waterproofing, the design of the wearing course is somewhat different. An inverted roof intended for foot or vehicular traffic requires one of the following three roof finishes, which are laid directly on the insulation.

Roof finishes in the form of concrete flags raised clear of the roof surface on proprietary mountings may not need to be laid to a fall, but this does depend on the National Technical Approval for the system. In this case we assume that most of the precipitation can drain away unhindered on the smooth surface of the insulation underneath and the thin film of water on the surface quickly drains through the joints between the flags and rapidly dries out again (Fig. D 2.31a, p. 107). The disadvantage of this form of construction is that the proprietary mountings supporting the flags can cause indentations in the insulation.

When using paving bricks embedded in a suitable material, a fall of at least 2.5 % at the water-

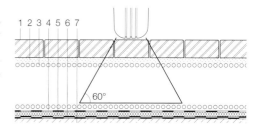

1 Wearing course: paving bricks, ≥ 12 cm up to 16 t, ≥ 14 cm up to 60 t
2 Base course: 3–5 cm chippings
3 Sub-base: min. 15 cm gravel
4 Filter membrane
5 Drainage layer
6 Separating/slip membrane
7 Stable supporting construction with suitable waterproofing

D 2.32

D 2.33

► ► ► Prevailing wind direction

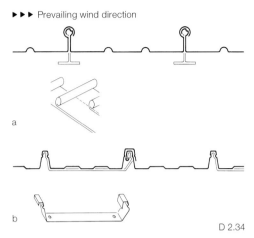

a

b

D 2.34

D 2.34 Longitudinal joints (side laps) between profiled
metal sheets
a Fixed with extruded aluminium sections
b Fixed with preformed sheet metal brackets
D 2.35 Fixed (left) and expansion (right) clips for standing
seam roofing laid manually
D 2.36 Roll-seam-welded standing seam with expansion
clip
D 2.37 Forms and customary sizes of profiled metal
sheets made from aluminium, copper (rare) and
galvanised steel
D 2.38 Waterproofing a glass roof with preformed seals
and clamping wing or cap (patent glazing)
D 2.39 Waterproofing a glass roof with silicone
D 2.40 Point-fixing of glass pane through drilled hole
D 2.41 Point-fixing of glass pane through the joint

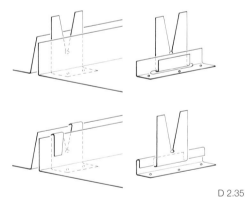

D 2.35

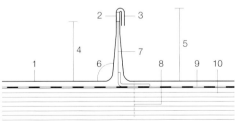

1 Sheet stainless steel
2 Weld seam
3 Folded-over edge of standing seam
4 Height to weld seam: approx. 16 mm
5 Height of material prior to folding over: approx. 30 mm;
   after folding: approx. 20 mm
6 Angle approx. 92°
7 Expansion clip
8 Stainless steel fastener
9 Sound-attenuating membrane
10 Decking

D 2.36

proofing level is vital. The joints between the
paving bricks and the bedding material must
be water- and diffusion-permeable, and therefore should not be sealed (Fig. D 2.31b, p. 107).
An inverted roof with in situ concrete slab on top
of the insulation also requires a minimum fall of
2.5%. Joints in the slab must be sealed to prevent the infiltration of water. In this case there is
no need to add surcharges to the U-value (see
"Surcharges on the U-value", pp. 57–58) because
as a rule no water can penetrate through to the
waterproofing material (Fig. D 2.31c, p.107).

Metal roofs

Engineers and architects have always been
looking for reliable methods of waterproofing
flat roofs and since the 1960s metal has been
used as well. Welded sheet stainless steel can
be employed on pitches from 0°, profiled metal
sheets on pitches from 1.5° or 2° (see "Sheet
metal roofing", p. 91).

Profiled metal sheets
The low pitches of flat roofs make them suitable
for profiled metal sheets made from steel, stainless steel, aluminium or titanium-zinc. The joints
between the individual sheets must lie above
the water run-off level.
Profiled metal sheets are laid in rows. In doing
so, the sheets are always laid with the wider
part of the profile at the bottom, to form the
water run-off level, and the side laps should be
arranged so that they face away from the prevailing wind direction (Fig. D 2.34). Depending
on the particular product, they are hooked
together, interlocked or overlapped. Transverse
joints should be avoided on metal flat roofs,
and the sheet lengths that can be supplied
these days generally enable this (Fig. D 2.37).
If transverse joints are, however, necessary,
then they must be welded. When laying directly
on timber or concrete, a textured separating
membrane must be laid in between to protect
the metal against aggressive constituents and
alkaline reactions. The separating membrane
also performs other functions: it acts as a run-
off layer for any condensation, provides a slip
plane to isolate against changes in length due
to expansion, and reduces the drumming noise
of rainfall.
Profiled metal sheets are clipped to special
brackets that interlock with the particular profile
(Fig. D 2.34). The brackets are fixed to the
loadbearing structure with rustproof fasteners.

Sheet stainless steel
Stainless steel in the form of strips or sheets,
350–670 mm wide in a gauge of 0.4 or 0.5 mm,
can be welded together to form a roof finish.
The material is produced in coils up to 250 m
long, but the means of transport used limits the
lengths that can be supplied to the building
site.
The strips or sheets are laid in rows. Bending
up the edges of the sheets creates standing

seams that are welded and folded together.
Roll-seam welding, which requires no filler
metal, is used for the joints (Fig. D 2.36). After
welding, the joint is folded over [13].
The long strip and sheet lengths available
obviate the need for any transverse joints in
most cases. Where they are unavoidable, they
must be welded together in a similar way to the
profiled sheets. A sound-attenuating separating
membrane beneath the stainless steel helps to
reduce the drumming noise of rainfall.
Expansion and fixed clips are used to attach
the stainless steel to the supporting structure
(Fig. D 2.36). Clips must be positioned according to the manufacturer's specification.
A welded sheet stainless steel roof is costly,
but will last almost forever, which means that
from both the ecological and economical view-
points, this is a sensible roofing material. Such
a finish is very useful for refurbishment projects
because the old roof covering can be left in
place.

Roof finishes
Metal roofs can be finished with planting or a
layer of gravel. However, with non-welded joints,
such finishes are only permitted when a build-up
of water is impossible and a rubber seal has
been fitted into the joint. On a sheet stainless
steel roof, a heavyweight finish makes clips
unnecessary. The main advantage of a roof
finish though is the fact that it reduces the temperature rise due to solar radiation and therefore the annoying loud cracking noises of the
metal as it expands and contracts with the temperature changes, an effect that is particularly
severe with sheet stainless steel.

	Profiled metal sheets		
	Aluminium	Copper	Galvanised steel
Profile height [mm]	50, 65	47	35, 65
Design width [mm]	250	457	305
	305		1000
	333		250
	400		333
	434		400
	500		500
	600		600
Gauge [mm]	0.7–1.2, customary 0.8	0.6	0.63–1.0, customary 0.75/0.88
Length [mm]	1200–7500	≤ 17000 (road) ≤ 30000 (rail)	≤ 50000
Weight [kg/m²]	3.1–5.2	7.9	6.7

D 2.37

Glass roofs

As glass roofs with a pitch of 0° have been possible for a number of years, glass represents yet another option for flat roofs (see "Glass", pp. 44–47 and p. 91).
Glass usually requires a slender loadbearing structure consisting of linear members and nodes or purlins, rafters and trusses. Such "minimal" forms of construction can be positioned either above or below the glazing. The panes of glass perform both the waterproofing and thermal insulation functions. The most important aspects of flat glass roofs are explained below, but the reader should consult specialist publications for more detailed information [14].

Design

When planning a glass roof, the "Technical Rules for the Use of Glazing on Linear Supports" (*Technische Regel für die Verwendung von linien-förmig gelagerten Verglasungen*) published by the Deutsches Institut für Bautechnik (DIBt, German Institute of Building Technology) requires that overhead glazing be made from a safety glass that prevents any fragments of glass from falling. Single glazing and the inner panes of insulating glass units must therefore be made from laminated safety glass. Alternatively, some other form of protection against falling glass may be employed, e.g. nets. Wired glass should not be used for overhead glazing because of the thermal stresses and the risk of corrosion.

Construction

In a glazed roof the glass is subjected to high temperatures – especially in the summer – because of the angle of incidence of the solar radiation. The following advice should therefore be followed in order to avoid unnecessary thermal stresses [15]:
- The panes of glass should be evenly ventilated to prevent a build-up of heat.
- Sunshades should be fitted no closer than 20 cm to the glass. It should be remembered that dark sunshades can heat up the glass to a greater extent than light-coloured fittings that reflect the light.
- Decoration painted on or attached to the panes of glass can cause uneven heating and hence thermal stresses.
- A thermal break is necessary between glazing bars and glazing wings/caps in order to prevent a temperature gradient between the perimeter and centre of each pane. The edge cover to the panes should therefore not exceed 20 mm. Adequate ventilation in the joints prevents, or least minimises, condensation, which must be able to drain away (Fig. D 2.38).
- Extreme shadows should be avoided because these can lead to temperature differences within the glass.
- As the air in the cavity of an insulating glass unit heats up, so the pressure on the glass also increases; therefore, the cavity of solar-control or low E glazing should not

be wider than 15 mm.
- Insulating glass units made up of solar-control glass/wired glass and wired glass/wired glass are unsuitable because glass and steel have different coefficients of thermal expansion, which can lead to the panes breaking as they heat up.
- Where additional thermal loads cannot be ruled out, then the panes affected must be made from a type of toughened glass.
- If an asymmetrical pane configuration is being used, the thinner and less stiff pane is much more likely to break. Therefore, with an aspect ratio > 2:1, the thinner pane must be made from a type of toughened glass. The inner pane of solar-control and low E glazing may not be more than 2 mm thinner than the outer pane.

Pitch

In practice it has been proved that glass roofs can be constructed with a pitch of 0°. At 2° precipitation can drain away properly, but leaves dirty streaks behind. Only with pitches of 7° and more does the self-cleaning effect become really evident. Only point fixings, which do not require any additional capping strips at the joints, are therefore suitable for shallow pitches.

Joints in patent glazing

In patent glazing the panes of glass are held in place with glazing wings or caps. The dimensions of the joint are specified in DIN 18545. The glazing edge cover in the case of non-toughened glass should be two-thirds of the depth of the rebate, but not more than 20 mm (Fig. D 2.38).
Glazing systems with open (drained) joints are used for roof glazing. Openings at the highest and lowest points ensure that air can circulate through these joints and compensate for any build-up of vapour pressure in the joint. If leaks lead to condensation, this must be able to drain away via these openings. The water run-off layer must lie below the sealing layer [16].

Sealing

In the transverse and longitudinal joints of a glass roof, seals alone prevent the ingress of water. The sealing materials must be compatible with each other and with the type of glass selected [17].
There are essentially two methods of sealing available these days:
- Preformed sealing strips and gaskets in conjunction with clamping strips (patent glazing wings/caps), which simultaneously provide the bearing for the glass as well as sealing against wind and rain. They provide a resilient bearing for the glass and compensate for the tolerances of the glazing units and glazing bars (Fig. D 2.38, p. 109) [18].
- Silicone, which is injected into the joint. A prestress guarantees contact with the sides of the joint. There should be no sealant in the space below the joint so that vapour pressure compensation and drainage are not hindered (Fig. D 2.39, p. 109).

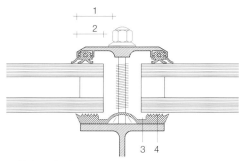

1 Rebate depth
2 Edge cover
3 Inner sealing level
4 Open (drained) joint, water run-off level

D 2.38

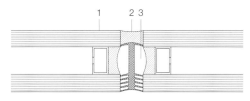

1 Insulating glass
2 Permanently elastic sealant
3 Open (drained) joint

D 2.39

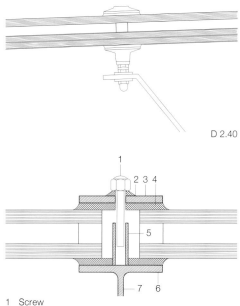

D 2.40

1 Screw
2 Sealing washer
3 Cover strip
4 Silicone
5 Screw sleeve
6 Supporting frame
7 Spacer piece to position frame clear of structure

D 2.41

109

Mounting

Two mounting methods are available at present [19]:

• Linear supports: the panes of glass are clamped in place by the lips of the seals fitted to the steel or aluminium patent glazing wings or caps, which are screwed to the glazing bars (Fig. D 2.38, p. 109). The wings/caps must be removable to allow repairs. Dust and dirt can collect at transverse glazing bars and therefore this form of mounting is only suitable for very shallow pitches when the glazing wings/caps are replaced by silicone at the transverse joints. Bevelled glazing wings/caps can be used for these joints but regular, intensive cleaning is required to remove the dirt or algae that still accumulates.

• Point supports: this method is better suited to flat roofs. The point fixings can be fitted through holes drilled in the glass (Fig. D 2.40, p. 109) or at the joints, between the panes of glass (Fig. D 2.41, p. 109). The fixing and sealing functions are separated in point-fixed glazing. The glazing units are fixed to the supporting structure with screws at the corners. The use of articulated fixings avoids introducing any additional stresses.

The latest developments favour fixings through the joints, which avoid the costly factory drilling and the associated resultant stresses in the glass.

Special aspects of glass roofs

Condensation may collect on the inner pane of a glass roof depending on the humidity of the air, the internal and external temperatures and the U-value of the glazing. At shallow pitches, this condensation drips off the glass. Condensation can be reduced by using panes with a low U-value.

In regions where heavy snowfall can be expected, it may be necessary to remove a layer of snow from a shallow-pitch roof. If this is to be carried out manually, accessible zones must be included (either areas without glazing or with glazing designed for maintenance foot traffic).

Another option for removing snow from glass is to melt it by using heated glass or glazing bars heated with hot water. However, both methods require a considerable energy input [20]. Generally speaking, glass flat roofs are very demanding in terms of their design and construction, which should be left to specialist companies.

Junctions and details

Special details to prevent water seeping below the waterproofing are required at the junctions between the waterproofing and other building components that do not form part of the roof surface and also around the perimeter of a roof. We distinguish between junctions with components firmly attached to the roof surface (rigid detail, e.g. parapet) and those junctions with components that can move with respect to the roof surface (non-rigid detail, e.g. penetration). The application categories according to DIN 18531-1 must be taken into account at junctions and details (see "Waterproofing", p. 86). There are two categories:

• Application category K 1: roof waterproofing materials satisfying normal requirements. Category K 1 requires a minimum pitch of 2%. Where the pitch is < 2%, waterproofing materials and details complying with category K 2 must be selected.

• Application category K 2: roof waterproofing that must satisfy more stringent requirements (e.g. a higher-value building usage, high-rise structures or roofs where access for maintenance is difficult). For this category falls of min. 2% at waterproofing level and min. 1% in valleys are compulsory. Junctions and details with difficult access are assigned to category K 2.

Waterproofing to roofs, or parts thereof, with a fall < 2% cannot be built according to the above application categories. Junctions employing sheet metal flashings integrated into flexible bitumen sheeting and those sealed solely with sealing compounds may only be used for application category K 1. They must be accessible for maintenance. For application category K 2 the flexible sheeting must be tucked into a groove and flashed, or protected by an apron or cover flashing (Fig. D 2.44). Joints and junctions at which excessive tensile or shear forces could occur are not permitted, e.g. a rigid joint in the waterproofing above a joint between two separate parts of the structure [21].

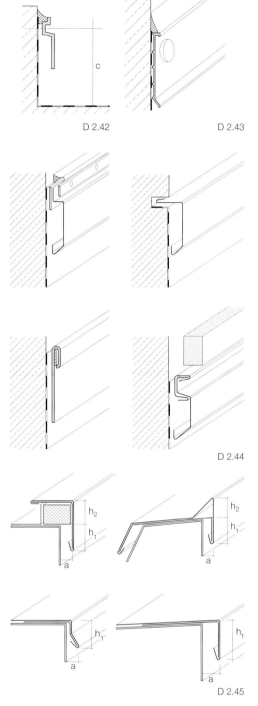

D 2.42

D 2.43

D 2.44

D 2.45

D 2.42 Skirting height c at building component rising above roof level
D 2.43 Clamping bar with permanently elastic seal
D 2.44 Various skirting flashing details
D 2.45 Roof edge flashing details produced manually on site
a Rainwater drip overhang
h_1 Overlap
h_2 Kerb
D 2.46 Two-part roof edge trim with continuous clamping bar
D 2.47 Parapet capping
a Rainwater drip overhang
h Overlap
c Skirting height
D 2.48 Minimum dimensions for sheet metal parapet cappings
D 2.49 Waterproofing detail at pipe penetration through plastic flexible sheeting

Junctions with components above roof level

Junctions with components that rise above the general roof level must be designed in such a way that they prevent water seeping behind the waterproofing even in the case of flooding. To do this, the waterproofing must continue up the side of the component for a sufficient height. The Flat Roof Directive specifies minimum values depending on the pitch of the roof: for a pitch ≤ 5° the minimum skirting height is 15 cm, for > 5° it is 10 cm, in each case measured above the topmost level of the roof finishes (e.g. layer of gravel, layer of planting) (Fig. D 2.42). A greater skirting height may be necessary in regions with heavy snowfall.

The skirting of flexible sheeting must be fixed at the top to prevent it slipping down, e.g. with mechanical fasteners. A linear fixing is generally employed, e.g. clamping bars or sections (Fig. D 2.43), or flashings welded to the waterproofing material, always on a flat substrate (application category K 1); the maximum distance between the fixings should not exceed 20 cm. Mechanical fasteners along the top edge are not usually necessary with liquid waterproofing materials, provided an adequate bond with the substrate can be guaranteed. The upper edge of the waterproofing material must be physically covered to make it rainproof and watertight. This is guaranteed by fixing apron or cover flashings (Fig. D 2.44) or prefabricated metal profiles to the component above roof level, which in Germany must be installed according to the *Fachregeln für Metallarbeiten im Dachdeckerhandwerk* (Metal Roof Covering Recommendations). Flashings and fixings exposed to the weather must be made from a corrosion-resistant material.

Where a skirting of waterproofing material is more than 50 cm high, it must be bonded or mechanically fastened to the vertical surface. Linear-type fixings such as bars or lines of fasteners are suitable here.

On a roof that is in regular use, the waterproofing material must be protected against mechanical damage where it is turned up a vertical surface. Sheet metal, stone or concrete flags, etc. can provide the necessary protection.

On a roof waterproofed with flexible bitumen sheeting, a bond enhancer will be required to ensure that the bitumen material adheres to the vertical surface. At least two plies of flexible bitumen sheeting must be provided at such junctions, and they must include a fillet, e.g. of insulating material, at the joint between roof surface and rising component in order to avoid excessive stresses or creases in the waterproofing material. The plies of waterproofing material at the junction must be bonded into the plies on the surface of the roof and must continue up the component, which could be vertical or at an angle, for the necessary height [22]. To comply with application category K 2, the waterproofing must be flashed or covered in some way.

Penetrations

Penetrations (e.g. vent pipes) must be planned and constructed like junctions. The joint between penetrating component and waterproofing is achieved with the help of a preformed metal flange, plastic collar, elastomeric or plastic flexible sheeting (Fig. D 2.49) or a liquid waterproofing material. Flanges and collars must be integrated into the waterproofing on the surface and the top end of the preformed element must be sealed to prevent ingress of water. The distances between penetrations and between penetrations and other components, e.g. walls, expansion joints or roof edges, should be at least 30 cm so that each detail can be properly constructed and permanently sealed. Critical here is the outer edge of a flange or collar.

Penetrations through the waterproofing for securing fall arrest systems, supports, masts and other roof-mounted items may be subjected to significant movements due to wind and other loads. The details at these penetrations should therefore be able to accommodate such movements. Sleeves with bonded flanges, collars or liquid waterproofing materials are all potential solutions for such fixing points. Each protective element must continue at least 15 cm above the top of the roof finishes and the top edge must be sealed to prevent ingress of

water. The bottom edge of a flange or collar must be integrated into the waterproofing material and must extend approx. 12 cm to each side [23].

Roof edge details

Special raised details are required at the edges of roofs, except adjacent to gutters. These details can take the following forms:
• Kerbs with capping flashings
• Kerbs with capping profiles
• Roof edge profiles

The detail along the edge of a roof can be formed manually on site by skilled workers (Fig. D 2.45) or can make use of preformed profiles (Fig. D 2.46). The Flat Roof Directive specifies general minimum values for the distance the waterproofing must extend above the level of the roof. For roof pitches ≤ 5° the minimum skirting height is 10 cm, and for > 5° it is 5 cm. These heights must be measured from the topmost surface of the roof finishes. Cappings and flashings must fall in the direction of the roof surface [24].

The Metal Roof Covering Recommendations specify the dimensions of such flashings for roof edge details. The clearance between a rainwater drip and the component behind must be min. 20 mm (Figs. D 2.47 and D 2.48). The reason for the folded edge and the clearance is to prevent dirty streaks on render, facing masonry, fair-face concrete, cladding, etc. [25]. A minimum clearance of 50 mm is necessary when using a copper capping. However, marks on walls below flashings and cappings cannot be entirely avoided.

Rooflights

A multitude of different prefabricated elements are available for admitting light, allowing smoke and heat to escape, and for ventilation (inlets or outlets). Rooflights enable daylight to reach the interior, but they can also function as smoke and heat vents, which are operated automatically in the event of a fire, or provide access to

D 2.46

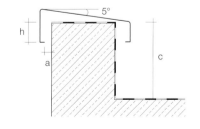

D 2.47

Building height [m]	Rainwater drip overhang a [mm]	Overlap h [mm]
< 8	20	50
8–20	30	80
> 20	40	100

D 2.48

D 2.49

111

Materials and form of construction	Colouring	U [W/m²K]	g⊥ [-]	τ$_{D65}$ [-]
Solid PMMA panel, single-leaf	clear	5.4	0.85	0.92
Solid PMMA panel, single-leaf	opal	5.4	0.80	0.83
Solid PMMA panel, double-leaf	clear/clear	2.7	0.78	0.80
Solid PMMA panel, double-leaf	opal/clear	2.7	0.72	0.73
Solid PMMA panel, double-leaf	opal/opal	2.7	0.64	0.59
Solid PMMA panel, double-leaf	clear, IR-reflective	2.7	0.32	0.47
Solid PMMA panel, triple-leaf	opal/opal/clear	1.8	0.64	0.60
Solid PC/PETG panel, single-leaf	clear	5.4	0.75	0.88

D 2.50

the roof. We distinguish between the following types of rooflight:
• Rigid rooflights with the infill element mounted permanently on a frame
• Opening rooflights with the infill element mounted on a separate frame attached to the main frame via hinges so that the infill element can be moved to allow ventilation or egress. Such rooflights do not necessarily have to be translucent, they can also be opaque.

Rooflights are available in single- and multi-leaf versions made from plastic, occasionally also glass (see "Rooflights", p. 97). These days, a triple-leaf rooflight is standard, apart from in unheated industrial buildings. Such rooflights achieve U-values of 2.5 W/m²K and can be installed in almost all situations (Fig. D 2.50). Despite this, condensation on the inside is still possible in interiors with higher temperatures and high levels of humidity, e.g. bathrooms, swimming pools. In such situations the thermal performance may have to comply with a higher specification. In the meantime, rooflights with up to four leaves have become available, which can have U-values below 2.0 W/m²K [26].
The junction with the roof is usually in the form of a mounting frame that is fixed to the loadbearing structure – either directly or via a tim-

ber frame – and integrated into the waterproofing. The larger the mounting frame, the greater is the risk of damage due to disparate thermal movements of the structure, waterproofing and mounting frame. The side length of the frame should therefore not exceed 2.5 m. The top edge of the mounting frame should extend at least 15 cm above the roof finishes. A timber frame – generally made from industrial-grade plywood – can be erected around the opening in the roof and the frame for the rooflight mounted on this (Fig. D 2.51). Several timber frames, one above the other, can be mounted on the roof if required in order to raise the mounting frame to suit the depth of thermal insulation. The waterproofing must continue up the side of the mounting frame for at least 15 cm above the roof finishes and must be fixed to the mounting frame. The fixing screws – max. spacing 30 cm (Fig. D 2.52) – attaching the mounting frame to the timber frame must be able to transfer all the loads on the rooflight back to the frame and hence to the roof structure. A factor of safety against wind uplift must be guaranteed.
If the mounting frame is sufficiently deep, then it can be attached directly to the loadbearing structure without the need for an intermediate timber frame. But here again, a factor of safety against wind uplift must be guaranteed and the skirting of waterproofing material should termi-

nate not less than 15 cm above the roof finishes. A fillet of insulating material should be fitted into the corner between insulation and frame when using thick waterproofing materials (especially flexible bitumen sheeting) in order to prevent the creasing that would otherwise occur at this point.
Irrespective of whether the mounting frame is mounted on a timber frame or directly on the roof structure, the rooflight is not fitted until the mounting frame has been integrated into the waterproofing. According to the Flat Roof Directive, the top edge of the mounting frame must be min. 15 cm above the finished roof surface, in the case of smoke and heat vents min. 25 cm.
Rooflights should only be installed in the middle of a roof because higher wind loads can be expected near edges and corners. DIN 1055-4 classifies the corner zone as one-quarter and the edge zone as one-tenth of the width of the building or twice the building height; the smaller figure governs.
The minimum clearance between a rooflight and a wall rising above the roof surface (made from incombustible materials and without windows) should be 2.5 m, or at least 5 m if there are openings in the wall.
In order to hinder the spread of fire, the distance between adjacent rooflights should not be less than 1.25 m; other dimensions are possible depending on the respective Federal State Building Regulations (LBO). The distance between smoke and heat vents should not be less than 4 m – but no more than 20 m – in order to rule out neighbouring vents influencing each other and also to guarantee good removal of smoke.
Rooflights installed for ventilation purposes should face in the direction opposite to the prevailing wind when open because this helps to prevent the ingress of rain and also creates a suction effect that aids the ventilation.
Riser frames – 10, 15 or 30 cm high – are available for refurbishment projects. These are placed over the existing mounting frame so that the original rooflight can be replaced after increasing the thickness of thermal insulation in the roof.

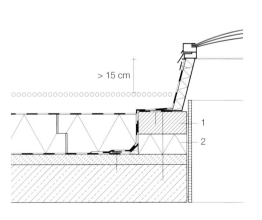

D 2.51

D 2.52

D 2.50 Reference values for thermal transmittance U, total energy transmittance g and light transmittance τ$_{D65}$ for rooflights to pre-standard DIN V 18599-4
D 2.51 Mounting frame on timber frame
  1 Timber frame
  2 Compression-resistant insulation or timber
D 2.52 Connecting the mounting frame to the structure
D 2.53 Example of the planning for a fall arrest system
D 2.54 Anchorage point penetrating waterproofing, with protective sleeve
D 2.55 Anchorage point that does not penetrate waterproofing
D 2.56 Horizontal safety system with movable anchorage point
D 2.57 Self-supporting safety barrier system

Safety

According to current German legislation, clients, or their appointed representatives (e.g. architects, engineers), are obliged to comply with the provisions of the Accident Prevention Regulations BGV C22, which cover building works and are drawn up by the Employers' Liability Insurance Association for the Construction Industry [27]. The requirements regarding safety are specified in cl. 12. For a fall height of 3 m or more, workplaces and traffic zones on roofs must include elements that prevent persons from falling (e.g. fall arrest systems, safety barriers). The safety precautions must also include the (temporary) construction and maintenance phases. The Flat Roof Directive also refers to safety during cleaning, maintenance and repair works [28]. On waterproofed roofs such safety systems must be considered at the design stage (e.g. anchorage points). In doing so, the requirements of the building regulations of the respective federal state must be taken into account. Likewise, if there is air-conditioning plant on the roof that requires regular maintenance, then the access routes to the plant must be considered at an early stage of the design. DIN 4426 specifies the technical safety requirements that must be complied with when designing and constructing permanent workplaces, traffic zones and other facilities on roofs that must be used for inspection and maintenance work, also brief repairs.

The fall height for the "free fall" should not exceed 2.0–2.5 m when designing safety features and systems. This stipulation forms the basis for the safety concept [29]. Fig. D 2.53 shows an example of such a concept. With a distance of 10 m between anchorage point and unprotected edge, the rope length needed to reach the corner is about 14 m. This results in a free fall of max. 4 m along the edges of the building. By installing an additional anchorage point where the arc intersects the diagonal, the free fall can be reduced to 1.2 m.

There are many different fixing elements that can be attached directly to the roof structure, e.g. by way of bolts or anchors, in order to guarantee an adequate level of safety. But all

of these penetrate the waterproofing (Fig. D 2.54). On green roofs and those finished with a layer of gravel, it is possible to install safety systems without penetrating the waterproofing. This is achieved by using the heavy roof finishes as ballast to hold the system components in place. To do this, a base plate with an anchorage point is laid on the waterproofed roof surface. A backup plate and a fleece to distribute the load of the ballast are then laid on top of this. The finished system obtains the necessary stability from the rooftop planting or layer of gravel. One key advantage of this method is its simplicity and flexibility when positioning the anchorage points. And as the waterproofing does not have to be penetrated, there is no risk of leaks or thermal bridges. Instead of a base plate, approved rotproof stainless steel or plastic nets or gratings can be incorporated in a sufficiently deep layer of gravel or green roof substrate (Fig. D 2.55). It is then easy to attach the anchorage points to such nets or gratings.

Another way of securing a fall arrest system is to provide a fixed stainless steel rope or rail that is connected to posts fixed to the roof structure (Fig. D 2.56). The person using the fall arrest system then clips the safety harness to this. A special rope grab allows the user to bypass fixed points without having to detach the harness, and shock absorbers at the ends of the system prevent overloads due to persons or materials.

Safety systems on large flat roofs or those in frequent use can take the form of safety barriers (Fig. D 2.57) or even walls, e.g. as an extension to a parapet. A multitude of different systems are available on the market.

Drainage

The precipitation falling on a roof must be collected and drained away via a system of outlets and pipes, unless some other method of dealing with rainwater has been envisaged. On a flat or shallow-pitch roofs it is particularly important to design the drainage system with adequate capacity in order to prevent an excessive build-up of water over the long-term. The systems used on pitched roofs generally drain the water to the outside, i.e. to the facades. However, this approach is used only rarely for flat roofs. As a rule, water on the roof surface is drained inwards, not outwards. Overflows must always be provided on roofs with internal drainage, and especially lightweight forms of construction (trapezoidal profile sheeting in particular). A serious build-up of water could occur on a flat roof with a skirting of waterproofing material or parapet around the perimeter in the case of clogged or overloaded outlets. Such a quantity of water could even overload the roof structure, and so the designer must take into account heavy rainfall (100-year return event) as well as

Anchorage point
D 2.53

□ Safe zone   ▨ Danger zone
4 m
Additional anchorage point
1.2 m | 2.8 m
10 m | 10 m
4 m   10 m

D 2.54

D 2.55

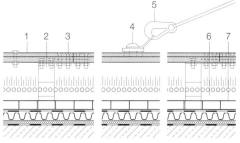

1 Rail
2 Rail mounting
3 Rail splice
4 Movable anchorage point
5 Connector
6 Substrate, ≥ 8 cm and ≥ 90 kg/m²
7 Grating

D 2.56

D 2.57

Location	Local peak rainfall (duration D = 5 min)		Area of roof surface to be drained [m²]					
			100		300		500	
	Design rain $r_{(5,5)}$ [l/s·ha]	100-year rain $r_{(5,100)}$ [l/s·ha]	Outlets [l/s]	Overflows (additional) [l/s]	Outlets [l/s]	Overflows (additional) [l/s]	Outlets [l/s]	Overflows (additional) [l/s]
Berlin	371	668	3.71	2.97	11.13	8.91	18.55	14.85
Dresden	323	602	3.23	2.79	9.69	8.37	16.15	13.95
Düsseldorf	316	607	3.16	2.91	9.48	8.73	15.80	14.55
Frankfurt am Main	329	601	3.29	2.72	9.87	8.16	16.45	13.60
Hamburg	266	463	2.66	1.97	7.98	5.91	13.30	9.85
Hannover	328	652	3.28	3.24	9.84	9.72	16.40	16.20
Kiel	230	426	2.39	1.87	7.17	5.61	11.95	9.35
Magdeburg	308	583	3.08	2.75	9.24	8.25	15.40	13.75
Munich	353	633	3.53	2.80	10.59	8.40	17.65	14.00
Stuttgart	446	858	4.46	4.12	13.38	12.36	22.30	20.60

D 2.58

Type of surface	Coefficient of discharge C
Impermeable surfaces, e.g.	
• Roof surfaces	1.0
• Concrete surfaces	1.0
• Ramps	1.0
• Paved surfaces with sealed joints	1.0
• Asphalt surfaces	1.0
• Paved surfaces with grouted joints	1.0
• Loose gravel on roofs	0.5
• Rooftop planting:	
- intensive planting	0.3
- extensive planting > 10 cm deep	0.3
- extensive planting < 10 cm deep	0.5
Partly permeable and poorly draining surfaces, e.g.	
• Concrete paving bricks laid in sand or slag, surfaces with flags	0.7
• Paved surfaces with proportion of joints > 15 %, e.g. 10 × 10 cm and smaller	0.6
• Waterbound surfaces	0.5
• Children's playgrounds with partial paving	0.3
• Sports facilities with drainage:	
- synthetic surfaces, artificial grass	0.6
- hard surfaces	0.4
- grass surfaces	0.3
Permeable surfaces with insignificant drainage or without drainage, e.g.	
• Parks and areas of vegetation, ballast and slag, rounded gravel, also with partially paved surfaces such as...	0.0 / 0.0
- garden paths with waterbound finish	
- driveways and parking spaces paved with "grasscrete" blocks	

D 2.59

D 2.60

D 2.61

the design rainfall intensity in order to avoid overloading the drainage system and causing flooding. The layout and sizing of the drainage system must therefore be given due attention at the design stage. All parts of a drainage system must be permanently accessible for inspection and maintenance.

Design

The design rainfall intensity $r_{5,5}$ is used for designing the rainwater downpipes, common horizontal discharge pipes and house drains. The design rainfall intensity is the amount of rainfall expected at the site, which is based on statistical information. In Germany rainfall figures can be obtained from the local authorities or the German Weather Service (Deutscher Wetterdienst, DWD). Furthermore, the DWD's "KOSTRA-DWD 2000" method allows the rainfall to be determined for any place in Germany. In this approach the design rainfall intensity $r_{(D, T)}$ is the quantity of precipitation that falls on an area of one hectare per second for a certain length of time. Here, D is the duration of the rainfall and T is the average return period. The duration of the rainfall is assumed to be that amount that falls in five minutes (D = 5 min). When designing roof surfaces, a return period of five years is assumed (T = 5 a), for an overflow 100 years (T = 100 a). Fig. D 2.58 shows the design rainfall intensities and the 100-year figures for a number of German cities.

The design of the outlets and overflows is carried out according to DIN 1986-100, also DIN EN 1253-1 and DIN EN 12056-3. Special measures, e.g. positioning outlets at points of maximum deflection, will be necessary on roof surfaces without falls.

The layout and sizing work embraces the following tasks:
• Calculating the rainwater flow rate
• Positioning the outlets
• Designing the overflows

Calculating the rainwater flow rate
According to DIN 1986-100, the rainwater flow rate Q is calculated from the rainfall intensity, coefficient of discharge and effective roof area as follows:

$$Q = r_{(D, T)} \cdot C \cdot A / 10\,000$$

where:
Q   rainwater flow rate [l/s]
$r_{(D, T)}$ design rainfall intensity [l/s·ha]
D   duration of rainfall [min]
T   return period [a]
C   coefficient of discharge [-]
A   effective precipitation area (corresponds to plan area of projected roof surface) [m²]

The coefficient of discharge C taken from DIN 1986-100 specifies the ratio of the surface water run-off to the total discharge for various surfaces (Fig. D 2.59). The higher the coefficient of discharge, the lower is the amount of rain seeping away. Taking this into account is only permitted when determining the flow rate based on the design rainfall intensity $r_{(5, 5)}$.

Positioning the outlets
The number of outlets required depends on the geometry of the roof, the flow rate and the capacity of the roof outlets and the drainage system. The following criteria must be considered when positioning the outlets:
- Every low point due to the form of construction must be provided with at least one roof outlet (including the low points that occur under load).
- If all the roof outlets are located along one linear low point without any appreciable differences in level, the maximum spacing of the outlets should not exceed 20 m. In the case of low points not in a straight line and with differences in level, the outlets should be positioned closer together in order to prevent an accumulation of rainwater.
- At least two gutter outlets (or one gutter outlet and one overflow) should be provided on a flat roofs or parts thereof surrounded by parapets or walls.

DIN 1986-100 contains the following equation for calculating the minimum number of roof outlets:

$$n_{DA} = Q/Q_{DA}$$

where:
n_{DA} minimum number of outlets, rounded up to the next whole number [-]
Q rainwater flow rate from one roof surface or part thereof [l/s]
Q_{DA} flow capacity of chosen roof outlet depending on depth of water (pressure head) at roof outlet [l/s]

The minimum values for the flow capacity Q_{DA}, depending on the pressure head, are given in Fig. D 2.62. The respective flow capacity must be verified by the manufacturer of the drainage system. For outlets built on site, the flow capacity must be calculated depending on the pressure head according to DIN EN 12056-3. The pressure head is the flooding at which the outlet exhibits its maximum performance. Gratings and guards, which prevent outlets becoming clogged by gravel or leaves, reduce the potential flow capacity of an outlet. In such cases the flow capacity must be reduced by 50 % unless this has already been taken into account in the manufacturer's figures. Trace heating is advisable in regions with frequent periods of freezing weather, especially for internal gutters and downpipes; such heating can prevent a drainage system becoming blocked with ice.
All outlets in rooftop gardens must be fully accessible for inspection and maintenance. Furthermore, guards and gratings to prevent soil, leaves, gravel, etc. blocking outlets are essential if a fully functioning drainage system is to be guaranteed (Figs. D 2.60 and D 2.61).

Designing the overflows
The drainage system to a flat roof must be able to drain a heavy rain event, i.e. one that, statistically, occurs every 100 years, without the roof suffering any damage due to overload, ingress of water, etc. This is where the overflows are important, which must be able to handle the difference between the design rainfall intensity and the 100-year return rain. Together, the drainage and overflow systems must be able to drain the 100-year return rain anticipated for the location of the building for a period of five minutes. In the case of buildings requiring greater protection (e.g. hospitals, theatres, museums), designers are recommended to size the overflow for the 100-year return rain $r_{(5,100)}$. The overflow system also protects the roof against flooding in the event of a clogged outlet. To do this, additional outlets are installed that are raised above the minimum flooding level of the roof drainage system and therefore only drain water when it rises above a certain level on the roof surface (Fig. D 2.64). Overflows can be provided in the following forms:
- Additional roof outlets with weir element (raised inlet) draining freely onto the ground
- Drainage via parapet outlets (outlets directly in the parapets draining into external downpipes)
- Openings in parapets (pipes, spouts)
- Lowering parts of the parapet to the lowest flooding depth

When draining via pipes, the respective flow capacity must be considered.

Sizing
DIN 1986-100 contains the following equation for calculating the minimum flow capacity of the overflow system:

$$Q_{overflow} = (r_{(5, 500)} - r_{(D, T)} \cdot C) \cdot A/10000$$

where:
$Q_{overflow}$ minimum flow capacity of overflow [l/s]
$r_{(5, 100)}$ 5-minute rainfall intensity expected once in 100 years [l/(s·ha)]
$r_{(D, T)}$ design rainfall intensity [l/s·ha]
D duration of rainfall [min]
T return period [a]
C coefficient of discharge [-]
A effective precipitation area [m²]

The invert of the overflow must lie above the pressure head required for the roof outlet used (Fig. D 2.64, p. 115). The addition of the pressure heads at the roof outlet and the overflow result in the maximum depth of flooding to be expected on the roof. The depth of flooding must be checked with the structural engineer because it also has an effect on the load on the roof in the event of very heavy rainfall. The load per unit area resulting from the flooding on the roof above the drainage low point (roof outlet) may not exceed the permissible load for the roof structure.

Nominal sizes of roof outlet pipes		Gravity drainage		Siphonic drainage	
DN/OD[1]	DN/ID[2]	Min. flow rate [l/s]	Pressure head h [mm]	Min. flow rate [l/s]	Pressure head h [mm]
40		–	–	2.5	
	40	–	–	3.0	55
50		0.9	35	4.0	
	50			6.0	
63		1.0	35	–	
75		1.7	35	12.0	
	70				55
80		2.6	35	14.0	
	75				
90		–	–	18.0	
110		4.5	35	22.0	55
	100				
125		7	45	–	–
	125			–	–
160		8.1	45	–	–
	150			–	–

[1] DN/OD outside diameter
[2] DN/ID inside diameter

D 2.62

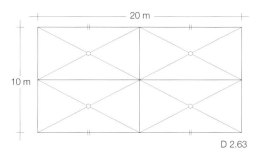

D 2.63

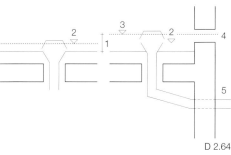

D 2.64

D 2.58 Rainfall intensity figures for a number of German cities (design and 100-year return rainfall) and drainage figures (standard and overflow) depending on area of roof surface to be drained
D 2.59 Coefficients of discharge C for determining rainwater flow rate, taken from DIN 1986-100
D 2.60 Roof outlet with fixed and detachable flanges
D 2.61 Two different outlet types for a flat roof with gravel finish
D 2.62 Pressure heads required at roof outlet in order to achieve minimum flow to DIN 1986-100
D 2.63 Example of how a flat roof can be divided into zones each falling to an outlet
D 2.64 Flooding depth for overflow drainage
    1 Pressure head required at overflow (difference in height between 3 and 2)
    2 Pressure head at roof outlet
    3 Maximum flooding depth
    4 Overflow opening in parapet, above 2
    5 Overflow outlet draining through facade

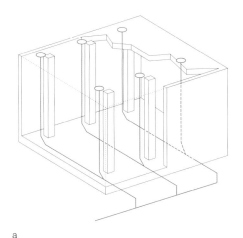

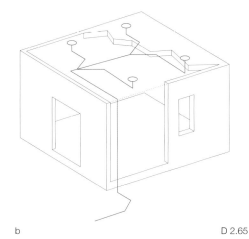

a

b

D 2.65

Downpipes

The flow capacities for vertical rainwater downpipes depend on their internal diameter and degree of filling according to DIN EN 12056. Internal rainwater downpipes must be able to withstand the pressure that can occur if the pipe becomes blocked. Insulation (lagging) to prevent condensation and to reduce noise levels is essential for rainwater pipes within buildings. Rainwater pipes should never be cast into load-bearing concrete components.

Inspections and maintenance every six months at least are essential if a functioning drainage system is to be guaranteed. Such work should be carried out in the autumn.

We distinguish between gravity and siphonic drainage systems.

Gravity drainage

The downpipes and discharge pipes are only partially filled with water in this type of drainage system (Fig. D 2.65a). Downpipes are generally sized assuming they are one-third full. Every roof outlet is connected to the house drain via its own downpipe. Reducing the pipe diameter in the direction of flow is not permitted with gravity drainage.

If the flow capacity of the downpipes is inadequate or a smaller pipe cross-section is desired, then a roof drainage system that assumes that pipes are filled (siphonic drainage) is required.

Siphonic drainage

The downpipes and discharge pipes are assumed to be filled with water in this type of drainage system (full-bore flow) (Fig. D 2.65b). This means that the entire water column between roof outlet and house drain is hydraulically effective, the difference in height is used to generate a powerful flow. The flow capacity is therefore much higher than a similar gravity drainage arrangement. In order to fill the downpipe completely, several roof outlets must be connected to the house drain via a common downpipe. Connections between individual pipes must be airtight in a siphonic drainage system. DIN 1986-100 is used for designing such systems.

Care and maintenance

Flat roofs need a certain amount of regular and proper care and maintenance if they are to function reliably over the long-term. In this respect, the German Roofing Contractors Association (ZVDH) recommends that roof coverings, waterproofing and other components be inspected and maintained at regular intervals. Inspections should include, for instance, a visual assessment of exposed materials for changes caused by external influences. Maintenance should include ensuring the proper functioning of system components, built-in items and customary components such as drainage systems, vents, flashings, edge details, etc. Accumulations of dirt and debris must be removed. Concluding an inspection and maintenance contract is a sensible precaution.

When changes or damage are detected in good time, then the work required to rectify the problems and their consequences is frequently reduced – and the costs, too. Regular care and maintenance can considerably prolong the lifetime of a roof [30].

D 2.65 Flat roof drainage
a Gravity drainage via vertical pipes
b Siphonic drainage via horizontal pipes
D 2.66 Roof construction for energy efficiency upgrade
D 2.67 Ventilated roof for drying out the thermal insulation

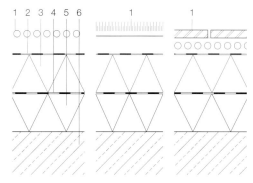

1 Ballast (water- and diffusion-permeable)
2 Separating membrane (diffusion-permeable)
3 New thermal insulation (XPS) with shiplap joints
Existing roof:
4 Waterproofing (still functioning)
5 Insulation
6 Loadbearing structure

D 2.66

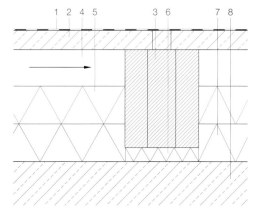

1 Waterproofing
2 Decking to support waterproofing
3 Beams
4 Air space
5 New insulation
6 New insulation
Existing roof:
7 Insulation
8 Reinforced concrete slab

D 2.67

Refurbishment

There are two main reasons why a flat roof may need to be refurbished:
- Inadequate thermal insulation
- Damage to the roof

Energy efficiency upgrade

The advantage of refurbishing a roof purely because of an inadequate layer of thermal insulation is that there has presumably been no damage to the roof so far. The work can be carried out without affecting the existing roof construction. One popular solution is to install an additional layer of thermal insulation made from extruded polystyrene (XPS) on the existing, intact waterproofing, which then creates an inverted roof (Fig. D 2.66). If the parapet around the perimeter of the roof is high enough, the work involved is minimal. However, a detailed inspection must be carried out first to establish the condition of the waterproofing.

This approach enables rooftop terraces and green roofs to be created, but the additional loads on the loadbearing structure must be assessed.

Repairing moisture damage

If the waterproofing has been damaged, then it is possible that water has infiltrated the construction. Moisture within the construction can cause a whole series of problems: damp insulation has a higher thermal conductivity, which reduces the insulating effect, and water can damage the loadbearing structure (e.g. rotting of wood or corrosion of metals), which if allowed to continue could lead to structural failure. There are two ways of dealing with this problem:
- Drying out the roof
- Removal and rebuilding

Drying out the insulation

It is only possible to dry out the insulation when the moisture content does not exceed approx. 10 % by vol. and when the loadbearing structure is not affected. However, it generally takes a considerable length of time to dry out the insulation, and it is a very costly procedure. The roof construction shown in Fig. D 2.67 is suitable when precipitation has infiltrated a flat roof and the existing roof construction is to be retained. Here, the existing unventilated roof is converted into a ventilated roof. To do this, the existing waterproofing is perforated or removed and beams are laid on top to form a new decking for the new waterproofing. The additional insulation required is laid in the resulting voids between the beams. Lightweight concrete slabs are preferred for the upper decking, supported on the concrete beams. This arrangement guarantees an adequate dead load and a secure base for the new roof structure. It also allows the insulation to dry out slowly, with the moisture being dissipated via the new air space. Good ventilation is essential and the dimensions of the air space must be much larger than those specified for normal situations (see p. 102).

Another advantage of this method is that a more or less fully functional roof remains in place during the entire refurbishment work. Temporary waterproofing is therefore superfluous.

This method of refurbishment is unsuitable for repairing moisture damage that may have affected loadbearing members (e.g. in a timber structure).

Individual vents

There are venting systems available that help to dry out a ventilated roof construction; the individual vents supplement the existing ventilation effect. This method involves installing individual vents at various points on the roof surface; they pass through the waterproofing to reach the air space below. The efficiency of such systems is, however, disputed because the diffusion flow in the roof is essentially vertical (following the temperature gradient) and transverse distribution of the moisture takes place to a limited extent only. Furthermore, the existing air flow may be diverted, with undesirable consequences. Practical experience has shown that the radius of influence of these individual vents is only about 1–2 m [31].

Drying through the waterproofing

It is also possible to allow damp insulation to dry out *through* the waterproofing. To do this, the existing waterproofing is removed or perforated before laying a new diffusion-permeable layer of insulation and a waterproofing material with a moderate diffusion resistance on top. Waterproofing materials with s_d values between 10 and 20 m are available for this. A dark-coloured waterproofing material should be preferred to improve the drying-out effect. No gravel, chippings, etc. should be laid on top.
Drying takes place only slowly, however; a period of several years is very likely. The drying options and the drying time should be checked prior to planning the refurbishment work (see "Hygrothermal simulation tools", pp. 70–73).

References

[1] Zentralverband des Deutschen Dachdeckerhandwerks; Hauptverband der Deutschen Bauindustrie e.V. (pub.): Fachregel für Abdichtungen – Flachdachrichtlinie. Cologne, 2008, p. 26
[2] ibid., p. 27
[3] Zentralverband des Deutschen Dachdeckerhandwerks (pub.): Merkblatt Wärmeschutz bei Dach und Wand. Cologne, 2008, p. 13
[4] ibid. [1], section 2.4.4, p. 15
[5] DIN 4108-2 Thermal protection and energy economy in buildings – Part 2: Minimum requirements for thermal insulation. Berlin, 2003
[6] Lohmeyer, Gottfried: Flachdächer – einfach und sicher. Konstruktion und Ausführung von Flachdächern aus Beton ohne besondere Dichtungsschicht. Düsseldorf, 1993
[7] Bundesverband der deutschen Zementindustrie e.V. (pub.): Flachdächer aus Zement. Zement-Merkblatt Hochbau. Hannover, 1999
[8] Optigrün: http://www.dachbegruenung-ratgeber.de/dachbegruenung, position as of 15 Feb 2009
[9] Buttschardt, Tillmann: Extensive Dachbegrünungen und Naturschutz. Karlsruhe, 2001;Liesecke, H. J.: Begrünung von Well-und Trapezprofilen mit einem Verbundschaumstoff. In: Dach & Grün, 1/2003, pp. 28ff.
[10] Forschungsgesellschaft Landschaftsentwicklung Landschaftsbau e.V. (FLL; pub.): Richtlinie für die Planung, Ausführung und Pflege von Dachbegrünungen – Dachbegrünungsrichtlinie. Bonn, 2008, p. 34
[11] Riegler, Rosina: Fachgerechte Ausführung und Sanierung von Flachdächern und Gründächern. Merching, 2009, section 6.2.1, p. 3
[12] ibid. [10], p. 11
[13] Euro Inox (pub.): Technischer Leitfaden: Dächer aus Edelstahl Rostfrei. Luxembourg, 2004, p. 22
[14] Schittich, Christian at el.: Glass Construction Manual. 2nd ed. Munich/Basel, 2007
[15] Schunck, Eberhard et al.: Roof Construction Manual. Munich/Basel, 2003, p. 165
[16] ibid., p. 166
[17] Bundesinnungsverband des Glaserhandwerks (pub.): Technische Richtlinie des Glaserhandwerks. No. 1: Dichtstoffe für Verglasungen und Anschlussfugen. Düsseldorf, 2009
[18] ibid. [15], p. 166
[19] ibid. [15], p. 168
[20] ibid. [15], pp. 168f.
[21] ibid. [1], p. 33
[22] ibid. [1], pp. 35f.
[23] ibid. [1], p. 37
[24] ibid. [1], p. 38
[25] Zentralverband des Deutschen Dachdeckerhandwerks (pub.): Fachregeln für Metallabdeckungen im Dachdeckerhandwerk. Cologne, 2006, p. 61
[26] http://www.fvlr.info/lik_dachdecker.htm, position as of 23 Feb 2010
[27] Berufsgenossenschaft der Bauwirtschaft (pub.): Unfallverhütungsvorschrift BGV C22 (formerly VBG 37): Bauarbeiten (Prävention Hochbau). Berlin, 1997
[28] ibid. [1], p. 10
[29] ABS – Absturzsicherung mit System, product information
[30] Zentralverband des Deutschen Dachdeckerhandwerks (pub.): Grundregel für Dachdeckungen, Abdichtungen und Außenwandbekleidungen. Cologne, 2008, p. 14
[31] Spilker, Ralf; Oswald, Rainer: Flachdachsanierung über durchfeuchteter Dämmschicht. Aachen, 2003, p. 7

Part E   Construction details

Fig. E   "Mountain dwellings", Copenhagen (DK), 2008,
BIG – Bjarke Ingels Group

Construction details

Eberhard Schunck

The drawings of the construction details make use of the following symbols – based on the Flat Roof Directive – to differentiate between the individual layers in the roof construction. (The drawings in Part F "Case studies" are in accordance with the customary DETAIL standards.)

Bitumen waterproofing

Vapour barrier

Synthetic waterproofing

Vapour barrier, synthetic

Liquid waterproofing

Loose gravel

Chippings

Protective membrane/filter fleece

Synthetic protective membrane

Facing (laminated)

Insulation

Acoustic blanket/board

Sealing tape, permanently elastic

Weld seam

Slip plane

Bond enhancer/primer

All construction details on pp. 122–147 are drawn to a scale of 1:10.

Up until the 1960s there were no uniform rules for the construction of flat roofs. But the publication of the so-called *Flachdachrichtlinie* (Flat Roof Directive) by the German Roofing Contractors Association and the German building industry in 1962 changed all that. The following forms of construction are based on that directive, further technical resources and the author's own experience.

Criteria

In the preparation of these construction details, with their various materials and sequences of layers, it was necessary to check the following issues:
- Can water infiltrate the construction from above or from the side?
- Can heat escape from inside to outside or hot external temperatures penetrate from outside to inside?
- Can moisture from the interior infiltrate the insulation or other parts of the construction?
- Is the construction protected from the sun, wind and mechanical damage?
- Are the materials compatible with each other (i.e. will not damage each other)?
- Can potential movements damage the materials and can the effects of movements be rendered harmless?
- How does the construction process affect the sequence of layers and the geometry of the design?

There are several ways of optimising the design with the help of the above questions. Those options are to be found in the choice of materials, the choice of waterproofing system or the methods of using them. The form of construction is always determined by the respective situation and the type of use, with the local climate always playing a major role. However, the size and height of the structure are also very important. Sometimes even the manual skills and experience available locally may have an effect on the design. The form of construction chosen must be systematically examined for weak spots where heat, water and moisture could penetrate from outside to inside, or inside to outside, infiltrate the construction and impair its functioning or service life.

Selecting the construction details

Damage does not usually occur over the surface, unless the problem is defective bonding or a fault in the waterproofing material. Rather, damage tends to occur at edges, junctions, joints, interfaces between different materials and different trades, and wherever the roof construction is penetrated (i.e. "damaged"). The details on the following pages therefore represent a selection that embraces all areas of the roof. These areas were considered in conjunction with the forms of construction and

materials commonly available for flat roofs. The waterproofing materials were varied within the forms of construction in order to present the widest possible choice of flat roof designs for the widest possible range of applications. But as it is not possible to deal with every potential case, solutions must be transferred from one form of construction to another.
Ventilated roofs have not been included because they frequently lead to problems which in each case must be solved by a building physics consultant (see "Ventilated flat roof", p. 102)

Stipulations

The standard of thermal insulation chosen corresponds to the requirements of the German *Energiesparverordnung* (EnEV, Energy Conservation Act) for the Central European climate zone, which results in a thickness of about 20 cm. Most details have been drawn without showing the falls that are recommended as a safety measure in the Flat Roof Directive. A screed laid to falls is only shown where the need for a fall is particularly important (e.g. with the insulation below the waterproofing) or there is a fall in the roof structure (e.g. helicopter landing pad) or in the insulation (e.g. refurbishment). Metal and glass roofs have been drawn with a 2° fall. As the compatibility issues of synthetic roofing materials vary considerably, their use in conjunction with protective membranes also varies (see "Incompatibilities between materials", p. 97). All flashings and cappings to parapets have been drawn with an overlap of 5 cm below the relevant joint, as recommended in the various German trade association publications for buildings up to 8 m high (see "Metal roofs", p. 108).
The levels marked on the drawings are those from which the dimensions given in the Flat Roof Directive should be measured to the uppermost layer of the construction.

Overview of forms of construction	Roof edge without overhang	Roof edge with overhang	Junction with wall	Penetration	Drainage	Joint	Individual rooflight	Continuous rooflight
1 Waterproofing above insulation	1.1.1 1.1.2 p. 122	1.2.1 1.2.2 p. 123	1.3 p. 124	1.4 p. 125	1.5.1 1.5.2 p. 124	1.6.1 1.6.2 p. 125	1.7 p. 126	1.8 p. 126
2 Waterproofing below insulation	2.1 p. 127	2.2 p. 127	2.3 p. 128	2.4 p. 128	2.5 p. 129	2.6 p. 129	2.7 p. 129	
3 Impermeable concrete slab	3.1 p. 130	3.2 p. 130	3.3 p. 131		3.5 p. 131	3.6 p. 131		
4 Green roof	4.1 p. 132	4.2 p. 132	4.3 p. 133	4.4 p. 133	4.5 p. 133			
5 Roof for foot traffic	5.1 p. 134	5.2 p. 134	5.3 p. 135	5.4 p. 135	5.5 p. 136	5.6 p. 136		
6 Roof for vehicular traffic	6.1 p. 137	6.2 p. 137	6.3 p. 138		6.5 p. 138			
7 Metal roof	7.1.1 7.1.2 p. 139	7.2.1 7.2.2 p. 140	7.3 p. 140					
8 Glass roof	8.1 p. 141	8.2 p. 141	8.3 p. 142	8.4 p. 142				
9 Refurbishment	9.1 p. 143	9.2 p. 144	9.3 p. 145	9.4 p. 146	9.5 p. 146		9.7 p. 147	

1 Waterproofing above insulation

1.1.1 Roof edge without overhang
· **Bitumen waterproofing**

Generally, at least two plies of bitumen water-proofing are required (built-up roofing). In this detail the waterproofing is protected by a layer of loose gravel. At the parapet a sheet metal skirting protects the waterproofing that would otherwise be exposed to UV radiation and mechanical damage. The prescribed dimension from the top of the protective layer (gravel) to the clamping bar must be adhered to (see "Junctions with components above roof level", p. 110). Any of the customary insulating materials approved for flat roofs may be used here. Whereas built-up roofing and thicker waterproofing materials benefit from the use of a triangular fillet of insulating material to prevent creases and cracking, this is not necessary under the single-ply vapour barrier. The parapet capping is fixed indirectly to the timber board underneath via a sheet metal clip that is screwed directly to the solid parapet.

1.1.2 Roof edge without overhang
· **Parapet overflow**
· **Bitumen waterproofing**

Special outlets, with horizontal and vertical flanges, are available for forming overflows at parapets. If such an outlet is used, the distance of 30 cm from the inside face of the parapet – as specified in the Flat Roof Directive – does not apply. Both the vapour barrier and the waterproofing should be clamped between the flanges when using products with double flanges. Rainwater may be discharged via a spout, as shown here, or into a downpipe, which in turn discharges onto the ground where the water can seep away. An overflow spout should project at least 30 cm, preferably further, beyond the facade.

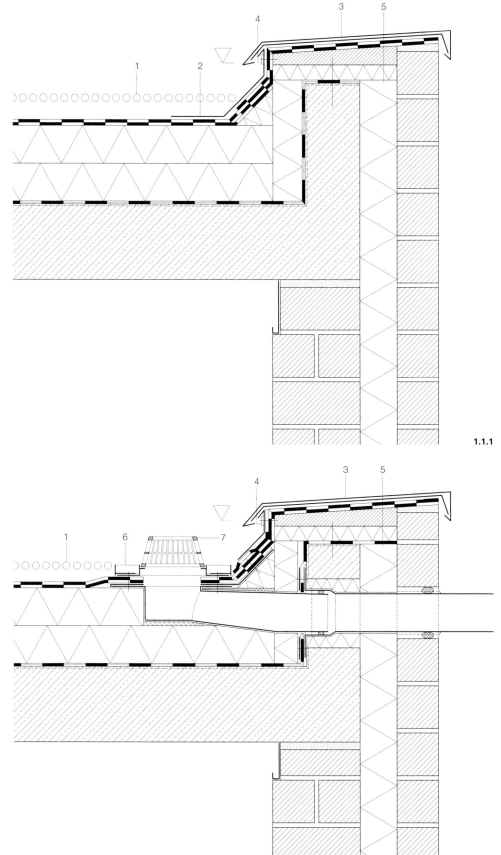

1.1.1

1.1.2

1	Loose gravel		Bitumen waterproofing,
	Bitumen waterproofing,		2-ply
	2-ply		Timber board
	Rigid plastic foam		Rigid plastic foam
	board		board
	Vapour barrier	4	Clamping bar
	Primer	5	Rigid plastic foam
	Reinforced concrete		board
2	Sheet metal skirting	6	Weir element
3	Parapet capping	7	Dome grating
	Sheet metal clip		

1.2.1 Roof edge with overhang
· **Synthetic waterproofing**

Synthetic waterproofing materials may in some instances be incompatible with the thermal insulation, which means that a protective membrane between the two will be necessary (note the manufacturer's recommendations). As synthetic waterproofing materials are normally single-ply products, they must be protected against being damaged by the loose gravel finish. And a separating membrane between vapour barrier and loadbearing structure may also be necessary, depending on the roughness of the surface. Synthetic waterproofing materials that are UV-resistant may be turned up the parapet without the need for any protection (but only on unused roofs). In principle, their tendency to shrink means that they will need to be fixed to the structure with mechanical fasteners along all edges. If they have a self-adhesive backing, then a clamping bar along the inside of the parapet may be unnecessary. This detail shows a precast concrete coping; the joints between the individual coping sections must be carefully sealed and regularly maintained.

1.2.2 Roof edge with overhang
· **Concealed gutter**
· **Synthetic waterproofing**

UV-resistant synthetic waterproofing materials may be used without any protective covering provided they are secured against wind uplift. To do this, the waterproofing requires mechanical fasteners in the form of concealed linear or point fixings.
The gutter consists of a single continuous sheet metal flashing with multiple folds. It extends from the waterproofing on the surface of the roof to the top of the parapet and is lined with a synthetic waterproofing material. As this is an internal gutter, the waterproofing must continue uninterrupted underneath. A suitable fall and two overflows will protect the roof construction against saturation.

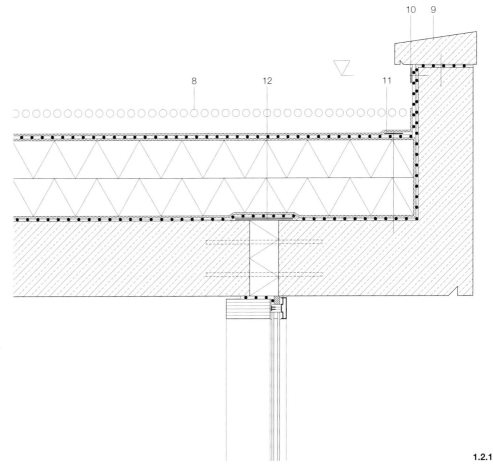

1.2.1

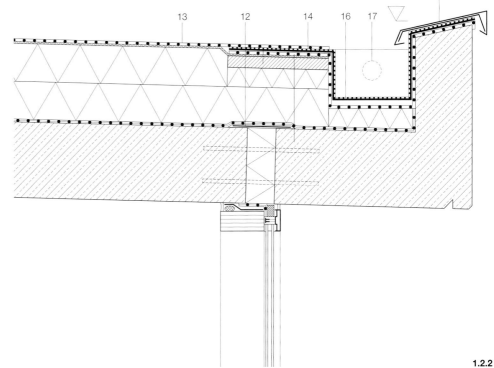

8	Loose gravel		Rigid plastic foam board
	Protective membrane		
	Synthetic waterproofing		Vapour barrier
	Protective membrane		Reinforced concrete
	Rigid plastic foam board	14	Sheet metal flashing
	Protective membrane		Synthetic waterproofing
	Vapour barrier		Protective membrane
	Reinforced concrete		Timber board
9	Precast concrete coping		Linear or point fixing
10	Clamping bar	15	Parapet capping
11	Linear fixing		Sheet metal clip
12	Bridging plate	16	Sheet metal gutter
13	Synthetic waterproofing	17	Overflow
	Protective membrane		

1.2.2

1.3 Junction with wall
  · Liquid waterproofing

The thermal insulation must have a facing (e.g. a bitumen facing in the example shown here) in order to provide a sufficiently stable surface for applying a liquid waterproofing material. Liquid waterproofing does not require a triangular fillet of insulation in the corner, nor a clamping bar at the top of the skirting. A cover flashing fixed directly to the external wall is a reliable way of protecting against water that is blown beneath the outer leaf (in this case clay brickwork). Any water draining down the face of the inner leaf of clay brickwork can drain away via the open joint. The timber moulding shown here could be replaced by a metal Z-section.

1.5.1 Gravity drainage
  · Plastic roof outlet
  · Synthetic waterproofing

Plastic outlets are fitted with a plastic flange that is firmly attached to the outlet body. This flange is laid on the vapour barrier and welded to this. The body of the outlet can be cast into the concrete slab in many instances. A sleeve of thin felt ensures minimal movement and absorbs any condensation that might occur. The extension piece, which compensates for the depth of insulation, has a fixing flange that is laid on the waterproofing and welded to this. Here again, synthetic waterproofing materials must be fixed to the structure (linear or point fixings) in order to avoid pulling the outlet out of the roof if the roof finish should shrink.

1.5.2 Siphonic drainage
  · Stainless steel roof outlet
  · Synthetic waterproofing

A stainless steel roof outlet has been chosen to illustrate the considerably more effective siphonic drainage. A slit limits the quantity of water entering the outlet. The synthetic waterproofing shown here must be UV-resistant and – as ever – fixed in place by adhesive or mechanical fasteners. With this type of outlet, both the vapour barrier and the waterproofing must be clamped between the flanges of the outlet (fixed and detachable). As a pipe blockage has a greater effect with siphonic drainage than with gravity drainage, careful design and workmanship is very important.

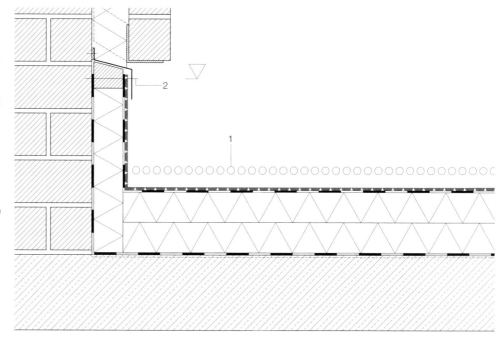

1.3

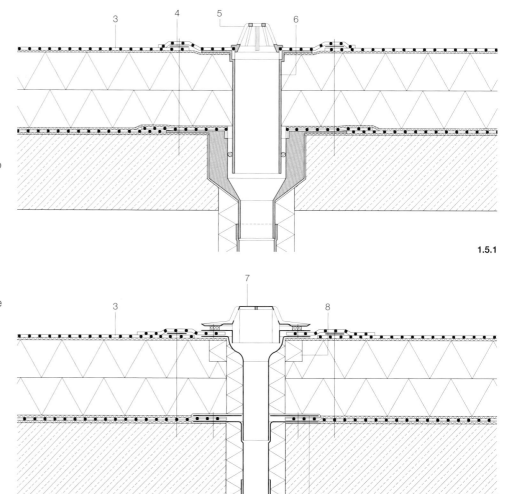

1.5.1

1.5.2

1	Loose gravel
	Liquid waterproofing
	Facing
	Rigid plastic foam board
	Vapour barrier
	Primer
	Reinforced concrete
2	Clamping bar
3	Synthetic waterproofing
	Protective membrane

	Rigid plastic foam board
	Protective membrane
	Vapour barrier
	Protective membrane
	Reinforced concrete
4	Linear fixing
5	Dome grating
6	Extension piece

1.4 Penetration
· Bitumen waterproofing

A pipe with a flange fixed to the vapour barrier has been selected here to show one of the many options possible with a roof penetration. A sleeve is slipped over this and connected to the waterproofing. The plies of waterproofing material are bonded above and below the flange of the sleeve. A drip flashing protects the joint between pipe and sleeve. The pipe lagging in the interior is essential.

1.6.1 Joint
· Synthetic waterproofing

The detail here shows a joint that must accommodate horizontal movement. To do this, two sheet metal kerbs raise the top of the joint the recommended 15 cm above the top of the waterproofing. An additional strip of the UV-resistant synthetic waterproofing material is bonded over the joint between the waterproofing on the roof and the sheet metal kerb.

1.6.2 Joint
· Liquid waterproofing

This detail is essentially identical to the one above except that different materials have been used for the waterproofing and the vapour barrier. The advantage of bitumen for the vapour barrier is that penetrations in the form of screws can be sealed better. Liquid waterproofing does not require any fixing at the top of the skirting because it forms a stronger bond with the sheet metal than is the case with synthetic or bitumen flexible sheeting.

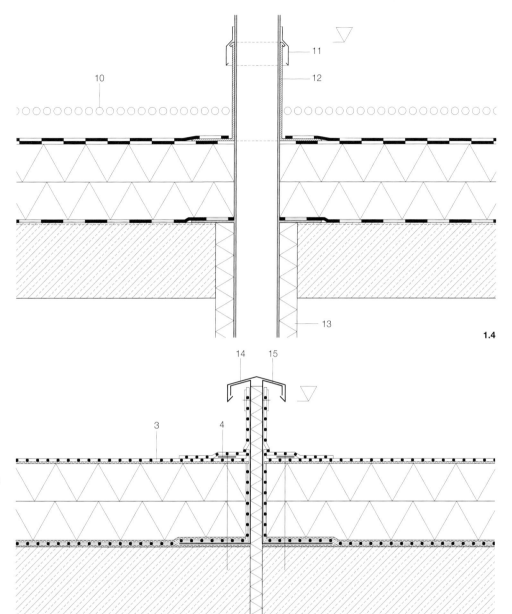

1.4

1.6.1

7	Stainless steel outlet	12	Sleeve
8	Lagging	13	Lagging
9	Clamping flange	14	Sheet metal capping
10	Loose gravel	15	Sheet metal kerb
	Bitumen waterproofing, 2-ply	16	Liquid waterproofing
	Rigid plastic foam board		Facing
	Vapour barrier		Rigid plastic foam board
	Primer		Vapour barrier
	Reinforced concrete		Primer
11	Drip flashing		Reinforced concrete

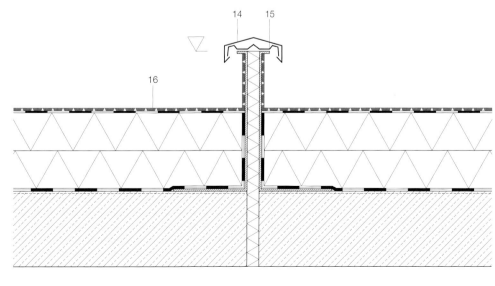

1.6.2

125

1.7 Individual rooflight
· Bitumen waterproofing

The detail for integrating an individual rooflight into a flat roof is not dissimilar to the detail at the junction with a wall. In order to avoid additional sheet metal skirtings around the rooflight, it is best to use a UV-resistant flexible bitumen sheeting with a slate granule finish as the topmost ply. The plies of waterproofing material surrounding the rooflight are bonded alternately with the waterproofing material on the surface of the roof. One problem when using bitumen waterproofing materials is the temperatures that occur when using a naked flame (felt torching). The actual rooflight must therefore be removed before beginning the work on the junction. Alternatively, a liquid waterproofing material can be used around rooflights.

1.8 Continuous rooflight
· Synthetic waterproofing

The problem at the junction with panes of glass in the roof tends to be the transition from the glass to the kerb rather than from the horizontal to the vertical waterproofing. An insulated sheet steel frame plus a sheet steel Z-section form the support along the edge of the stepped insulating glass. Preformed plastic seals and permanently elastic sealant ensure an airtight junction. A coating or printing along the bottom edge of the glass prevents an excessive temperature rise here and therefore reduces the risk of differential stresses damaging the pane of glass or the hermetic edge seal.

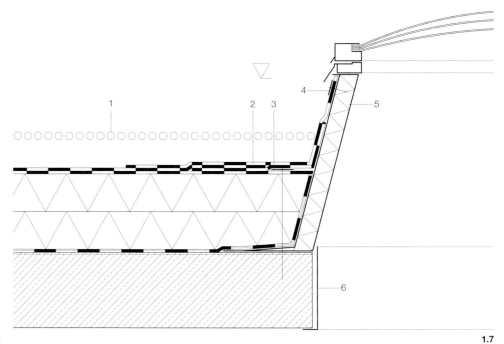

1.7

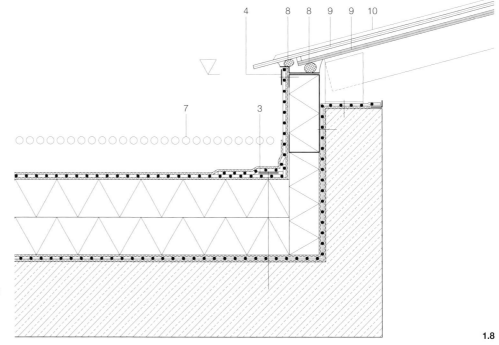

1.8

1	Loose gravel	
	Bitumen waterproofing, 2-ply	
	Rigid plastic foam board	
	Vapour barrier	
	Primer	
	Reinforced concrete	
2	Bitumen waterproofing, 2-ply, slate granule finish to top ply	
3	Linear fixing	
4	Clamping bar	
5	Mounting frame	
6	Lining	

7	Loose gravel
	Protective membrane
	Synthetic waterproofing
	Protective membrane
	Rigid plastic foam board
	Protective membrane
	Vapour barrier
	Protective membrane
	Reinforced concrete
8	Preformed sealing strip
9	Insulating glass
10	Glazing bar

2 Waterproofing below insulation

2.1 Roof edge without overhang
· Synthetic waterproofing

Inverted roofs require more ballast, which is why a loose gravel finish on such a roof is generally deeper than on roofs where the waterproofing is above the insulation (approx. 10 cm gravel for 20 cm insulation). If the protective membrane above the insulation can at same time act as a water run-off layer, then it is not necessary to add a surcharge to the U-value. Owing to the very hard, sharp edges of the extruded polystyrene insulating boards, a triangular fillet of insulation must be laid beneath the waterproofing and a protective membrane installed as well. Chamfering the edges of the insulating boards obviates the need for a fillet. The waterproofing on the roof continues over the top of the parapet and is clamped beneath a bar along the top of the facade. In principle, it is possible to position the linear fixing just above the waterproofing level further down the parapet. A sheet metal skirting protects the insulation against mechanical damage and UV radiation.

2.2 Roof edge with overhang
· Liquid waterproofing

This parapet is made from stiffened sheet metal that is fixed to the loadbearing structure. The splices in the parapet capping, which in this case extends the full depth of the parapet on the outside, require good support and must be sealed with an appropriate tape. The sharp edges of the XPS insulating boards make it necessary to include a triangular fillet of insulation, even below robust liquid waterproofing. The insulated joint between the cantilevering precast concrete element and the roof slab must be bridged over with a suitable material (e.g. sheet metal, plastic) to form a stable base for the liquid waterproofing.

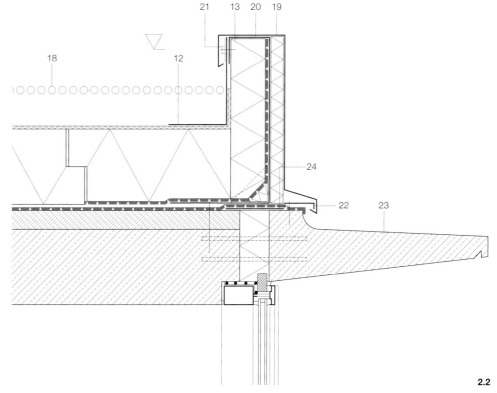

2.1

2.2

11	Loose gravel	
	Filter fleece	
	Extruded polystyrene (XPS) board	
	Protective membrane	
	Synthetic waterproofing	
	Protective membrane	
	Screed laid to falls	
	Reinforced concrete	
12	Sheet metal skirting	
13	Extruded polystyrene (XPS) board	
14	Parapet capping	
	Sheet metal clip	
	Synthetic waterproofing	
	Timber board	
15	Hollow facade section	

16	Clamping bar	
17	Insulated panel	
18	Loose gravel	
	Filter fleece	
	Extruded polystyrene (XPS) board	
	Liquid waterproofing	
	Screed laid to falls	
	Reinforced concrete	
19	Parapet capping	
20	Stiffened sheet metal parapet	
21	Clamping bar	
22	Sheet metal clip	
23	Precast concrete element	
24	Bridging plate	

2.3 Junction with wall
· Bitumen waterproofing

The two-ply bitumen waterproofing is reinforced with a third ply in the corner. Fixing the top edge of the skirting with a sheet metal rail does create a minimal thermal bridge, the effect of which is hardly reduced by the two plies of bitumen waterproofing on the wall. This also applies to the rail at the bottom of the trapezoidal profile wall cladding, which must be fixed to the wall via a sealing tape.

2.4 Penetration
· Synthetic waterproofing

In this solution, fabricated on site, it is essential to ensure that the pipe is firmly attached to the loadbearing structure. After cutting the waterproofing to suit, a collar is formed around the pipe. The gap between the collar and the waterproofing on the surface of the roof must be closed off with an additional strip of waterproofing material. A hose clip secures the top edge of the collar.

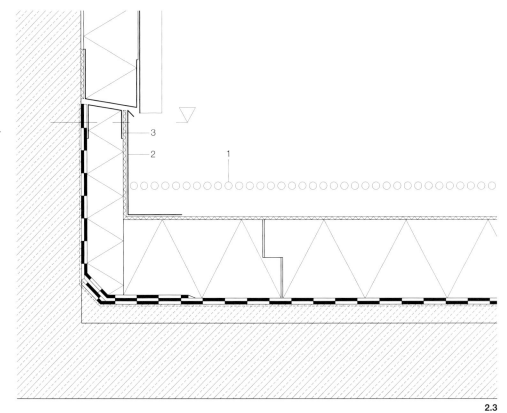

2.3

1	Loose gravel		Extruded polystyrene
	Filter fleece		(XPS) board
	Extruded polystyrene		Protective membrane
	(XPS) board		Synthetic waterproofing
	Bitumen waterproofing,		Protective membrane
	2-ply		Screed laid to falls
	Primer		Reinforced concrete
	Screed laid to falls	5	Collar of waterproofing
	Reinforced concrete		material
2	Sheet metal skirting	6	Hose clip
3	Fixing rail	7	Linear fixing
4	Loose gravel	8	Lagging
	Filter fleece		

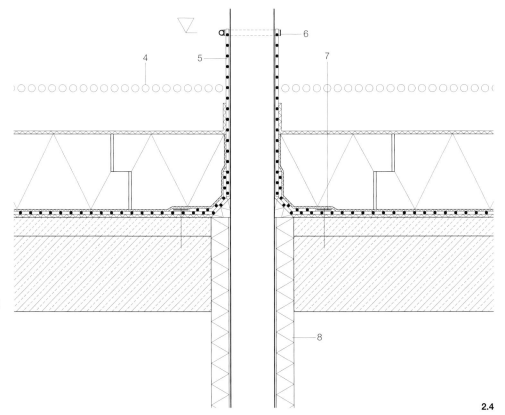

2.4

2.5 Drainage
· Liquid waterproofing
The roof outlet shown here schematically is fixed to the loadbearing structure by casting it and its fixing flange into the screed (which is laid to falls). A sleeve of thin felt ensures that a minimum amount of movement is possible and also absorbs any condensation. The liquid waterproofing and its reinforcing fleece are laid over the flange. An extension piece bridges the distance between the waterproofing (i.e. the top of the body of the outlet) and the top of the insulation.

2.6 Joint
· Polymer-modified bitumen waterproofing
The joint, which in this case has to accommodate horizontal movement, is raised above the water run-off level by means of a strip of XPS insulation and a preformed foam strip. Two additional strips of polymer-modified bitumen waterproofing are applied over the joint. A loop of sufficient length must be included in the vapour barrier material so that it, too, can accommodate the horizontal movement.

2.7 Individual rooflight
· Synthetic waterproofing
Raising the flange of the rooflight mounting frame to the level of the top of the roof insulation is achieved with the help of a timber frame which is in turn supported on incompressible thermal insulation. The junction between frame and roof is made watertight with a strip of waterproofing material that extends up to the top of the rooflight mounting frame. A separating membrane between the two plies of waterproofing acts as a slip plane to accommodate minor movement of the rooflight; it also covers the heads of the mounting frame fixing screws. A clamping bar secures the waterproofing to the top of the mounting frame. Any waterproofing material that remains uncovered must be UV-resistant (or else it must be covered, too). Here again, linear fixings for the synthetic waterproofing are required on all sides of the rooflight.

9 Loose gravel
    Filter fleece
    Extruded polystyrene (XPS) board
    Liquid waterproofing
    Screed laid to falls
    Reinforced concrete
10 Dome grating
11 Extension piece
12 Roof outlet
13 Preformed sealing strip
14 Extruded polystyrene (XPS) board
15 Elastomeric bitumen waterproofing
16 Synthetic waterproofing
17 Clamping bar
18 Mounting frame
19 Lining
20 Rigid plastic foam board

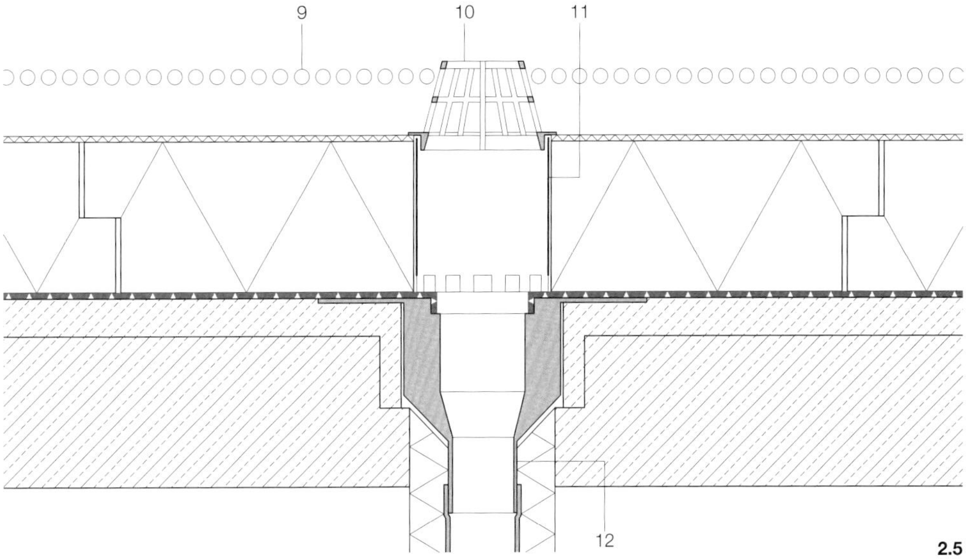

2.5

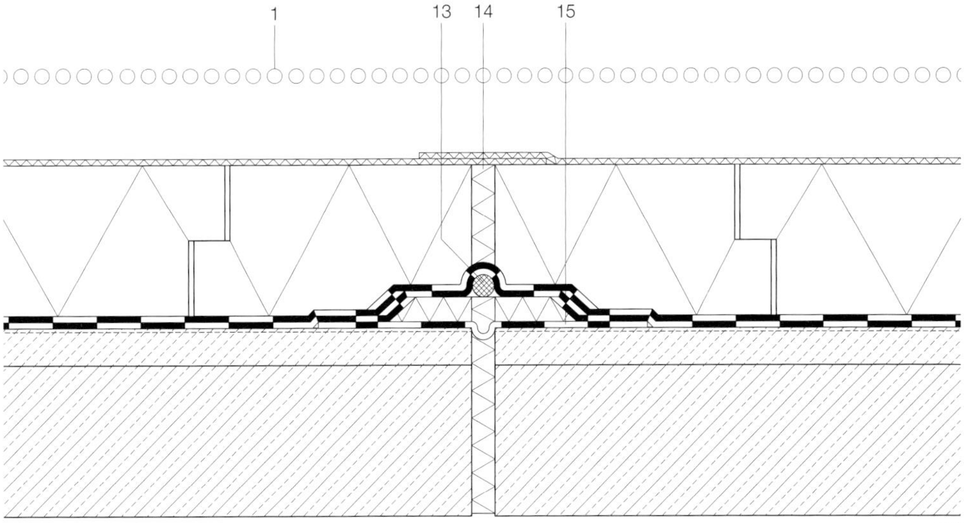

2.6

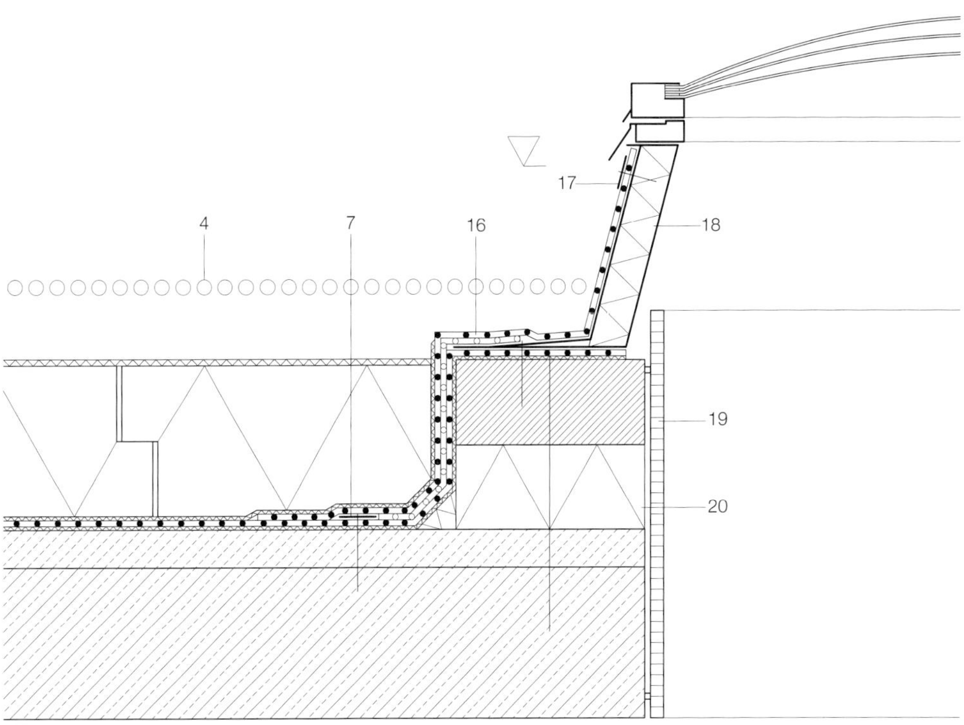

2.7

3 Impermeable concrete slab

3.1 Roof edge without overhang
· Insulation above slab

The parapet must be cast monolithically with the slab in one pour. To do this, the inside face of the parapet formwork is suspended from the outside face. The insulation on top of the power-floated slab protects the concrete against the effects of the weather and represents the best solution for insulation to a slab of impermeable concrete. When the protective membrane above the insulation forms a water run-off layer, moisture below the insulation is only minimal and so it is not necessary to reduce the U-value. Incidentally, the remarks for the details for waterproofing below insulation apply here as well (pp. 127–129). A slip plane on the ring beam protects the slab against restraint forces during the curing phase and also accommodates movement (e.g. due to temperature fluctuations) in the service condition.

3.2 Roof edge with overhang
· Insulation below slab

With the insulation attached on the inside, the roof slab is cold at low external temperatures and must be thermally isolated from the interior of the building. A slab supported at individual points allows insulation to be inserted between slab and ring beam. The slab must be finished with a layer of gravel in order to minimise temperature fluctuations and erosion. A wearing course for foot or vehicular traffic or a substrate for plants (a root barrier is required) achieve the same result. The kerb itself can be flashed with sheet metal. A smart vapour barrier attached to the underside of the mineral wool internal insulation protects the insulation against moisture in the room. This vapour barrier also allows any residual moisture that may find its way through the concrete to dry out towards the inside. Attaching the suspended ceiling to a grid minimises the number of hanger penetrations through the vapour barrier. The outer leaf of facing brickwork must be isolated from the movements of the concrete roof slab.

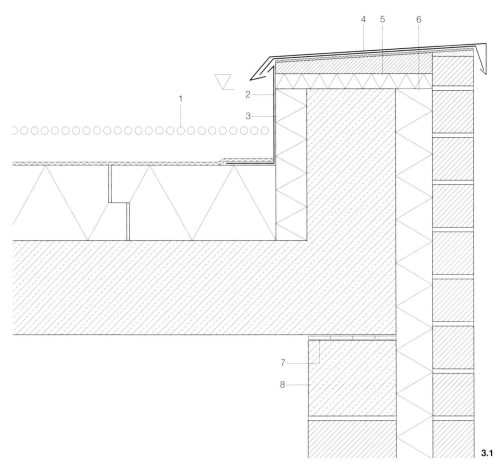

3.1

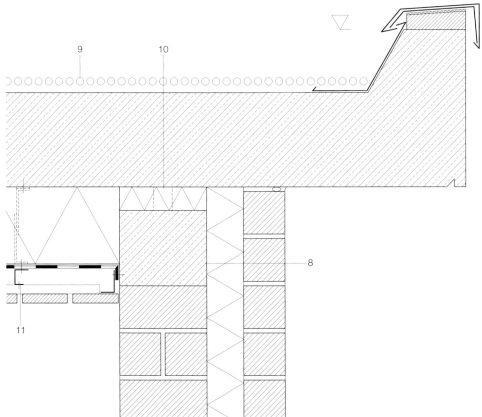

3.2

1	Loose gravel	6	Mineral fibre board
	Filter fleece	7	Slip plane
	Extruded polystyrene	8	Ring beam
	(XPS) board	9	Loose gravel
	Impermeable concrete		WU-Impermeable
	slab		concrete slab
2	Sheet metal skirting		Mineral fibre board
3	Extruded polystyrene		Smart vapour barrier
	(XPS) board	10	Slip plane as point
4	Parapet capping		support
	Sheet metal clip	11	Grid to support
	Protective membrane		suspended ceiling
	Timber board		Suspended ceiling
5	Rigid plastic foam board		

3.3 Junction with wall
· Insulation below slab
The main danger with this detail is that water can infiltrate behind the brickwork via the joints. It is therefore especially important to ensure good waterproofing at the junction between the cover flashing on the wall and the sheet metal finish to the concrete kerb. An elastic sealing tape placed in the joint where the two overlap can help to prevent water infiltration. The use of liquid waterproofing instead is another good option.

3.5 Drainage
· Insulation above slab
Various manufacturers have found simple solutions for outlets in concrete slabs. They consist of a metal pipe surrounded by a so-called waterstop flange. Both are cast into the impermeable concrete slab without the need for a sleeve of thin felt. Above this there is an extension piece, topped by a simple dome grating, to accommodate the thick insulation.

3.6 Joint
· Insulation above slab
The joint is waterproofed in this case by raising it above the water run-off level (i.e. the top of the concrete slab). The loops of material laid over the preformed plastic strip permit horizontal movement at the joint. Several layers of fleece are necessary when using liquid waterproofing. The insulation on top of the kerb is raised above the water run-off level and the heavy concrete coping prevents the insulation floating during heavy rainfall. Sheet metal skirtings, held in place by the weight of the gravel on the horizontal leg, protect the insulation against the weather and mechanical damage.

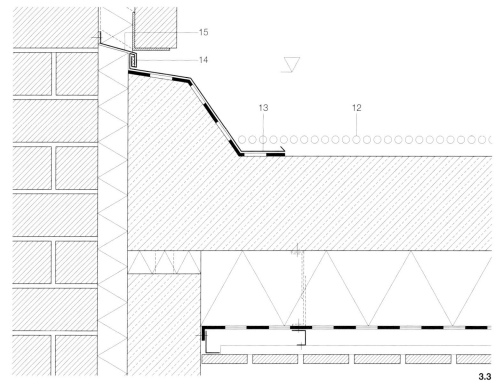

3.3

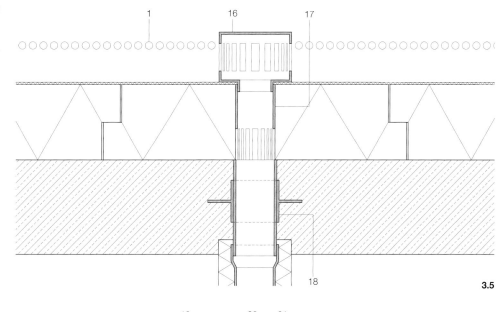

3.5

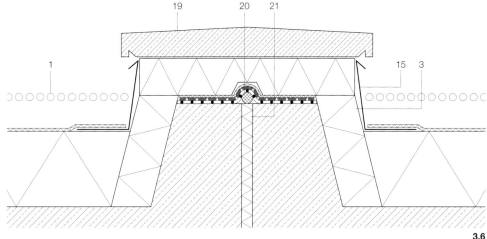

3.6

12	Loose gravel	19	Precast concrete coping
	Impermeable concrete slab		Extruded polystyrene (XPS) board
	Mineral fibre board		Protective membrane
	Smart vapour barrier		Liquid waterproofing
13	Cover flashing		Synthetic waterproofing
14	Elastic sealing tape	20	Preformed sealing strip
15	Cover flashing		
16	Dome grating		
17	Extension piece		
18	Waterstop flange	21	Mineral fibre blanket

4 Green roof

4.1 Roof edge without overhang
· Extensive planting
· Waterproofing below insulation
· Liquid waterproofing

On green roofs it is advisable to place the waterproofing below the insulation where it is protected against mechanical damage. Liquid waterproofing materials may not contain any solvents when laid on a timber loadbearing structure. When selecting the drainage layer and the substrate, it is vital to ensure that there is sufficient ballast to prevent the drainage layer from lifting; it may even be necessary to use "grasscrete" paving blocks. The filter fleece should extend at least to the top of the drainage layer if the drainage layer is in the form of a loose fill material.

4.2 Roof edge with overhang
· Extensive planting
· Waterproofing above insulation
· Bitumen waterproofing

As the vapour barrier must be welded, a separating membrane is necessary between the edge-fixed timber roof elements and the vapour barrier. The desire to leave a timber parapet exposed conflicts with the regulations and guidelines for metal roof coverings, which call for a 15 cm high skirting of waterproofing alongside any component rising above roof level. This detail does not comply with this requirement because the overhang is relatively small and the sheet metal finish falls to the outside.

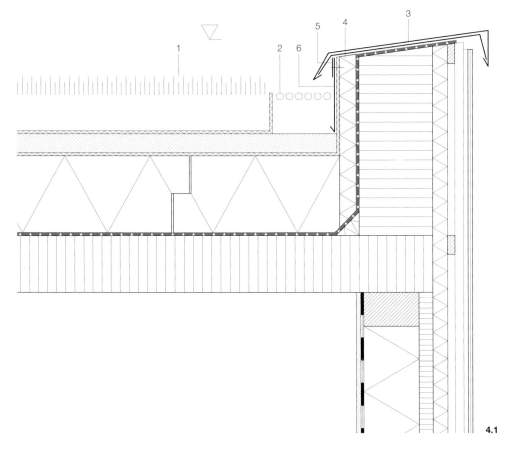

4.1

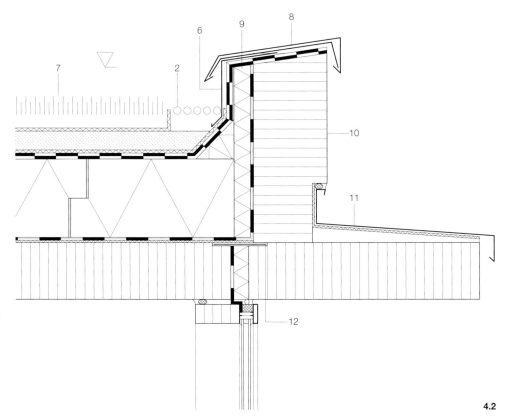

1	Planting		Bitumen waterproofing,
	Substrate		2-ply
	Filter fleece		Rigid plastic foam
	Drainage layer		board
	Filter fleece		Vapour barrier
	Extruded polystyrene		Separating membrane
	(XPS) board		Edge-fixed timber roof
	Liquid waterproofing		element
	Edge-fixed timber roof	8	Parapet capping
	element		Sheet metal clip
2	Loose gravel		Bitumen waterproofing,
3	Parapet capping		2-ply
	Sheet metal clip	9	Rigid plastic foam
	Liquid waterproofing		board
4	Extruded polystyrene	10	Parapet, glued laminated
	(XPS) board		timber
5	Clamping bar	11	Sheet metal finish
6	Sheet metal skirting		Protective membrane
7	Planting		Timber battens
	Substrate		Edge-fixed timber roof
	Filter fleece		element
	Drainage layer	12	Bridging plate
	Protective membrane		

4.2

4.3 Junction with wall
- **Extensive planting**
- **Waterproofing above insulation**
- **Synthetic waterproofing**

In principle, multiple layers of insulation must be laid with their joints offset. Certain constituents of timber can make it necessary to include a separating membrane in the case of some synthetic waterproofing materials. The prescribed skirting height of 15 cm is measured from the top of the substrate or concrete flags. The proximity of the flags makes it necessary to protect the skirting of waterproofing material with sheet metal. If the synthetic waterproofing material is fixed to the vertical timber elements instead of to the horizontal roof elements, the screws do not need to be so long.

4.4 Penetration
- **Waterproofing above insulation**
- **Synthetic waterproofing**

Pipes penetrating the roof must always be firmly attached to the roof structure. In this detail the penetration is integrated into both the vapour barrier and the waterproofing with a collar fabricated on site. A hose clip secures the top edge of the collar at the prescribed height of 15 cm above the roof finishes. The mechanical fasteners for the synthetic waterproofing material must be guaranteed on all sides.

4.5 Drainage
- **Waterproofing above insulation**
- **Bitumen waterproofing**

Roof outlets, too, must be firmly attached to the roof structure. This detail shows an outlet and an extension piece, both with double flanges (one fixed, one detachable) for clamping the waterproofing and vapour barrier materials. To ensure good access for maintenance, an area measuring approx. 50 × 50 cm around the dome grating should be kept clear of substrate and plants.

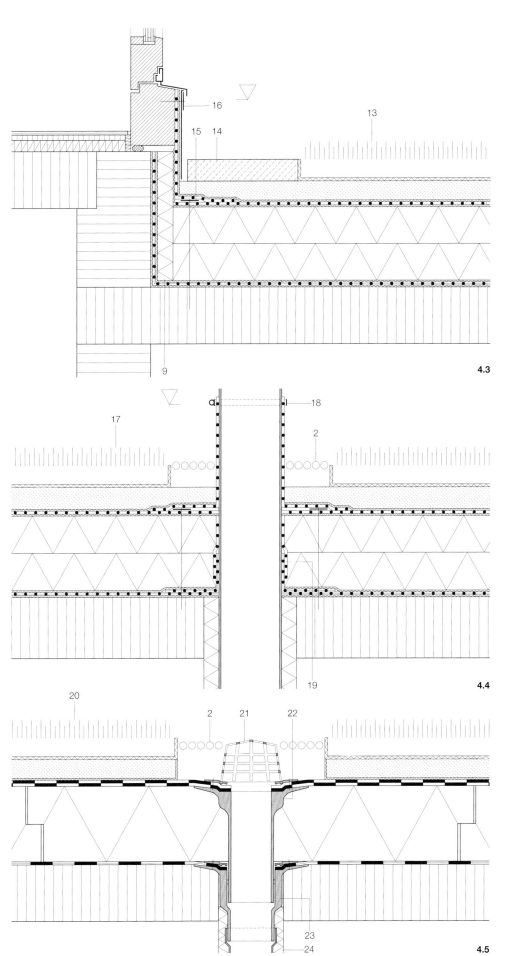

13	Planting		Separating membrane
	Substrate		Vapour barrier
	Filter fleece		Separating membrane
	Drainage layer		Edge-fixed timber roof
	Protective membrane		element
	Synthetic waterproofing	18	Hose clip
	Separating membrane	19	Pipe collar
	Rigid plastic foam board	20	Planting
	Separating membrane		Substrate
	Vapour barrier		Filter fleece
	Separating membrane		Drainage layer
	Edge-fixed timber roof		Protective membrane
	element		Bitumen waterproofing,
14	Concrete flags		2-ply
15	Linear fixing		Rigid plastic foam
16	Clamping bar		board
17	Planting		Vapour barrier
	Substrate		Edge-fixed timber roof
	Filter fleece		element
	Drainage layer	21	Dome grating
	Protective membrane	22	Extension piece
	Synthetic waterproofing	23	Body of roof outlet
	Separating membrane	24	Lagging
	Rigid plastic foam board		

5 Roof for foot traffic

5.1 Roof edge without overhang
· **Waterproofing above insulation**
· **Bitumen waterproofing**

A protective membrane over the waterproofing is necessary because of the bedding of angular (i.e. sharp) chippings for the stone flags. A sheet metal skirting protects the bitumen flexible sheeting at the parapet. It is easy to offset the joints in the insulation on the roof surface: simply lay the lower layer of insulation up to the inside face of the concrete parapet and start the upper layer of insulation from the face of the vertical insulation to the inside of the parapet. This method is suggested in the Flat Roof Directive.

5.2 Roof edge with overhang
· **Balustrade**
· **Waterproofing below insulation**
· **Synthetic waterproofing**

If the thermal insulation is incompatible with the synthetic waterproofing material, then a separating membrane must be laid between the two. The balustrade in this detail consists of laminated safety glass spanning between steel plates connected to the precast concrete coping. The insulated joint between the roof slab and the cantilevering section requires a stable cover to support the materials above. A screed laid to falls is necessary on the roof slab.

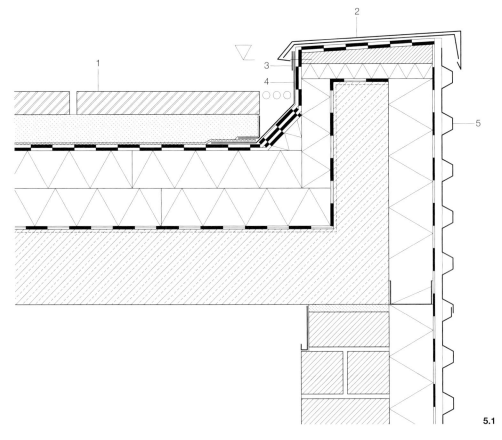

5.1

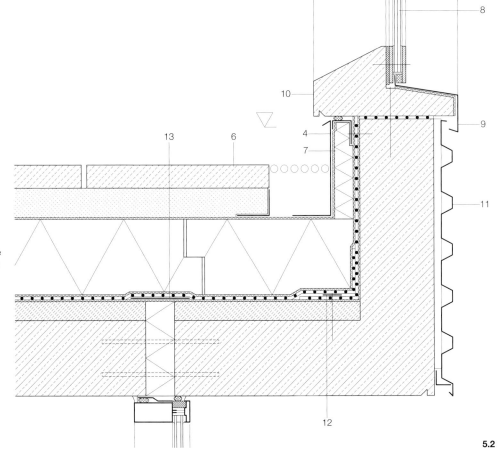

1   Stone flags
    Chippings
    Protective membrane
    Bitumen waterproofing,
    2-ply
    Rigid plastic foam
    board
    Vapour barrier
    Primer
    Reinforced concrete
2   Parapet capping
    Sheet metal clip
    Bitumen waterproofing,
    2-ply
    Timber board
    Rigid plastic foam
    board
3   Clamping bar
4   Sheet metal skirting
5   Trapezoidal profile
    metal cladding
    Facing
    Mineral fibre board

6   Concrete flags
    Chippings
    Filter fleece
    Extruded polystyrene
    (XPS) board
    Separating membrane
    Synthetic waterproofing
    Protective membrane
    Screed laid to falls
    Extruded polystyrene
    (XPS) board
7   Extruded polystyrene
    (XPS) board
8   Laminated safety
    glass
9   Sheet metal sill
10 Precast concrete
    coping
11 Trapezoidal profile
    metal cladding
12 Linear fixing
13 Bridging plate

5.2

134

5.3 Junction with wall
· Liquid waterproofing

A skirting of liquid waterproofing does not necessarily need to be fixed with a clamping bar; the protective sheet metal is simply riveted to the panel. The skirting height required is measured from the top edge of the roof finishes, but if it is measured from the invert of the drainage channel, then the channel must be connected directly to the drainage system and further conditions specified in the Flat Roof Directive must be complied with, e.g. the sill in the facade must overhang the roof finishes by min. 5 cm. The insulation in roofs intended for foot traffic should always exhibit an adequate compressive strength, and that of the insulation adjacent to drainage channels should be even higher.

5.4 Penetration
· Bitumen waterproofing

The waterproofing at this detail has been improved by using a post base section with a diameter larger than that of the post itself. The base plate is embedded in the vapour barrier with the help of a strip of elastomeric bitumen waterproofing material. A sheet metal sleeve is slipped over the base section and integrated into the roof waterproofing. The advantage of using sheet metal is that it is much less vulnerable to damage. The post itself is bolted to the base section.

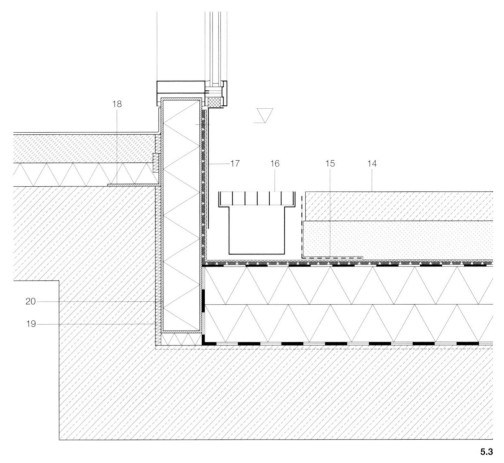

5.3

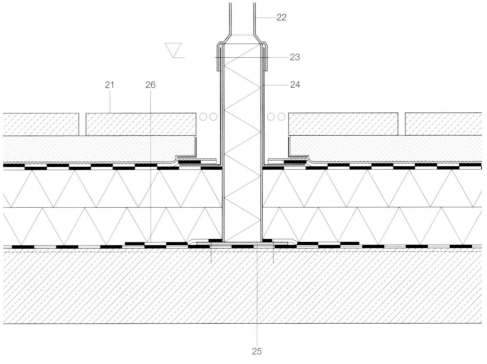

14	Concrete flags	21	Concrete flags
	Chippings		Chippings
	Protective membrane		Protective membrane
	Liquid waterproofing		Bitumen waterproofing,
	Bitumen facing		2-ply
	Rigid plastic foam		Rigid plastic foam
	board		board
	Vapour barrier		Vapour barrier
	Primer		Primer
	Reinforced concrete		Reinforced concrete
15	Mesh to prevent loss	22	Post
	of chippings	23	Bolt
16	Drainage channel	24	Sheet metal sleeve
17	Sheet metal skirting	25	Steel post base
18	Fixing angle	26	Elastomeric bitumen
19	Acoustic board		waterproofing
20	Panel filled with cellular glass		

5.4

5.5 Drainage
· Synthetic waterproofing

Roof outlet components must be securely fixed to the roof construction. The extension piece within the insulation must be integrated into the waterproofing via a flange, just like the body of the outlet is integrated into the vapour barrier. Both vapour barrier and waterproofing must be fixed in place with mechanical fasteners. The second extension piece (with grating for foot traffic) that raises the outlet to the level of the concrete flags, compensates for the depth of the flags plus bedding and is height-adjustable. When using synthetic waterproofing materials, proprietary mountings for the concrete flags can be considered instead of a bedding of chippings.

5.6 Joint
· Liquid waterproofing

A secure joint is ensured by raising it above the level of the water run-off layer (i.e the waterproofing). The vapour barrier and the waterproofing can accommodate transverse movements with the help of additional loops of material. The liquid waterproofing above the preformed foam strip is applied in several coats.

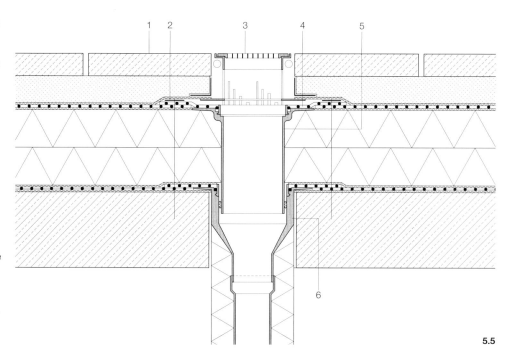

5.5

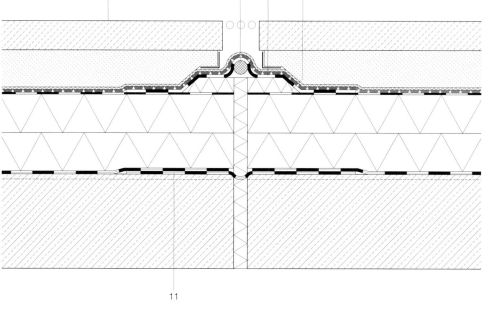

	1	Concrete flags		7	Concrete flags
		Chippings			Chippings
		Protective membrane			Protective membrane
		Synthetic waterproofing			Liquid waterproofing
		Separating membrane			Facing
		Rigid plastic foam board			Rigid plastic foam board
		Separating membrane			
		Vapour barrier			Vapour barrier
		Separating membrane			Primer
		Reinforced concrete			Reinforced concrete
	2	Linear fixing		8	Preformed foam/ rubber strip
	3	Grating for foot traffic			
	4	Collar of synthetic water- proofing		9	Bitumen waterproofing
				10	Rigid plastic foam board
	5	Extension piece			
	6	Body of outlet		11	Vapour barrier

5.6

136

6 Roof for vehicular traffic

6.1 Roof edge without overhang
· Synthetic waterproofing
· Composite waterproofing (compact roof)

A roof intended to be used as a helicopter landing pad should always include so-called composite waterproofing below the topmost layer of waterproofing, i.e. boards of cellular glass insulation fully bonded with hot bitumen and all joints filled with hot bitumen. The first ply of waterproofing material should be made from polymer-modified bitumen; a bitumen-compatible synthetic material is laid on top of this. A screed is required to protect the waterproofing. The fall essential for this type of roof can be provided by the concrete slab of the landing pad itself. An incompressible acoustic board ensures the acoustic decoupling of the landing pad from the rest of the structure.

6.2 Roof edge with safety net
· Liquid waterproofing
· Composite waterproofing (compact roof)

Liquid waterproofing is used for the top ply in this detail. The safety net required for this helicopter landing pad may not project above the height of the parapet.

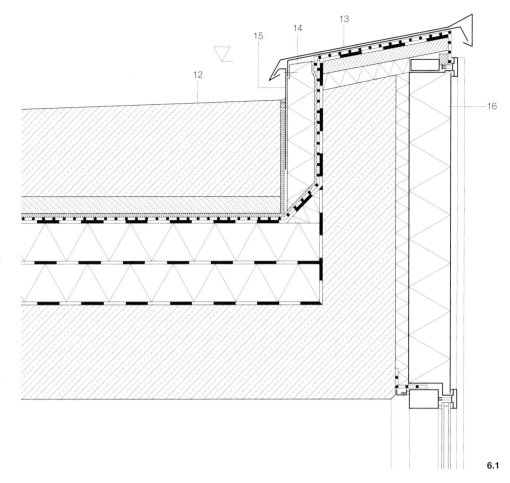

6.1

12 Reinforced concrete slab laid to falls
Screed
Incompressible acoustic board
Synthetic waterproofing
Polymer-modified bitumen waterproofing
Composite waterproofing: cellular glass, 2 layers, bonded with bitumen
Bitumen waterproofing
Reinforced concrete
13 Parapet capping
Sheet metal clip
Synthetic waterproofing
Polymer-modified bitumen waterproofing
Timber board
Rigid plastic foam board
14 Extruded polystyrene (XPS) board
15 Sheet metal skirting
16 Insulated panel

17 Reinforced concrete slab laid to falls
Screed
Incompressible acoustic board
Liquid waterproofing
Synthetic waterproofing
Composite waterproofing: cellular glass, 2 layers, bonded with bitumen
Vapour barrier
Primer
Reinforced concrete
18 Parapet capping
Sheet metal clip
Liquid waterproofing
Timber board
Rigid plastic foam board
19 Wire mesh
20 Loadbearing member
21 Bracket

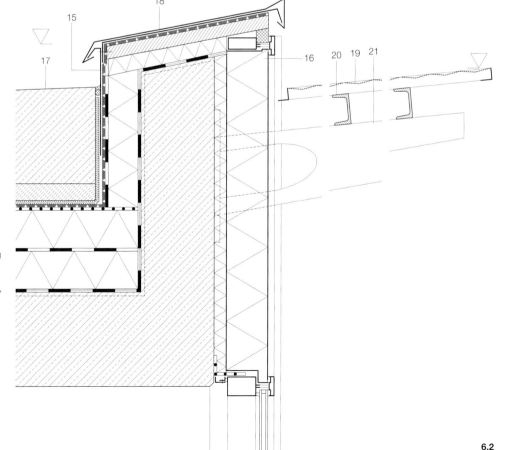

6.2

6.3 Junction with wall/facade
· Bitumen waterproofing
· Composite waterproofing (compact roof)

Several measures are necessary in order to achieve a level transition between helicopter landing pad and interior, as laid down in the Flat Roof Directive. These measures include a direct connection between drainage channel and drainage system, a water run-off level laid to falls, protection against driving rain, waterproofing material connected to the door frame via a flange detail, and possibly additional waterproofing to the interior with separate drainage (not shown in this detail). The drainage channel is bedded on a block of non-impermeable in situ concrete.

6.5 Drainage
· Bitumen waterproofing
· Composite waterproofing (compact roof)

The roof outlets used in helicopter landing pads must comply with the highest loading category, which means using cast iron. The body of the outlet must be firmly screwed to the structure and enclosed in lagging; the multi-part extension piece is cast into the concrete landing pad itself. The outlet should be wrapped in a synthetic felt to allow some movement but at the same time absorb any condensation.

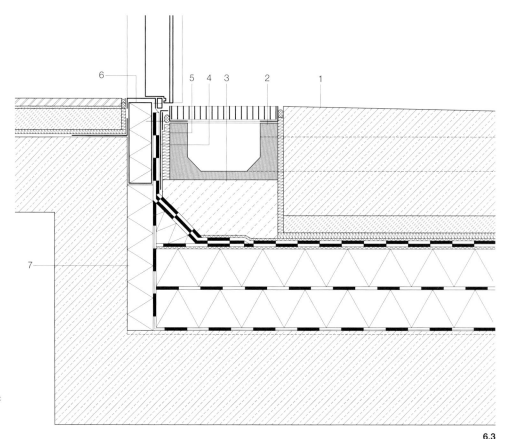

6.3

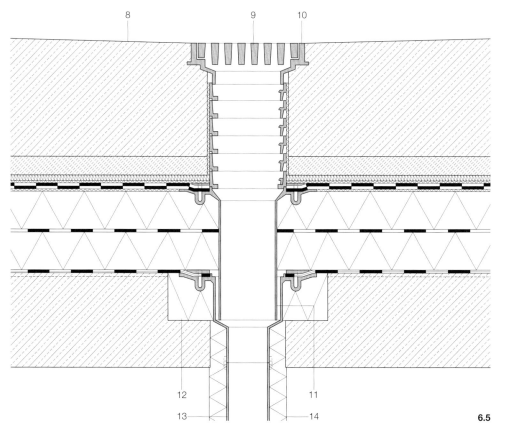

1	Reinforced concrete slab		laid to falls
	laid to falls		Screed
	Screed		Incompressible
	Incompressible acoustic		acoustic board
	board		Protective membrane
	Protective membrane		Bitumen waterproofing,
	Bitumen waterproofing,		2-ply
	2-ply		Separating membrane
	Separating membrane		Composite water-
	Composite waterproofing:		proofing: cellular glass,
	cellular glass, 2 layers,		2 layers, bonded with
	bonded with bitumen		bitumen
	Vapour barrier		Vapour barrier
	Primer		Primer
	Reinforced concrete		Reinforced concrete
2	Drainage channel	9	Cast iron grating
3	Permeable in situ con-	10	Grating frame
	crete	11	Extension piece
4	Sheet metal skirting	12	Block of insulation
5	Acoustic blanket	13	Flat roof outlet with
6	Insulated panel		fixed and clamping
7	Rigid plastic foam board		flanges
	Reinforced concrete slab	14	Lagging

6.5

7 Metal roof

7.1.1 Roof edge without overhang
· Aluminium standing seam sheets

A two-layer system is available for very deep layers of insulation on metal roof sheets; rails are mounted on the first layer to which the clips are attached. The ends of the standing seams are welded together and to the sheet metal skirting at the parapet. Expansion elements are required to permit transverse movement of the sheet metal skirting. And a compressible insulating material must be installed to enable longitudinal movement of the sheets. The insulation from the parapet to the front edge of the standing seam sheets plus the entire lower layer of insulation must be incompressible. Even if a backing plate is used to protect the insulation from the heat of welding, the top layer of insulation should still be incombustible. The parapet in this detail is formed by stiffened sheet metal. The troughs of the loadbearing trapezoidal sheets are covered with sheet metal which is connected to the stiffened sheet metal parapet via a sealing tape.

7.1.2 Roof edge without overhang
· Welded sheet stainless steel

On this timber structure the parapet is made from a glued laminated timber beam. The sheet stainless steel is welded at the transverse seams and to the parapet. Adequate tolerances must be available to accommodate the longitudinal expansion. A sound-attenuating board has been laid beneath the stainless steel to reduce the drumming noise of rainfall.

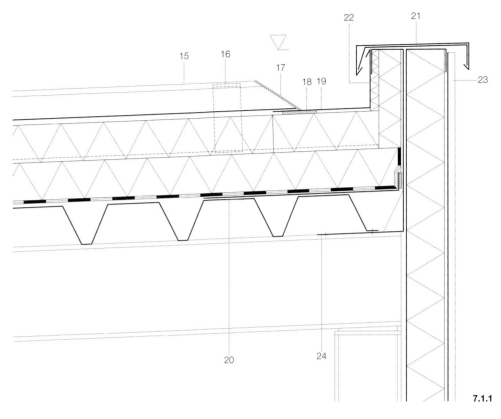

7.1.1

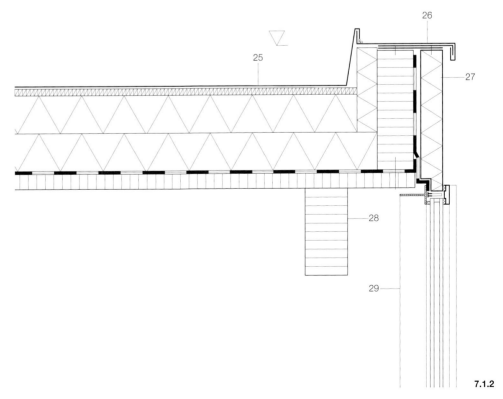

7.1.2

15 Aluminium standing seam sheets
   Mineral fibre board, 2 layers, top layer incombustible, bottom layer incompressible
   Vapour barrier
   Trapezoidal profile sheets
   Loadbearing section
16 Clip for standing seam
17 Weld seam
18 Protective membrane
19 Sheet metal skirting
20 Sheet metal covering to troughs
21 Parapet capping
   Sheet metal clip
22 Mineral fibre board
23 Trapezoidal profile metal cladding

Mineral fibre board
24 Stiffened sheet metal parapet
25 Stainless steel sheet
   Acoustic board
   Rigid plastic foam board
   Vapour barrier
   Wood-based board product
26 Sheet stainless steel capping
   Sheet metal clip
27 Insulated panel
   Vapour barrier
   Glued laminated timber parapet
   Rigid plastic foam board
28 Beam
29 Post

7.2.1 Roof edge with overhang
· Gutter
· Aluminium standing seam sheets

The eaves board is fixed to brackets anchored in the aluminium standing seam sheets. A short cantilever is possible with these sheets, but longer cantilevers require round steel bars to be inserted into the standing seams as reinforcement. The gutter bracket is held in place with the help of a clip.

7.2.2 Roof edge with overhang
· Sheet steel
· R-girder composite deck

This patented composite construction consists of two so-called R-girders (lattice beams) made from 4 mm thick galvanised sheet steel strip, which determine the depth of the construction and are thermally isolated from each other, plus stiffeners to prevent buckling. Filled with insulating material, the whole construction performs several functions: waterproofing (with epoxy resin coating), loadbearing, thermal insulation and vapour barrier; it also forms the soffit. The gutter outlet is at the bottom of the gutter and there is a overflow to the side. The columns supporting this lightweight roof structure are integrated into the facade.

7.3 Junction with wall
· Profiled aluminium sheets

A wall junction profile that can be bent to suit and can provide a skirting the prescribed height above the water run-off level enables a wall junction parallel to the standing seams. The vapour barrier is clamped to the wall with a loop of material and a sealing strip in order to accommodate relative movements between the metal roof structure and the wall.

1 Profiled metal sheet
  Mineral fibre board,
  2 layers, bottom layer
  incompressible
  Vapour barrier
  Trapezoidal profile
  sheet
2 Mounting rail
3 Bracket
4 Eaves board
5 Clip for standing seam
6 Preformed sealing strip
7 Gutter
8 Gutter bracket
9 Trapezoidal profile
  metal cladding
10 Mineral fibre board
  Frame
11 Sheet metal covering
  to troughs
12 Beam
13 Column
14 Galvanised sheet
  steel
  Mineral fibre board
  Galvanised sheet
  steel
15 R-girder (lattice beam)
16 Stiffener
17 Cover flashing
18 Sheet metal skirting
19 Sealing strip

7.2.1

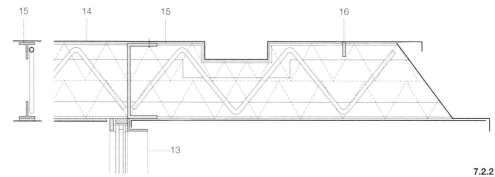

7.2.2

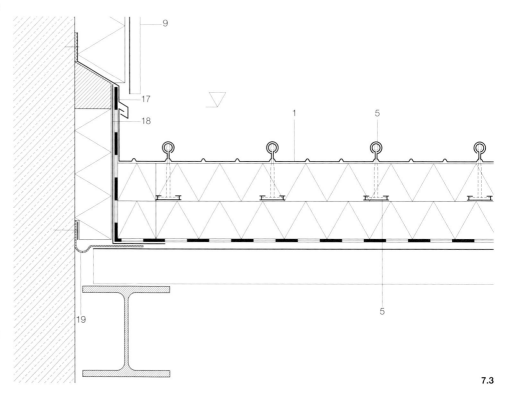

7.3

8 Glass roof

8.1 Roof edge without overhang (verge)
· **Linear fixing**

Glass has also been used for flat roofs in recent years because it can be laid without transverse joints at pitches as shallow as 2% without suffering any damage. However, intensive cleaning and maintenance are required because the self-cleaning effect only starts to function at pitches of approx. 15.5% (7°). The insulating glass unit with an outer pane of toughened glass and inner pane of laminated safety glass is clamped directly to the I-section with sealing gaskets. U-shaped patent glazing caps are used in this example. The corner joint at the verge is stepped and sealed with silicone. The roof glazing is supported on the vertical facade glazing, which is clamped to the vertical I-section posts with the same patent glazing caps.

8.2 Roof edge with overhang (gable)
· **Gutter**
· **Point fixing**

The panes in this detail are fixed with point fixings in drilled holes. However, it would also be possible to use point fixings fitted in the longitudinal joints between the panes. Each pair of fixings is mounted on a bracket which transfers the loads back to the structure; all longitudinal joints are sealed with a preformed strip and an elastic compound. The same brackets are used for the facade glazing as well, but the fixings here are flush with the outer surface of the glass and they include an articulated joint in the plane of the glazing. This method of fixing permits a certain amount of movement which results in lower stresses in the glass. The joint between roof glazing and facade glazing is stepped; the length of the glass cantilever depends on the thickness of the glass (up to 30 cm is possible). The internal valley gutter must be laid to a fall of at least 0.5% and must include overflow outlets.

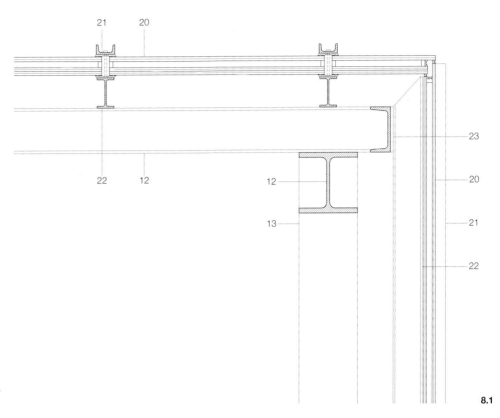

8.1

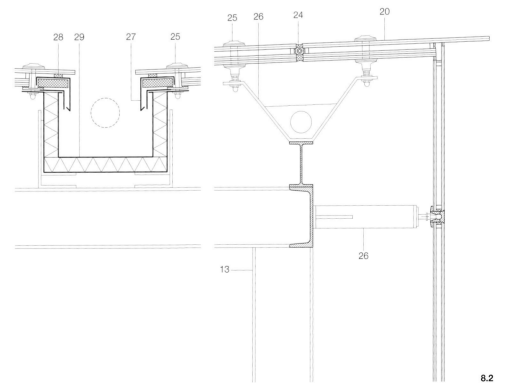

20	Insulating glass: tough. glass + lam. safety glass	24	Preformed sealing strip
21	Patent glazing cap	25	Point fixing
22	Glazing bar	26	Bracket
23	Steel channel edge rail	27	Drip flashing
		28	Spacer
		29	Gutter

8.2

8.3 Junction with side wall
· Linear fixing

The junction with a side wall is achieved with the help of a multi-ply composite panel (left) supported between the patent glazing bar (clamped in place by the glazing cap) and an angle fitted to the wall. The sheet metal skirting, extending the prescribed height above the glazing, is also clamped to the glazing bar; a certain amount of vertical movement is possible here. A multi-ply composite panel in the shape of a gutter (right) could be used here as an alternative, which would improve the insulation.

8.4 Penetration und junction with wall at top of slope
· Linear fixing

The detail at the penetration is similar to the wall junction detail (8.3 left). A sheet metal pipe collar is clamped in a panel between two pairs of glazing bars. This collar rises the recommended height above the glazing. A drip flashing, held in place with a hose clip, covers the top of this collar.

The junction with the wall at the top of the slope is similar to the detail for side wall without gutter (8.3 left).

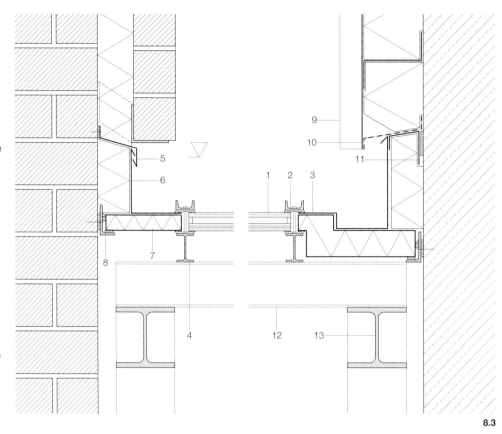

8.3

1	Insulating glass	10	Insect screen
2	Glazing cap	11	Fixing rail
3	Multi-ply composite panel in gutter form	12	Purlin
		13	Steel beam
4	Glazing bar	14	Glazing cap
5	Cover flashing	15	Hose clip
6	Sheet metal skirting	16	Drip flashing
7	Multi-ply composite panel	17	Steel angle
		18	Column
8	Steel angle		
9	Trapezoidal profile metal cladding		

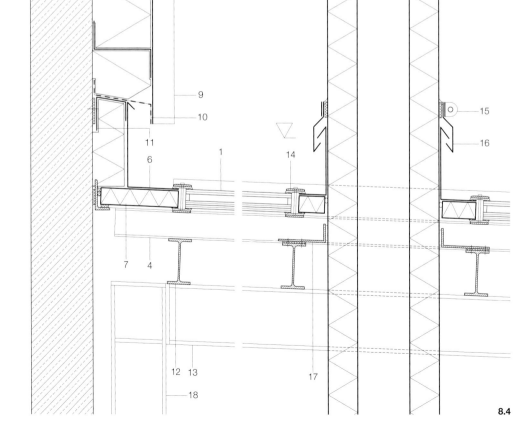

8.4

9 Refurbishment

The examples of refurbishment details shown here presume that the layer of insulation and the waterproofing to the roof are still functioning properly. If that is not the case, both must be removed and the roof rebuilt from the roof structure upwards according to the current regulations. As many older flat roofs were built without a fall, designers are advised to use tapered insulation products for their refurbishment projects. Each detail shows the roof construction prior to (black) and after refurbishment (red).

9.1 Roof edge without overhang
· **Waterproofing above insulation**
· **Bitumen waterproofing**

The parapet height required to accommodate the new depth of insulation and the fall can be built up on the existing parapet very simply with additional layers of insulation and/or timber. To do this, the outside face of the parapet must be flashed with sheet metal which is fixed to the timber at the top and held in place with a sheet metal clip at the bottom. The additional insulation in the cavity of the external wall can be provided in the form of a blown, moisture-resistant material, e.g. finely ground igneous rocks (e.g. basalt) with a water-repellent coating, expanded clay beads or lightweight silicate foam.

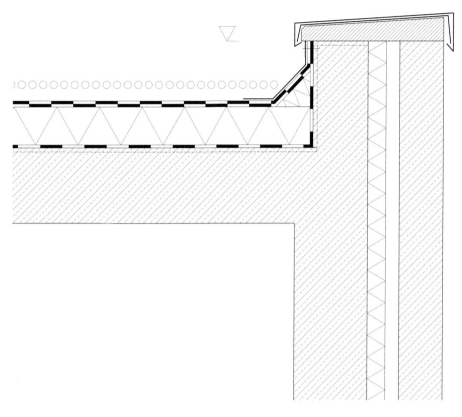

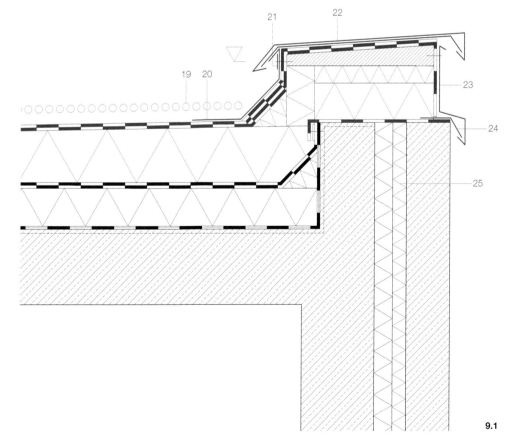

19	Loose gravel	21	Clamping bar
	Bitumen waterproofing, 2-ply	22	Parapet capping Sheet metal clip
	Rigid plastic foam bds., tapered to create falls		Bitumen waterproofing, 2-ply
	Bitumen waterproofing, 2-ply		Timber board
	Mineral fibre board		Rigid plastic foam board
	Vapour barrier	23	Sheet metal parapet
	Primer		facing
	Reinforced concrete	24	Sheet metal clip
20	Sheet metal skirting	25	Blown insulation

9.1

9.2 Roof edge with overhang
  · **Waterproofing between layers of insulation**
  · **Synthetic waterproofing**

Where the waterproofing is still in a good condition, refurbishment in the form of waterproofing between two layers of insulation represents an interesting option (sandwich roof). The new waterproofing material is attached to the existing one and clamped at the prescribed height above the top of the gravel. To do this, a sheet metal frame must be fixed to the timber and the new parapet capping folded over this. The energy efficiency upgrade to the patent glazing facade is in this case achieved simply by replacing the double glazing with triple glazing.

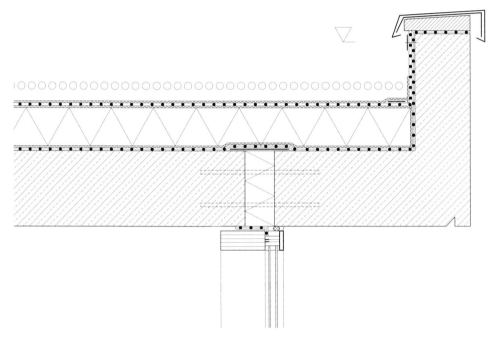

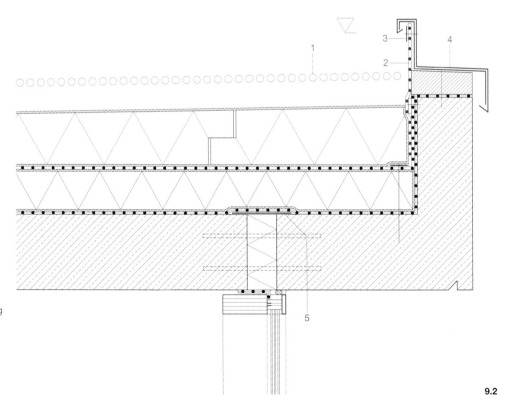

1	Loose gravel		Vapour barrier
	Protective membrane		Reinforced concrete
	Extruded polystyrene	2	Synthetic waterproofing
	(XPS) board, tapered to	3	Clamping bar
	create falls	4	Parapet capping
	Separating membrane		Sheet metal frame
	Synthetic waterproofing		Timber board
	Separating membrane		Vapour barrier
	Rigid plastic foam board	5	Bridging plate
	Separating membrane		

9.2

9.3 Junction with wall
· **Waterproofing above insulation**
· **Liquid waterproofing**

The existing waterproofing and insulation must be removed locally in order to attach the new skirting bracket to the roof structure. The bracket is required to support the new skirting of waterproofing material. The void between the old skirting and the new bracket is filled with insulating material. The energy efficiency upgrade to the external wall – additional insulation plus trapezoidal profile metal cladding – does change the external appearance but offers the chance to create a rainproof covering to the outer leaf of clay brickwork and thus exclude water from the cavity.

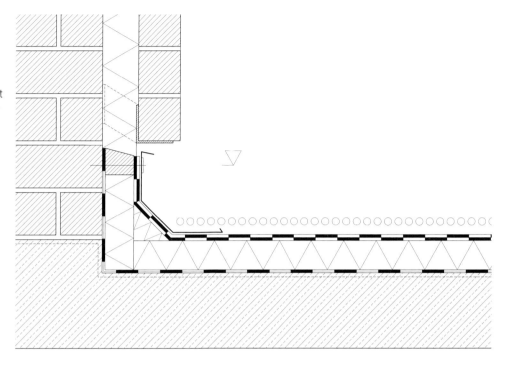

6	Loose gravel		Vapour barrier
	Liquid waterproofing		Primer
	Bitumen facing		Reinforced concrete
	Rigid plastic foam board boards, tapered to create falls	7	Trapezoidal profile metal cladding
			Mineral fibre board
	Bitumen waterproofing, 2-ply	8	Cover flashing
	Rigid plastic foam board	9	Stiffened support bracket
		10	Mineral wool

9.3

145

9.4 Penetration
- **Waterproofing above insulation**
- **Bitumen waterproofing**

The junction with the new waterproofing is achieved with a new sleeve that is slipped over the old one. The flange is integrated into the layers of bitumen waterproofing. Liquid waterproofing represents a less costly option.

9.5 Drainage
- **Waterproofing above insulation**
- **Synthetic waterproofing**

Various manufacturers offer outlets for refurbishment projects, which raise the outlet to the level of the new roof. These anti-backflow products are simply inserted into the existing outlet and integrated into the new waterproofing.

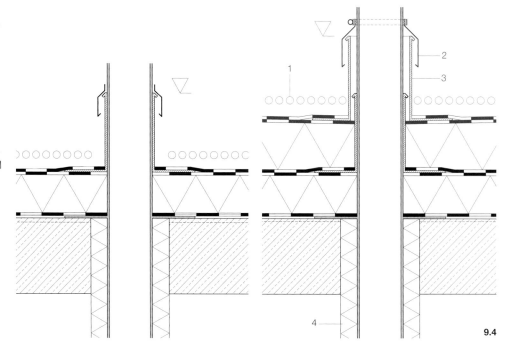

9.4

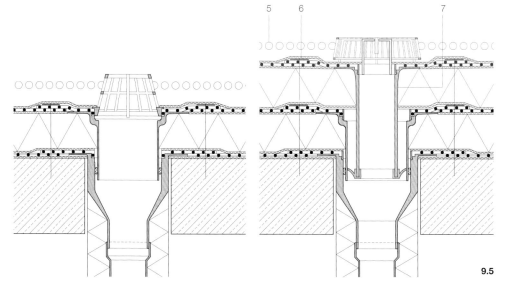

9.5

1	Loose gravel	Synthetic waterproofing
	Bitumen waterproofing, 2-ply	Separating membrane
	Rigid plastic foam board	Rigid plastic foam board
	Bitumen waterproofing, 2-ply	Separating membrane
	Rigid plastic foam board	Synthetic waterproofing
	Vapour barrier	Separating membrane
	Primer	Rigid plastic foam board
	Reinforced concrete	Separating membrane
2	Drip flashing	Vapour barrier
3	Pipe sleeve	Separating membrane
4	Lagging	Reinforced concrete
5	Loose gravel	6 Linear fixing
	Protective membrane	7 Extension piece

146

9.7 Individual rooflight
- **Waterproofing above insulation**
- **Liquid waterproofing**

The existing waterproofing and insulation must be removed locally so that the new vapour barrier can be connected to the existing one. The new depth of insulation is achieved by increasing the depth of the timber frame. Normally, an energy efficiency upgrade project will also include replacing the old rooflights with new two- or three-leaf models which include a mounting frame with improved insulation.

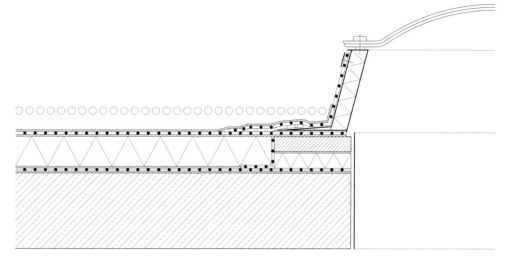

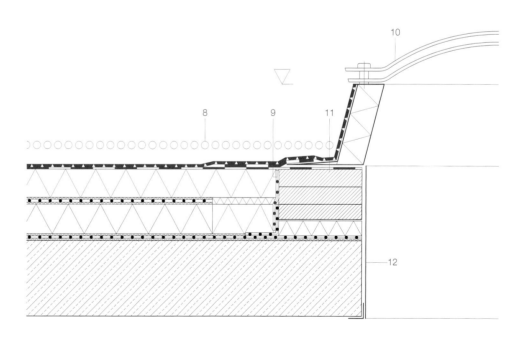

9.7

8	Loose gravel		Vapour barrier
	Liquid waterproofing		Separating membrane
	Facing		Reinforced concrete
	Rigid plastic foam board	9	Vapour barrier
	Separating membrane	10	Double-leaf rooflight
	Synthetic waterproofing	11	Timber frame to suit
	Separating membrane		depth of insulation
	Rigid plastic foam board	12	Lining
	Separating membrane		

Part F   Case studies

Fig. F   MAXXI Museum, Rome (I), 2009,
Zaha Hadid Architects

Example 01

High-bay racking warehouse

Lüdenscheid, Germany, 2001

Architects:
Schneider + Schumacher, Frankfurt am Main
Project team:
Gunilla Klinkhammer, Robert Binder, Nadja
Hellenthal, Alexander Probst, Till Schneider
Structural engineers:
Posselt Consult, Übersee

The design for this high-bay racking warehouse
in Lüdenscheid has been kept very simple, but
the architecture far surpasses that of an average
industrial building. The brief, adding a pallet
racking warehouse to the two existing production
buildings, has been solved in a functional but
convincing way. The client is an international
manufacturer of lighting systems and so light
plays an important role in the facade concept.
The basic idea was to create different facades
for day and night. During the day the cladding
to the building – profiled glass elements with
their flanges pointing outwards – reacts to the
changing patterns of daylight. At night, the
fluorescent tubes installed in vertical rows behind
the facade evoke an image of the barcodes
that mark the products in the warehouse.
The storage racking doubles as a steel frame
for the building itself. Only the staircase and
the plant rooms are in solid construction. Steel
sections are used for the roof structure too, and
trapezoidal profile sheets for the loadbearing
deck. The synthetic waterproofing requires no
further protection. The perimeter kerb is set
back from the edges, which makes the roof
construction less obtrusive. Six continuous
rooflights along the length of the warehouse
ensure a good level of daylight even deep within
this 30 × 74 m building.

· Plastic waterproofing without protective finish
· Roof overhang with steel beams
· Continuous rooflights

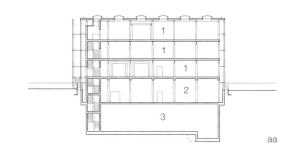

aa

Sections · Plan
Scale 1:750

1 Order-picking
2 Plant room
3 Sprinkler system
   control room
4 Bridge to dispatch
   department
5 High-bay racking
   warehouse
6 Entrance
7 Deliveries
8 WC

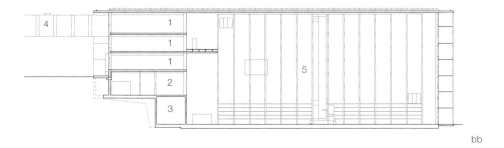

bb

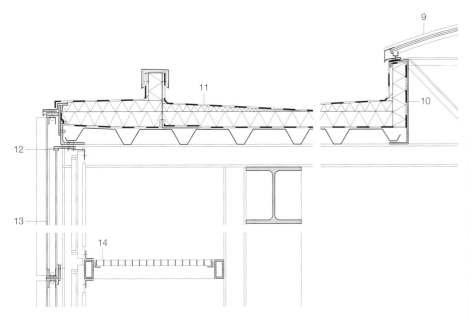

Vertical sections
Scale 1:20

9   Rooflight, 16 mm twin-wall polycarbonate
    sheet
10 Steel flat, folded, 4 mm
11 Roof construction:
    2 mm polyolefin waterproofing
    80–120 mm mineral wool thermal insulation,
    tapered to create falls
    vapour barrier, 80 × 307 mm trapezoidal
    profile sheet
    HEA 120 steel sections, HEA 300 steel sections
12 Sheet aluminium, folded

13 Wall construction:
    profiled glass, 262 × 60 × 7 mm
    profiled glass, 262 × 41 × 6 mm
    HEB 120 steel sections, galvanised
14 Maintenance walkway, 100 × 50 mm steel hollow
    section frame, 30 mm open-grid flooring
15 Insulated panel: 2 mm sheet steel,
    38 mm thermal insulation, 2 mm sheet steel
16 Insulating glass: 2 No. 6 mm lam. safety glass +
    16 mm cavity + 10 mm tough. safety glass, vertical
    joints sealed with silicone, 60 × 20 mm aluminium
    capping strip to horizontal joint, aluminium post-
    and-rail facade, 150 × 60 mm aluminium hollow
    sections, HEM 400 steel columns

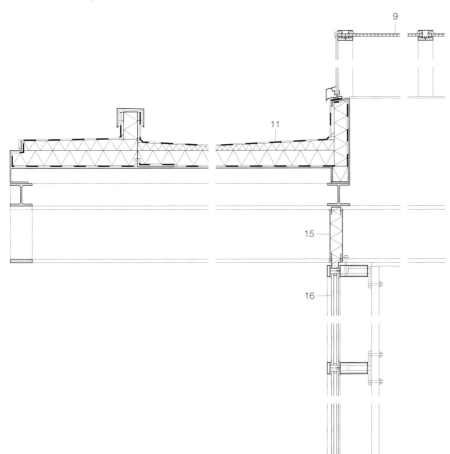

Example 02

Concert hall

Raiding, Austria, 2006

Architects:
Atelier Kempe Thill, Rotterdam
Andre Kempe, Oliver Thill
Project team:
Saskia Hermanek, Sebastian Heinemeyer,
Cornelia Sailer, David van Eck, Andre Boucsein,
Takashi Nakamura, Kingman Brewster, Frank
Verzijden
Structural engineers:
Vasko Woschitz Engineering, Eisenstadt
Building physics consultants:
Bölcskey + Scherpke, Vienna

A concert hall in memory of Franz Liszt has been
built in his birthplace, the village of Raiding not
far from Austria's border with Hungary The
building is totally tailored to his music and has
a capacity of about 600 – almost the entire
population of this little village! The minimalistic
design with its white cladding is hidden behind
the walls and trees of a small park. Two large
openings, one 13 m wide, the other 18 m, link
the foyer with the landscape on two sides. Only
a closer look reveals that conventional panes
of glass could not span this distance without
some form of intermediate support. And indeed,
it is not glass at all that bonds interior and exterior here, but rather 50 mm thick transparent
acrylic sheets.
The facade design takes its cue from the rendered facades of the locality; but the aim was
to create a more noble, more modern form –
without being too expensive. The design team
therefore opted for a cost-effective external
thermal insulation composite system with a
sprayed plastic finish that represents a new
variation on the traditional structure. The sprayed
polyurethane coating provides not only a decorative finish to the facade, but also forms the
waterproofing to the roof – a seamless, jointless
covering to the entire building in fact. Thanks to
its tear resistance and ability to bridge over
small cracks, it can accommodate the small
movements of the construction and forms a full
bond with the substrate which withstands
mechanical actions and prevents moisture
seeping underneath. Polyurethane coatings are
quite common on horizontal roof surfaces, but
on vertical surfaces wind suction represents a
considerable risk for the lightweight construction. The coating was therefore attached with
an especially strong adhesive. The thick, coarse
coat of render underneath was deliberately left
uneven to create an animated play of light and
shadow on the shiny surface.

• Sprayed polyurethane waterproofing
• No protective finish

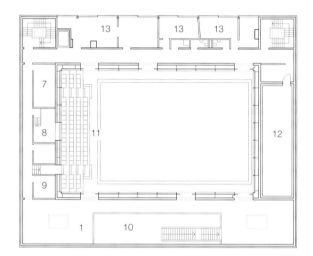

Plans · Sections
Scale 1:500

1 Foyer
2 Box office/bar
3 Cloakroom
4 Chamber music hall
5 Piano/chairs store
6 Small foyer
7 Storage
8 Stage equipment
9 Plant room
10 Void
11 Balcony
12 Plant and
   equipment yard
13 Dressing room

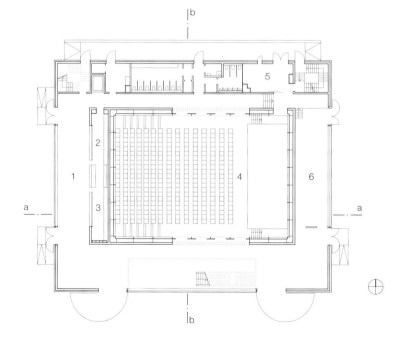

aa

bb

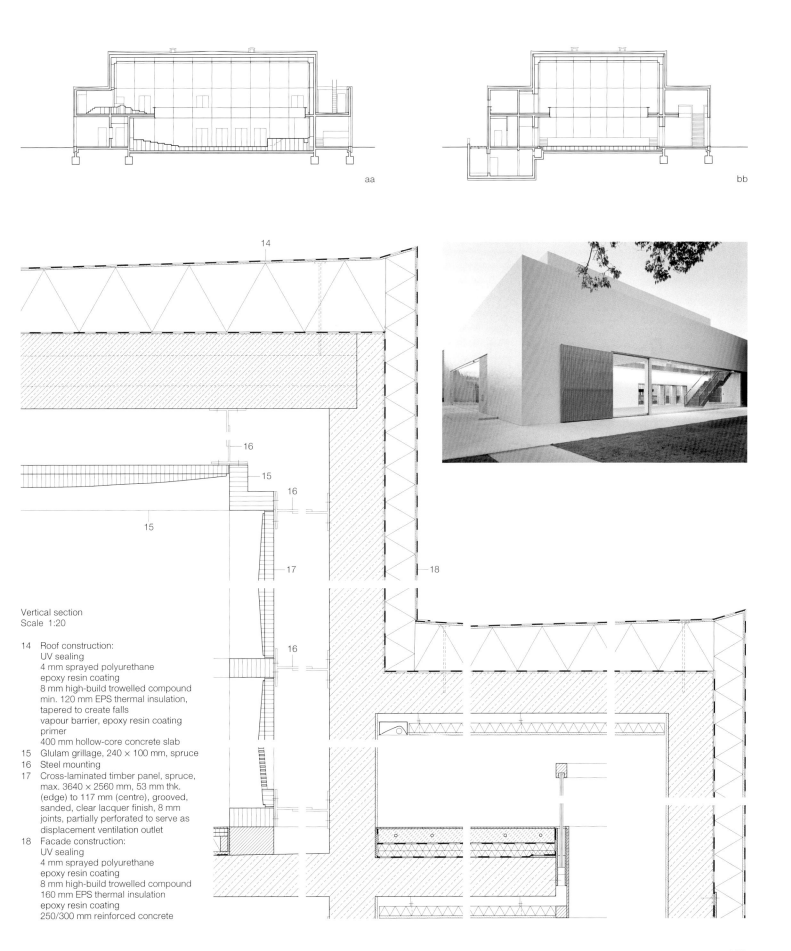

Vertical section
Scale 1:20

14 Roof construction:
UV sealing
4 mm sprayed polyurethane
epoxy resin coating
8 mm high-build trowelled compound
min. 120 mm EPS thermal insulation,
tapered to create falls
vapour barrier, epoxy resin coating
primer
400 mm hollow-core concrete slab
15 Glulam grillage, 240 × 100 mm, spruce
16 Steel mounting
17 Cross-laminated timber panel, spruce,
max. 3640 × 2560 mm, 53 mm thk.
(edge) to 117 mm (centre), grooved,
sanded, clear lacquer finish, 8 mm
joints, partially perforated to serve as
displacement ventilation outlet
18 Facade construction:
UV sealing
4 mm sprayed polyurethane
epoxy resin coating
8 mm high-build trowelled compound
160 mm EPS thermal insulation
epoxy resin coating
250/300 mm reinforced concrete

Example 03

America's Cup Building

Valencia, Spain, 2006

Architects:
David Chipperfield Architects, London
b720 Arquitectos, Fermín Vázquez, Barcelona
Project team:
Marco de Battista, Mirja Giebler, Jochen Glemser,
Regina Gruber, David Gutman, Melissa Johnston,
Andrew Phillips, Antonio Buendía, Peco Mulet,
Amparo Casaní, Lorena Lindberg, Magdalena
Ostornol, Sebastían Khourian
Structural engineers:
Brufau, Obiol, Moya & Associats S. L., Barcelona

"Veles e Vents" – sails and winds – is the name
that has been given to David Chipperfield's
radiant white landmark in the port of the Spanish
city of Valencia. Completed in a record time of
just 11 months after winning the design compe-
tition, in time for the qualifying regattas, the
building was a big attraction during the 2006
America's Cup itself and was able to present its
advantages to best effect: with public observation decks, bars, restaurants, lounges, VIP
facilities and management offices, it acted as
the central service point throughout this premier
event. From the urban planning viewpoint, this
building occupies a key position in the restruc-
turing of a port that has been vegetating for
many years and which now, step by step, is
being turned into a public space with an essentially commercial character.
With the help of straightforward, unambiguous
forms and a design reduced to just a few critical
details, the result is a building with a concise,
unadorned stature. The appearance is dominated by the use of just a few materials: white-
painted steel encasing the concrete loadbearing
structure, glass for the facades and spandrel
panels, dark hardwood for the floor finish on
the observation decks. A spectacular ramp
provides access to the first floor, which is open
to the public, whereas VIPs enter the building
via the foyer at ground floor level and are transported to the upper floors in lifts. The most
characteristic feature, however, is the cantilevering floors and roof, which function as visitor terraces and observation platforms and almost
seem as if they are floating in the air. They
screen the facade and the areas below against
the sun, but most importantly they frame the
view. The result is magnificent panoramic views,
lending a somewhat dramatic effect to what are
actually rather unspectacular views of port basins
and container terminals.

• Long cantilevering platforms
• Rooftop terraces

Plans • Sections
Scale 1:1250

1   Reception
2   Bar
3   Restaurant
4   Kitchen
5   Store
6   Access ramp
7   Boutique
8   Observation deck
9   Public viewing platform
10 Restaurant terrace
11 Wellness facilities
12 Club

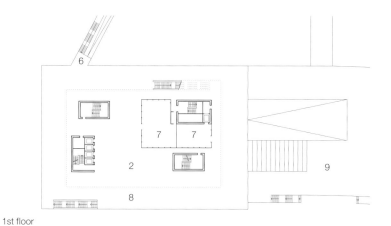

2nd floor

1st floor

Ground floor

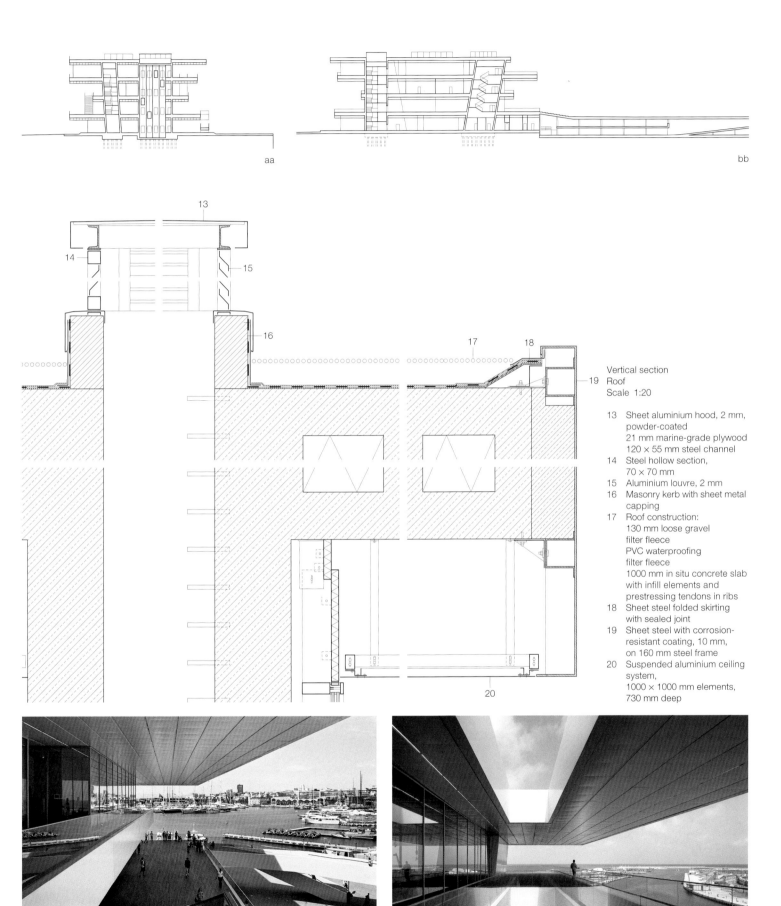

aa

bb

Vertical section
Roof
Scale 1:20

13 Sheet aluminium hood, 2 mm,
    powder-coated
    21 mm marine-grade plywood
    120 × 55 mm steel channel
14 Steel hollow section,
    70 × 70 mm
15 Aluminium louvre, 2 mm
16 Masonry kerb with sheet metal
    capping
17 Roof construction:
    130 mm loose gravel
    filter fleece
    PVC waterproofing
    filter fleece
    1000 mm in situ concrete slab
    with infill elements and
    prestressing tendons in ribs
18 Sheet steel folded skirting
    with sealed joint
19 Sheet steel with corrosion-
    resistant coating, 10 mm,
    on 160 mm steel frame
20 Suspended aluminium ceiling
    system,
    1000 × 1000 mm elements,
    730 mm deep

Example 03

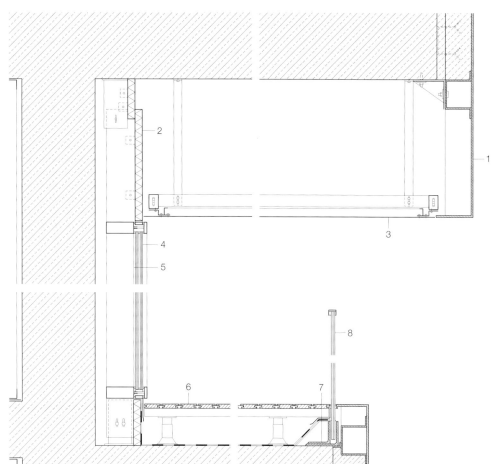

Vertical section
Observation deck
Scale 1:20

1 Sheet steel with
  corrosion-resistant
  coating, 10 mm, on
  160 mm steel frame
2 Insulated panel, 40 mm
3 Suspended aluminium
  ceiling system,
  1000 × 1000 mm elem-
  ents, 730 mm deep
4 Insulating glass: 2 No.
  6 mm lam. safety glass
  + 12 mm cavity + 2 No.
  8 mm lam. safety glass
5 Post-and-rail construc-
  tion, 60 × 150 mm
  steel hollow sections

6 Floor construction:
  22 mm hardwood floor-
  boards
  45 mm pine battens
  height-adjustable plastic
  pedestals
  waterproofing
  1000 mm in situ con-
  crete slab with infill
  elements and pre-
  stressing tendons in
  ribs
7 Steel angle,
  60 × 60 mm
8 Balustrade: 2 No.
  10 mm lam. safety
  glass with stainless
  steel handrail

MAXXI Museum

Rome, Italy, 2010

Architects:
Zaha Hadid Architects, London
Zaha Hadid and Patrik Schumacher, Rome
Project team:
Gianluca Racana (project manager),
Paolo Matteuzzi, Anja Simons, Mario Mattia
(site manager)
Structural engineers:
Anthony Hunt Associates, London
OK Design Group, Rome
Studio S.P.C. S.r.l., Giorgio Croci, Aymen
Herzalla, Rome (detailed design work)

It was in the spring of 2010 that MAXXI, Italy's
new national museum for the art and architecture
of the 21st century, opened on the site of a
former military barracks in the north of Rome,
not far from Pier Luigi Nervi's Palazzetto dello
Sport and Renzo Piano's Parco della Musica.
Rome had for a long time dedicated itself to the
past in particular and now it was time to establish
a venue for cultural innovation, in terms of both
building and content.
Interlinked elongated blocks begin at an existing
building and snake across the L-shaped site of
the former barracks, following the main lines of
the urban axis. This low-rise concrete sculpture
therefore fits surprisingly neatly into its surroundings. When viewed from the air, it is the dynamic
of the roof that particularly catches the eye. The
roof is divided up by glass ribbons that give the
impression of movement as they trace the shape
of the building below.
The gallery areas within the building reveal the
key ingredient of the design concept, which is
based on the elements so critical for such a
museum: lighting and wall surfaces. Concrete
walls, which are designed as deep beams
spanning 30 m without intermediate supports,
define the framework of the design. An internal
lining offers the works of art a neutral back-
ground and at the same time conceals all the
services necessary for the operation of the
museum. This leaves the soffits free for carefully shaped rooflights from which works of art
or partitions can be suspended as required. At
the same time, the rooflights ensure a good supply of daylight for the exhibition rooms below. The
soffit of the roof is defined by panels made from
glass fibre-reinforced concrete. They are reminiscent of rails and therefore emphasize the flow
of the elongated galleries.

· Continuous rooflights
· Roof as architectural element

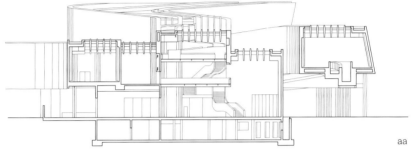

aa

Plan
Scale 1:1500
Section
Scale 1:750

1 Foyer
2 Reception
3 Café
4 Auditorium
5 Shop
6 Graphics
  collection
7 Temporary exhibitions
  (existing building)
8 Exhibition room 1

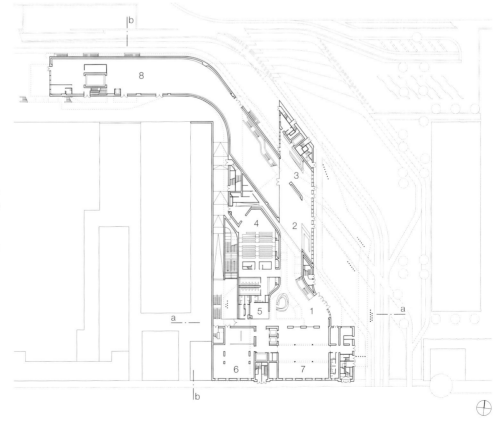

Example 04

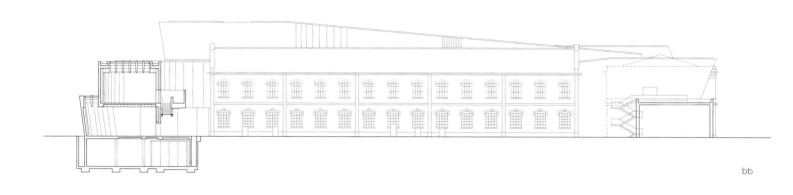

bb

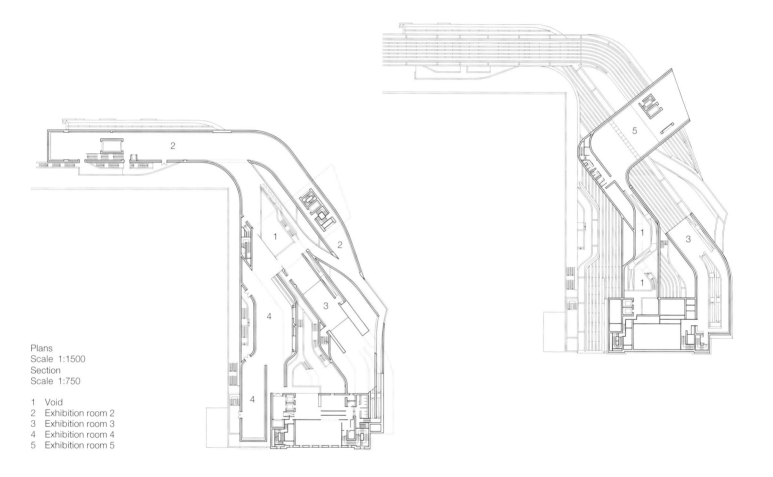

Plans
Scale 1:1500
Section
Scale 1:750

1 Void
2 Exhibition room 2
3 Exhibition room 3
4 Exhibition room 4
5 Exhibition room 5

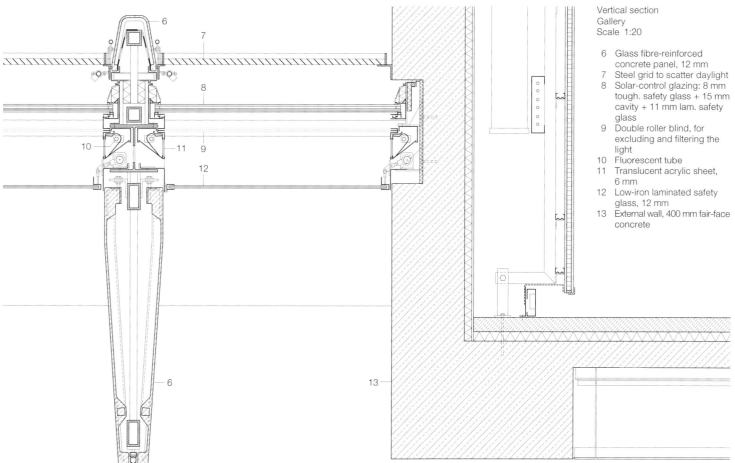

Vertical section
Gallery
Scale 1:20

6 Glass fibre-reinforced
concrete panel, 12 mm
7 Steel grid to scatter daylight
8 Solar-control glazing: 8 mm
tough. safety glass + 15 mm
cavity + 11 mm lam. safety
glass
9 Double roller blind, for
excluding and filtering the
light
10 Fluorescent tube
11 Translucent acrylic sheet,
6 mm
12 Low-iron laminated safety
glass, 12 mm
13 External wall, 400 mm fair-face
concrete

Example 04

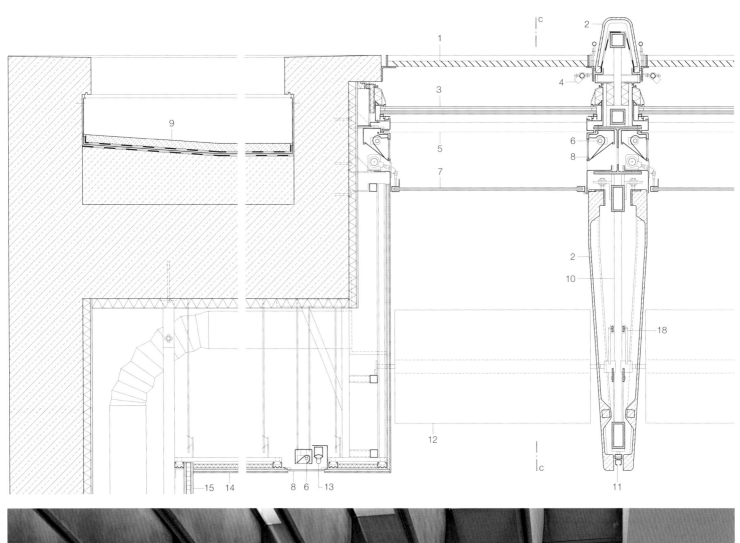

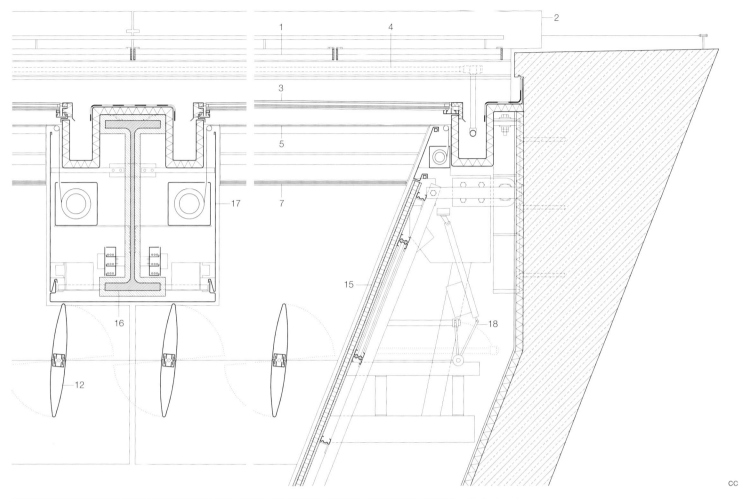

cc

Vertical sections
1st floor gallery
Scale 1:20

1 Glass fibre-reinforced concrete panel, 12 mm
2 Steel grid to scatter daylight, galvanised and painted
3 Solar-control glazing: 8 mm tough. safety glass + 15 mm cavity + 11 mm lam. safety glass
4 Automatic glass cleaning system
5 Blackout roller blind
6 Fluorescent tube
7 Translucent acrylic sheet, 6 mm, for scattering the daylight
8 Low-iron laminated safety glass, 12 mm, can be opened mechanically for cleaning and maintenance, 3 No. panes each 600 mm in aluminium frames (open joints for ventilation)
9 Roof construction:
screed laid to falls
waterproofing
sheet steel
waterproofing
screed laid to falls
10 Steel truss
11 Track for suspending works of art or partitions
12 Swivelling aluminium louvres for controlling incoming daylight
13 Fluorescent tube for emergency lighting
14 Ceiling:
5 mm sprayed acoustic plaster
12.5 mm perforated plasterboard
20 mm acoustic board
15 Lining:
12.5 mm glass fibre-reinforced plasterboard
25 mm MDF
12.5 mm glass fibre-reinforced plasterboard
galvanised steel framework
16 Steel beam, HEM 900, with fire-resistant coating
17 Coated sheet aluminium casing
18 Linear spindle drive for louvres, electric

Example 05

Private house

London, UK, 2007

Architects:
Adjaye Associates, London
David Adjaye
Project team:
Rashid Ali, Yohannes Bereket,
Candida Correa de Sa, Nikolai Delvendahl,
Cornelia Fischer-Ekhorn
Structural engineers:
Eurban Construction, London

This client's childhood dream of owning his own house came true when he managed to purchase one of the rare and expensive building plots in London's East End. The simple style of the new building clad in black-stained wood fits in with the workshops opposite, but its dimensions match those of the Victorian clay brick semi-detached houses of this neighbourhood. However, this house has one extra storey because the whole plot has been lowered. The private, sunken courtyard offers a certain degree of intimacy and allows the kitchen and dining area to be extended to the outside. Contrary to the typical British home, the bedrooms are at road level, i.e. the first floor in this house, and the living room is on the topmost floor. A hatch provides access to the platform on the flat roof, from where there is a view across the rooftops of London's East End.

The entire timber loadbearing structure was prefabricated in Germany. Two goods vehicles transported the individual components to London and the structure was erected within two days. The cladding of cedar wood planks treated with linseed oil was attached in situ. Ventilation flaps, doors and window openings are integrated flush with the dark cladding, thus accentuating the homogeneous overall impression. As the building is designed with loadbearing solid members on all sides, it was possible to position the large, frameless openings virtually anywhere, without having to consider spans and loadbearing supports. From the windows in the facades there is a view of the existing nearby trees; the openings in the roof frame the sky and allow the evening sunlight to reach deep into the interior of the house. So despite the introvert nature of this building, it still maintains sensitive contact with its surroundings.

• Rooftop platform
• Identical cladding on facade and roof

Section • Plans
Scale 1:250

1 Sunken courtyard
2 Kitchen
3 Dining
4 Studio
5 Bedroom
6 Storage
7 Entrance
8 Living room

aa

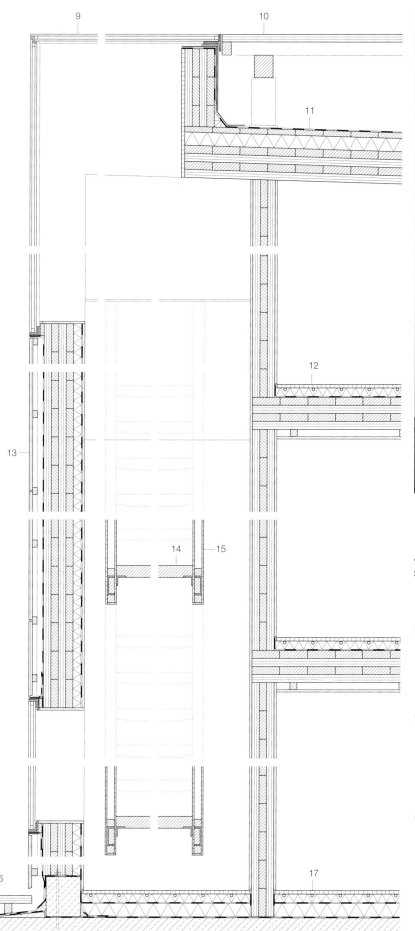

Vertical section
Scale 1:20

9 Insulating glass glued to
   stainless steel angle frame,
   50 × 50 mm
10 Roof construction:
   40 mm cedar wood planks,
   surface fluted, black stain finish
   50 × 70 mm cedar wood
   framing
   100 × 120 mm spruce beam
   spacer block
   rubber granulate bearing pad
11 Roof deck:
   waterproofing
   prefabricated solid timber
   deck laid to falls:
   30 mm decking
   70 mm hemp thermal insulation
   160 mm cross-laminated
   timber, planed spruce, white
   paint finish
12 Suspended floor construction:
   18 mm plywood, white paint
   finish
   50 mm insulation with underfloor heating
   separating membrane
   Prefabricated solid timber deck:
   170 mm cross-laminated timber,
   spruce
   38 × 38 mm softwood battens
   12.5 mm plasterboard

13 Wall construction:
   20 mm cedar wood planks,
   surface fluted, black stain
   finish, 5 mm open joints
   25 × 38 mm battens
   38 × 38 mm counter battens
   waterproofing
   Prefabricated solid timber wall:
   160 mm cross-laminated
   timber, spruce
   50 mm fire-resistant rigid foam
   insulation with integral vapour
   barrier
   12.5 mm plasterboard
14 Stair tread, 63 mm walnut,
   stained
15 Balustrade, 9 mm plywood,
   walnut veneer finish
16 Floor construction, terrace:
   40 mm cedar wood boards,
   surface fluted, black stain finish
   40 × 60 mm pretreated
   softwood battens
   rubber granulate bearing pad
   waterproofing
   250 mm reinforced concrete
   ground slab
17 Floor construction, ground floor:
   18 mm plywood, white paint
   finish
   50 mm insulation with
   underfloor heating
   separating membrane
   70 mm rigid foam insulation

Example 06

Production building and retail premises

Eitensheim, Germany, 1996

Architects:
Homeier + Richter, Munich
Assistant:
Thomas Bauer
Structural engineers:
Grad Ingenieurplanungen, Ingolstadt

Brandl, a mid-size metalworking company based in Eitensheim in Upper Bavaria, has extended its facilities in two directions. The floor area of the existing production building was doubled by simply continuing the proven loadbearing structure of steel lattice beams further out into the spacious site; a wide, continuous rooflight marks the boundary between the old and new parts of the production building. And facing the road, a two-storey extension was built to extend the retail zone. The lightweight steel roof here consists of loadbearing, long-span, thermally insulated sandwich elements. Plain sheet steel flanges form the tension and compression zones of the roof construction; these are stiffened to prevent buckling and are connected by way of thin shear-resistant webs. Combining loadbearing elements of steel results in long spans and also long cantilevers with a low structural depth because each individual component performs multiple tasks. The upper sheet metal flange is simultaneously loadbearing element and roof covering. The lower sheet metal flange forms the ceiling in the retail premises and enables indirect lighting via its reflective surface. In addition, it acts as a vapour barrier and includes integral heating elements whose radiant heat output has a positive effect on the interior climate. Cooling units can be installed as an alternative. The low structural depth of the roof due to the optimisation of thermal insulation and the ensuing low roof load enabled the columns in the interior to be kept slender and delicate.

This roof design is the result of close cooperation between architect, client and structural engineer, and is the subject of a patent application. The great advantage of the system lies in its high degree of prefabrication. Almost all the operations could be tested on the client's premises unaffected by the weather; a number of components can be manufactured in long production runs.

• Metal roof construction
• Heating elements

aa

Section · Plans
Scale 1:750

1 Retail premises
2 Showroom
3 Storage
4 Office
5 Deliveries
6 Staff amenities
7 Production building

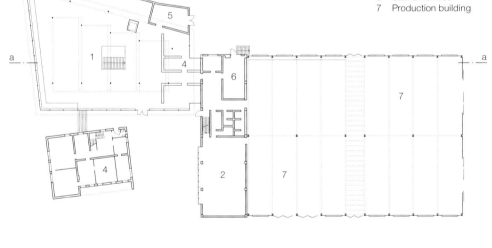

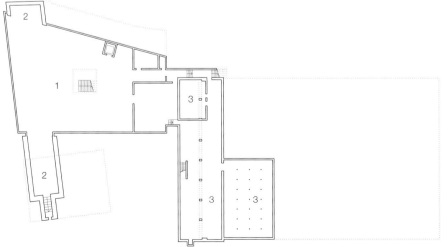

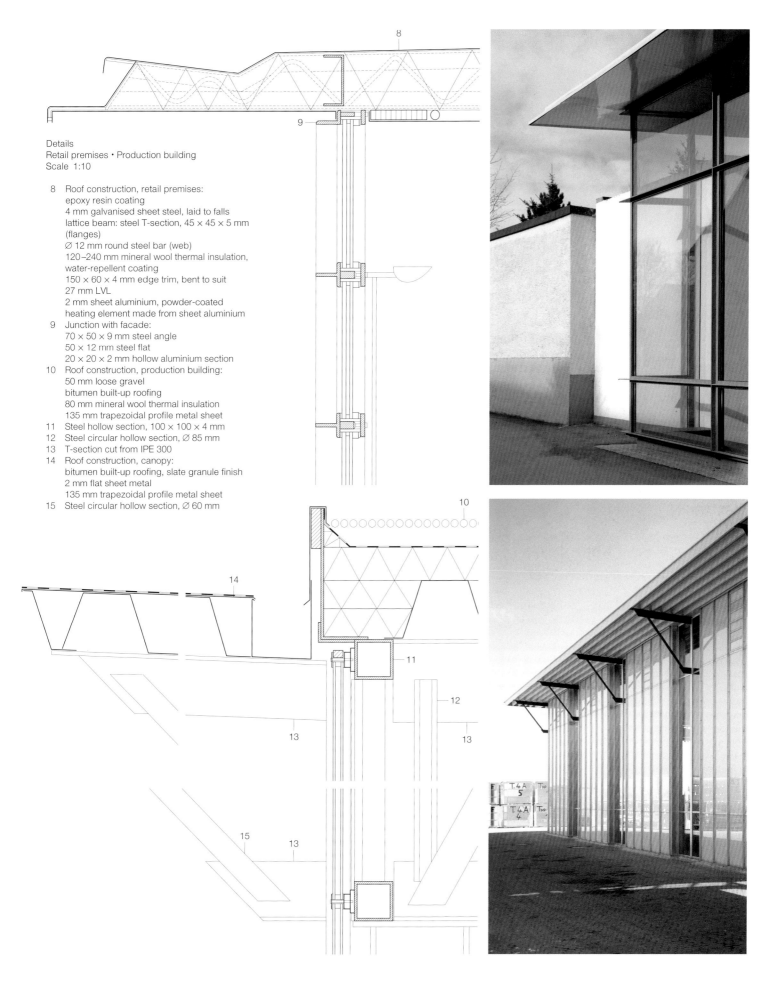

Details
Retail premises · Production building
Scale 1:10

8 Roof construction, retail premises:
epoxy resin coating
4 mm galvanised sheet steel, laid to falls
lattice beam: steel T-section, 45 × 45 × 5 mm
(flanges)
⌀ 12 mm round steel bar (web)
120–240 mm mineral wool thermal insulation,
water-repellent coating
150 × 60 × 4 mm edge trim, bent to suit
27 mm LVL
2 mm sheet aluminium, powder-coated
heating element made from sheet aluminium
9 Junction with facade:
70 × 50 × 9 mm steel angle
50 × 12 mm steel flat
20 × 20 × 2 mm hollow aluminium section
10 Roof construction, production building:
50 mm loose gravel
bitumen built-up roofing
80 mm mineral wool thermal insulation
135 mm trapezoidal profile metal sheet
11 Steel hollow section, 100 × 100 × 4 mm
12 Steel circular hollow section, ⌀ 85 mm
13 T-section cut from IPE 300
14 Roof construction, canopy:
bitumen built-up roofing, slate granule finish
2 mm flat sheet metal
135 mm trapezoidal profile metal sheet
15 Steel circular hollow section, ⌀ 60 mm

Example 07

Sports hall

Bietigheim-Bissingen, Germany, 2003

Architects:
Auer + Weber + Architekten, Stuttgart
Project team:
Felix Wiemken, Jürgen Weigl
Structural engineers:
Mayr + Ludescher, Stuttgart (concept)
Arge Müller & Merkle, Bietigheim-Bissingen

This new sports hall closes off the last open side of an ensemble of school buildings dating from the 1950s. The design blends in harmoniously with the rest of the complex: the ancillary rooms and the playing area itself are below ground level, which means that this large structure appears to be only one storey high. Glazing on three sides permits a view through from the road to the central open area between the school buildings, and vice versa. The main entrance is from the school side, through a glazed pavilion alongside the main building. The roof structure, which measures 45 × 25.5 m, is a steel-timber composite construction. In this variation on a monitor roof, the monitors have glazing on the north side only, but this still allows a generous amount of daylight to illuminate the playing area evenly from above. Sheet stainless steel has been used for the roof covering.

Six main beams of glued laminated timber span across the playing area. These form the closed sides of the monitors and are supported in the plane of the facade on steel circular hollow section columns, which also serve as rainwater downpipes. The secondary beams, in the shape of an elongated Z-frame, are made up of welded HEB steel sections. These span in the longitudinal direction between the main beams, forming a structural grid together with these. Stability in the transverse direction is guaranteed by the roof plate and the monitors; X-bracing behind the glazing and channel sections between the secondary beams help to withstand compressive forces in the transverse direction. Steel circular hollow sections along the perimeter walls brace the building in the longitudinal direction.

The ceiling – which seems to float between the monitors – consists of prefabricated timber box elements with a white glaze finish. These are suspended between the secondary beams and ensure a flush soffit.

- Stainless steel sheet roof covering
- Monitor-type roof

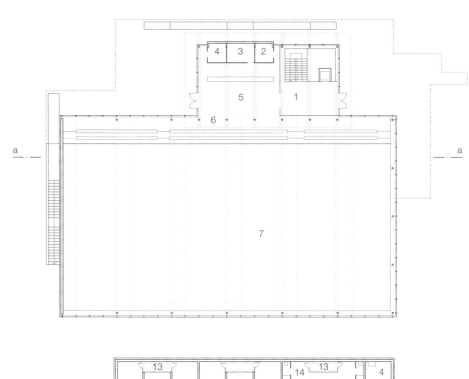

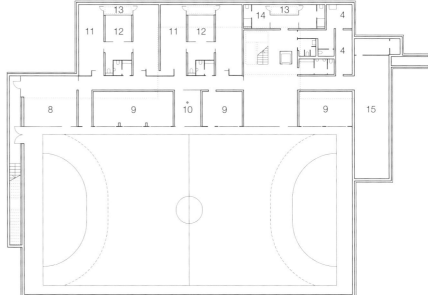

Location plan
Scale 1:3000
Plans · Longitudinal section
Scale 1:500

1   Entrance
2   Caretaker
3   Café storeroom
4   Storage
5   Café
6   Spectators area (seating for 148)
7   Void over playing area
8   Multi-purpose room
9   Equipment store
10 Umpires and officials
11 Changing room
12 Showers
13 Lightwell
14 Teachers' changing room
15 Plant room

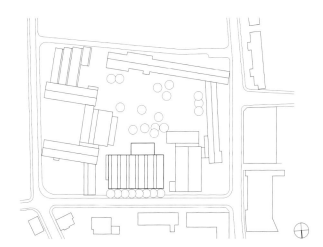

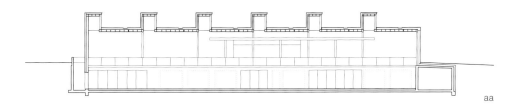

aa

Example 07

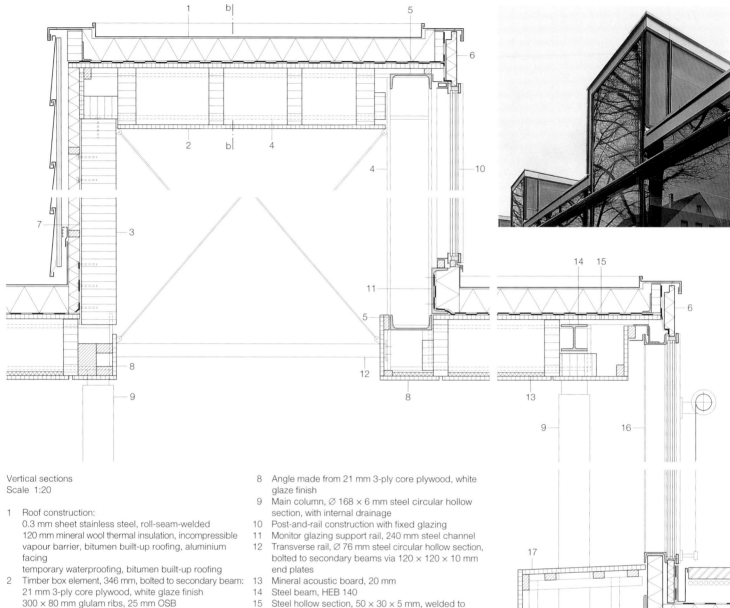

Vertical sections
Scale 1:20

1 Roof construction:
0.3 mm sheet stainless steel, roll-seam-welded
120 mm mineral wool thermal insulation, incompressible vapour barrier, bitumen built-up roofing, aluminium facing
temporary waterproofing, bitumen built-up roofing
2 Timber box element, 346 mm, bolted to secondary beam:
21 mm 3-ply core plywood, white glaze finish
300 × 80 mm glulam ribs, 25 mm OSB
3 Main beam: 1850 × 200 mm glulam, facing quality on inside, white glaze finish
4 Secondary beam: Z-frame of HEB 240 steel sections, rigid welded connections
5 Strut between secondary beams: 220 mm steel channel bolted via end plates
6 Panel, 2 mm sheet stainless steel and insulation
7 Secondary beam fixed to main beam via
1550 × 240 × 8 mm steel flat, with M20 bolts

8 Angle made from 21 mm 3-ply core plywood, white glaze finish
9 Main column, ⌀ 168 × 6 mm steel circular hollow section, with internal drainage
10 Post-and-rail construction with fixed glazing
11 Monitor glazing support rail, 240 mm steel channel
12 Transverse rail, ⌀ 76 mm steel circular hollow section, bolted to secondary beams via 120 × 120 × 10 mm end plates
13 Mineral acoustic board, 20 mm
14 Steel beam, HEB 140
15 Steel hollow section, 50 × 30 × 5 mm, welded to HEB for fixing facade with 2 steel angles
16 Facade construction:
IPE 120 steel post, post-and-rail construction with fixed glazing
sunblind, plastic-coated, silver, light-permeable
17 Impact-resistant lining: 21 mm 3-ply core plywood, white glaze finish

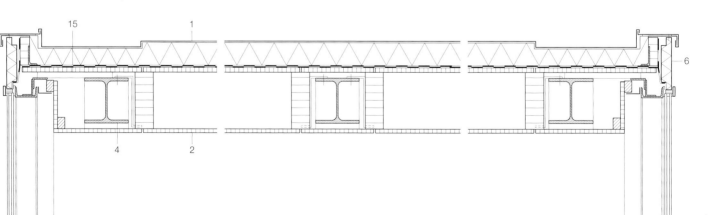

bb

Cité du Design

Saint-Étienne, France, 2009

Architects:
LIN Finn Geipel and Giulia Andi, Berlin/Paris,
with Cabinet Berger, Saint-Étienne
Project team:
Stefan Jeske, Philip König, Jacques Cadilhac,
Jan-Oliver Kunze, Judith Stichtenoth, François
Maisonnasse, Muriel Poncet, Marielle Gilibert,
Heiko Walth
Structural engineers:
Werner Sobek + Thomas Winterstetter, WSI,
Stuttgart/New York

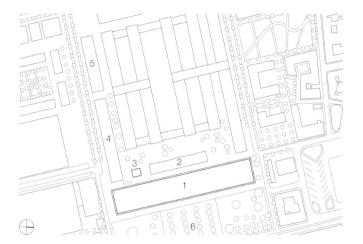

Location plan
Scale 1:5000

1 "Platine" design centre
2 Offices/management/
  biennale
3 Viewing tower
4 Technical studios
5 Teaching studios
6 Place d'Armes

A former armaments factory, now protected by
a conservation order, in the north of Saint-Étienne
provides the location for this new design centre
for the French region of Rhône-Alpes. To achieve
this change of use, three of the factory buildings
were converted into seminar rooms and workshops for the Art & Design Academy as well as
offices and apartments for guests. But the real
eye-catcher at the Cité du Design is the new
200 × 32 m multi-purpose building, dubbed the
"Platine" (plate). This flat, low structure unites
the key functions of the art and design complex:
exhibition areas, auditorium, media centre with
materials library, showroom, restaurant and
unheated glasshouse.
The steel space frame of the roof is reminiscent
of factory architecture and renders internal
columns unnecessary. The envelope, both
facade and roof, consists of about 14 000 triangular panels with a side length of 1.2 m in 11
different versions – from aluminium sandwich
panel to glazing with integral solar cells. These
panels are positioned to suit the various interior
requirements. They filter light, absorb or transform it into energy, and regulate the ventilation
and heating requirements as required. In addition, the envelope provides a field for experiments
with new technological developments and will
be equipped with photosynthesis modules in
the near future. The architects' concept includes
the replacement of facade and roof panels at
any time, to suit changing internal requirements
or to take into account innovative technologies.
Sensors have been installed to enable this climatic
experiment to be monitored for a period of two
years by the Environment & Energy Management
Agency, the French energy company EDF and
the Scientific Centre for Building Technology.

• Steel structure
• Adaptive envelope made up of modules with
  different functions

Example 08

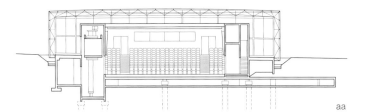

aa

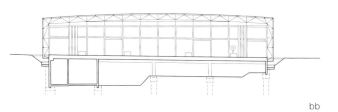

bb

Sections	1	Seminar room
Scale 1:500	2	Auditorium
Section · Plan	3	Exhibition
Scale 1:1000	4	Showroom, cafeteria
	5	Restaurant
	6	Glasshouse
	7	Media library

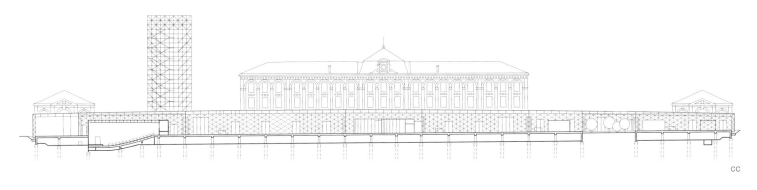

cc

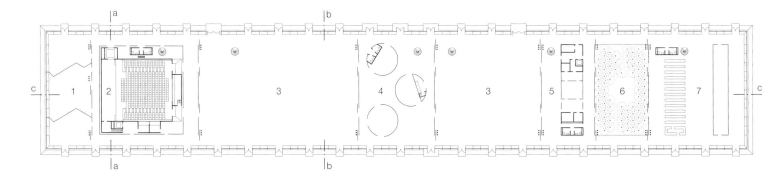

Vertical section
Scale 1:20

Details of modules
Scale 1:5

8 Opaque module:
   6 mm sheet aluminium,
   anodised
   50 mm mineral wool
   0.75 mm galvanised
   sheet steel
   50 mm mineral wool
   white felt
   1.5 mm perforated sheet
   aluminium
9 Opening module:
   2 No. 2 mm anodised
   sheet aluminium, with
   100 mm mineral wool in
   between
10 Transparent insulating
   glass module: 6 mm
   float glass + 16 mm
   cavity + 2 No. 5 mm
   lam. safety glass
11 Roof frame: 1200 mm
   equilateral triangle
   made from 80 × 50 mm
   steel hollow sections
12 Steel hollow section,
   180 × 60 mm
13 Smoke curtain, 6 mm
   single glazing (float glass)
   on aluminium angles
14 Transparent module:
   6 mm coated float glass
   + 16 mm cavity + 6 mm
   clear float glass
15 Facade frame: 1200 mm
   equilateral triangle
   made from 80 × 40 mm
   steel hollow sections
16 Curtain, glass-fibre fabric

A Opaque module: 2 mm anodised sheet aluminium, 50 mm
   mineral wool, 0.75 mm galvanised sheet steel, 50 mm mineral wool, white felt, 1.5 mm
   anodised sheet aluminium,
   perforated
B Coloured insulating glass
   module: 6 mm float glass +
   16 mm argon-filled cavity +
   2 No. 6.4 mm lam. safety
   glass with coloured foil
C Experimentation module:
   2 No. 5.5 mm lam. safety
   glass, aluminium frame,
   modules interchangeable
   (e.g. photosynthesis module)
D Photovoltaic module: 6 mm
   float glass, 2 mm resin coating with 12.5 × 12.5 mm
   monocrystalline cells + 4 mm
   float glass + 16 mm argon-
   filled cavity + 2 No. 5.5 mm
   lam. safety glass
E Insulating glass module with
   louvres: 6 mm float glass +
   24 mm cavity, aluminium
   louvres + 2 No. 6.4 mm lam.
   safety glass
F Insulating glass module for
   controlling incoming daylight: 8 mm float glass, silk-
   screen printing on underside
   + 6 mm float glass, silk-
   screen printing on top, electric drive for sliding inner
   pane + 16 mm cavity + 2
   No. 6.4 mm lam. safety glass

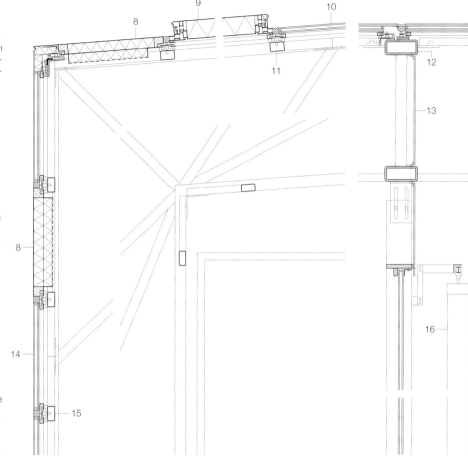

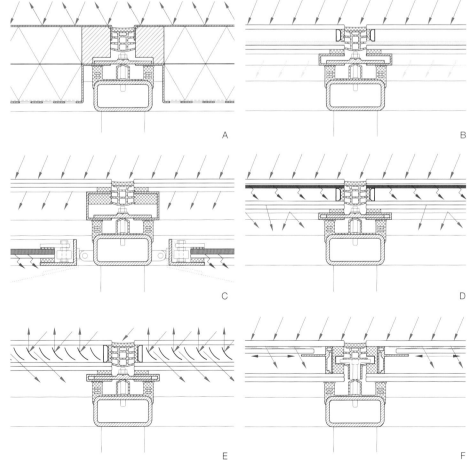

A

B

C

D

E

F

Example 09

Apartment building

Jona-Kempraten, Switzerland, 2004

Architects:
roos architekten, Rapperswil
Bernhard Roos
Structural engineers:
Horst Schuhmacher, Uetliburg (concrete)
Renggli AG, Schötz (timber)

This small housing development at the foot of a vineyard consists of two identical blocks each containing three maisonettes with rooftop terrace and two apartments with garden. Each apartment has a floor area of 180 m² and a flexible internal layout which can be determined by the occupants themselves. The large rooftop terraces of the maisonettes face north and south on one block, east and west on the other; the gardens of the ground-floor apartments include pergolas and covered seating areas. Car parking and storage rooms are located in the basement. Although the ground floor is divided into two units and the floors above into three, this is concealed from the observer by the simple, unsegmented facade. Parallel horizontal battens of Canadian red cedar clad the entire building and also form the balustrades to the rooftop terraces, which merge seamlessly into the main cladding. The large windows, seemingly precisely cut out of the facades to suit the interior layouts, are the only interruptions to the uniform wooden cladding.

Some of the soil excavated for the basement was reused on the green roofs. These are ventilated roofs (cold deck). Thermally insulated timber elements form a compact building envelope covering roof and external walls. Up to 40 cm of insulation has been installed in order to cut energy consumption. However, the basement, suspended floors and internal walls are all in concrete for structural or sound insulation reasons. "Cold" parts of the building (stair and lift shafts, basement parking) are rigorously separated from the "warm" parts (apartments) in order to avoid thermal bridges. These buildings comply with the Swiss "Minergie-P" standard, which is comparable with the German passive house standard.

· Ventilated green roof
· Rooftop terraces
· Passive house standard

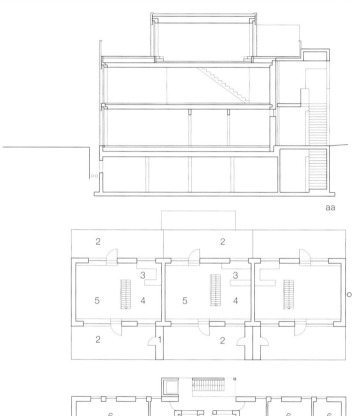

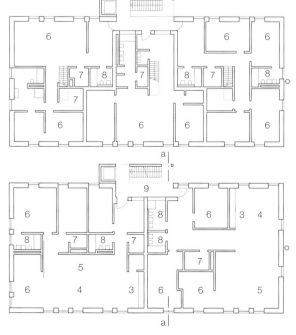

Section
Scale 1:250
Plans
Scale 1:400

1   Storage
2   Rooftop terrace
3   Kitchen
4   Dining
5   Living
6   Room
7   Boiler room
8   Bathroom
9   External stair to
    maisonette

aa

a

a

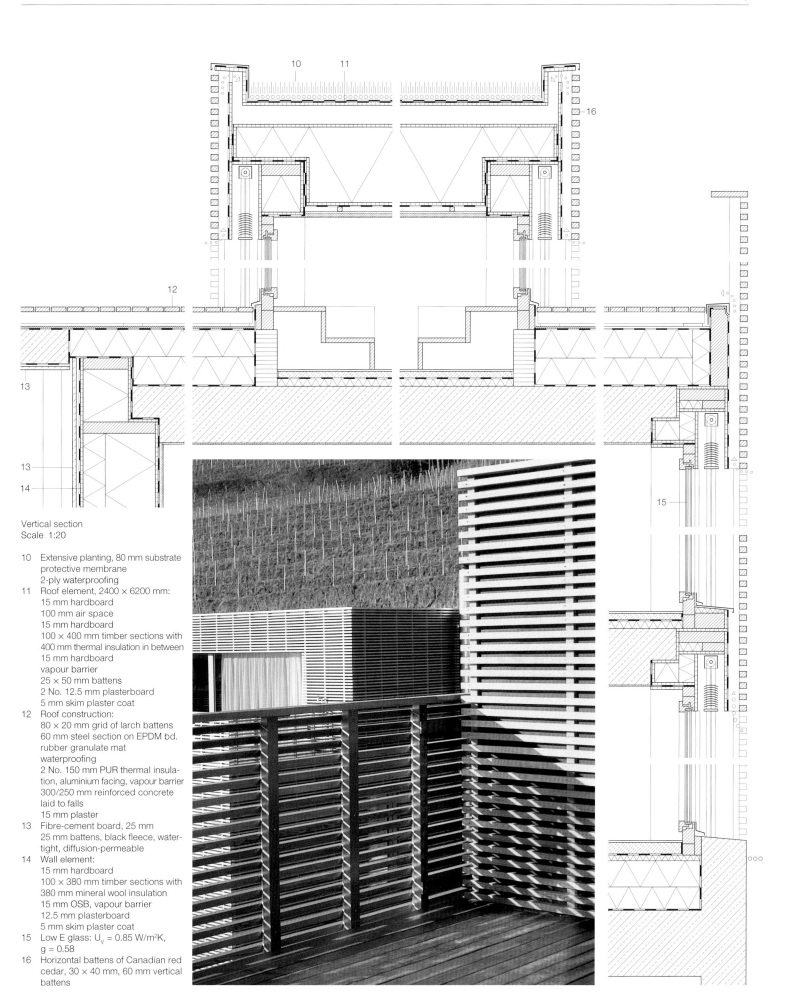

Vertical section
Scale 1:20

10 Extensive planting, 80 mm substrate
    protective membrane
    2-ply waterproofing
11 Roof element, 2400 × 6200 mm:
    15 mm hardboard
    100 mm air space
    15 mm hardboard
    100 × 400 mm timber sections with
    400 mm thermal insulation in between
    15 mm hardboard
    vapour barrier
    25 × 50 mm battens
    2 No. 12.5 mm plasterboard
    5 mm skim plaster coat
12 Roof construction:
    80 × 20 mm grid of larch battens
    60 mm steel section on EPDM bd.
    rubber granulate mat
    waterproofing
    2 No. 150 mm PUR thermal insula-
    tion, aluminium facing, vapour barrier
    300/250 mm reinforced concrete
    laid to falls
    15 mm plaster
13 Fibre-cement board, 25 mm
    25 mm battens, black fleece, water-
    tight, diffusion-permeable
14 Wall element:
    15 mm hardboard
    100 × 380 mm timber sections with
    380 mm mineral wool insulation
    15 mm OSB, vapour barrier
    12.5 mm plasterboard
    5 mm skim plaster coat
15 Low E glass: U_V = 0.85 W/m²K,
    g = 0.58
16 Horizontal battens of Canadian red
    cedar, 30 × 40 mm, 60 mm vertical
    battens

Example 10

Gallery

La Pizarrera, Spain, 2009

Architect:
Elisa Valero Ramos, Granada
Project team:
Leonardo Tapiz, Juan Fernández,
Jesús Martínez
Building services:
Luis Ollero

The shiny panes of glass placed in neat rows on the grass look like features of landscape architecture. It is hard to imagine that below them there is a private gallery containing important works from the Spanish art scene. Located just 10 km from the royal palace of El Escorial, this is the home of the art collector Plácido Arango. The two houses have now been joined by exhibition areas which will display his extensive collection.

To retain the oak trees in the garden, which are covered by a preservation order, the architects designed totally subterranean exhibition rooms – which at the same time fortuitously create a link between the two houses. The random plan form of the exhibition area is the outcome of the positions of the trees, the roots of which could not be damaged.

The greatest challenge of this project was supplying the exhibition with daylight. This was solved by including 45 virtually horizosntal rooflights, each measuring approx. 60 × 140 cm, which ensure uniform illumination of the works of art below. Steel frames are fitted into the openings in the flat roof in order to support the laminated safety glass flush with the grass, i.e. the rooftop planting. To protect against direct sunlight in summer, a system of aluminium louvres can be fitted into the openings temporarily.

• Green roof
• Rooflights flush with grass

Section · Plans
Scale 1:500

1 Living room
2 Office
3 Garden
4 Exhibition
5 Lobby
6 Kitchen
7 Plant room
8 Store
9 Goods lift

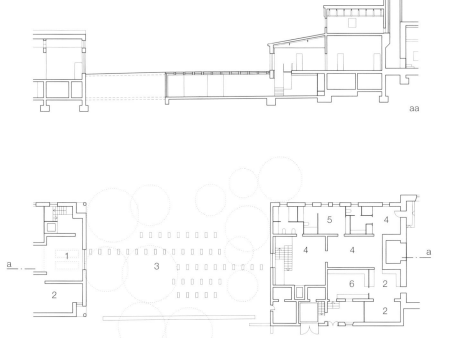

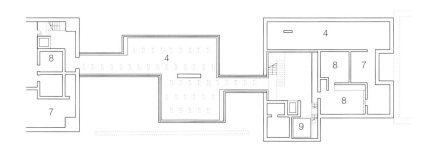

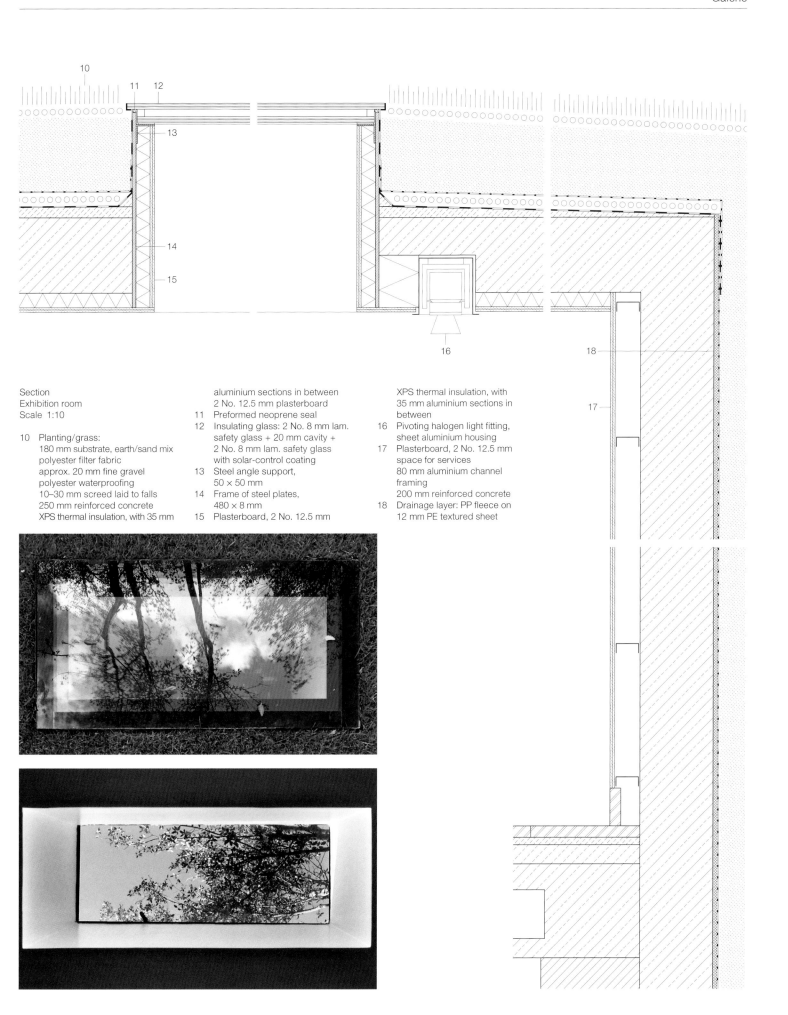

Section
Exhibition room
Scale 1:10

10 Planting/grass:
    180 mm substrate, earth/sand mix
    polyester filter fabric
    approx. 20 mm fine gravel
    polyester waterproofing
    10–30 mm screed laid to falls
    250 mm reinforced concrete
    XPS thermal insulation, with 35 mm

aluminium sections in between
2 No. 12.5 mm plasterboard
11 Preformed neoprene seal
12 Insulating glass: 2 No. 8 mm lam.
    safety glass + 20 mm cavity +
    2 No. 8 mm lam. safety glass
    with solar-control coating
13 Steel angle support,
    50 × 50 mm
14 Frame of steel plates,
    480 × 8 mm
15 Plasterboard, 2 No. 12.5 mm

XPS thermal insulation, with
35 mm aluminium sections in
between
16 Pivoting halogen light fitting,
    sheet aluminium housing
17 Plasterboard, 2 No. 12.5 mm
    space for services
    80 mm aluminium channel
    framing
    200 mm reinforced concrete
18 Drainage layer: PP fleece on
    12 mm PE textured sheet

Example 11

"Mountain dwellings"

Copenhagen, Denmark, 2008

Architects:
BIG – Bjarke Ingels Group, Copenhagen
Project team:
Jakob Lange, Hendrik Poulsen
Structural engineers:
Moe & Brødsgaard, Copenhagen

A new suburb of Copenhagen has been under development between the centre of the city and the airport since 1992. The plan is that this new district, Ørestad City, will become the centre of the Øresund region. The light rail rapid transit line, with driverless trains, directly alongside the development has been in operation since 2002. The "VM Bjerget" project (bjerg = mountain), an unconventional mix of parking and housing in the form of a stepped pyramid, marks the third phase of the development and is situated to the north of the "V" and "M" buildings. The client's brief specified a multi-storey car park over two-thirds of the site, housing over the other third. But instead of simply placing the buildings side by side, the architects decided to merge the two functions. The car park with 480 spaces covers the whole area of the plot and provides the foundation for the 80 apartments above. The boundary between public and private areas extends deep into the "mountain" – parking close to the apartments is possible even on the upper floors.
The apartments enjoy a good level of daylight and a roofscape with planting that varies depending on the season. There are 10 storeys of stepped-back apartments facing south, permitting views of the surroundings and the nearby town of Tårnby. Every apartment has a rooftop terrace finished with ipe wood floorboards and artificial grass. Projections at the corners and deep parapets guarantee privacy for the occupants. The wooden boards on the parapets conceal plant tubs in which ivy, clematis and honeysuckle have been planted. Irrigation is carried out by means of a central system of pipes. In contrast to the facade of the housing part of the development, the car park levels are clad with aluminium which is perforated in such a way that it reproduces the outline of the Mount Everest massif!

• Rooftop terraces
• Parapets with planting

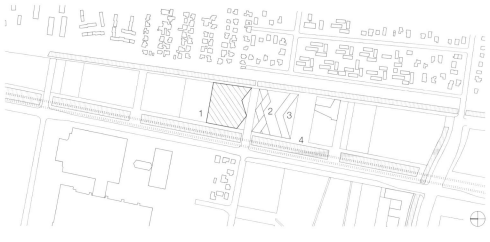

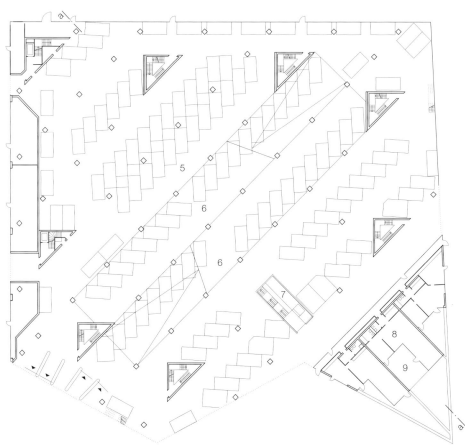

Location plan
Scale 1:8000
Plan · Section
Scale 1:750

1 "VM Bjerget"
2 "M" House
3 "V" House
4 Light rail rapid
  transit line
5 Car park
6 Ramp
7 Inclined lift
8 Apartment
9 Terrace

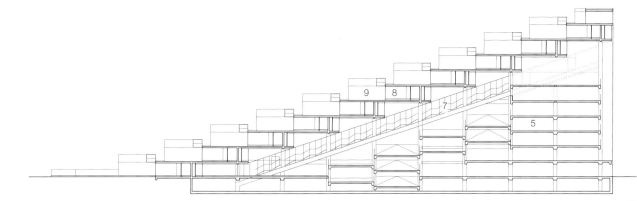

aa

Example 11

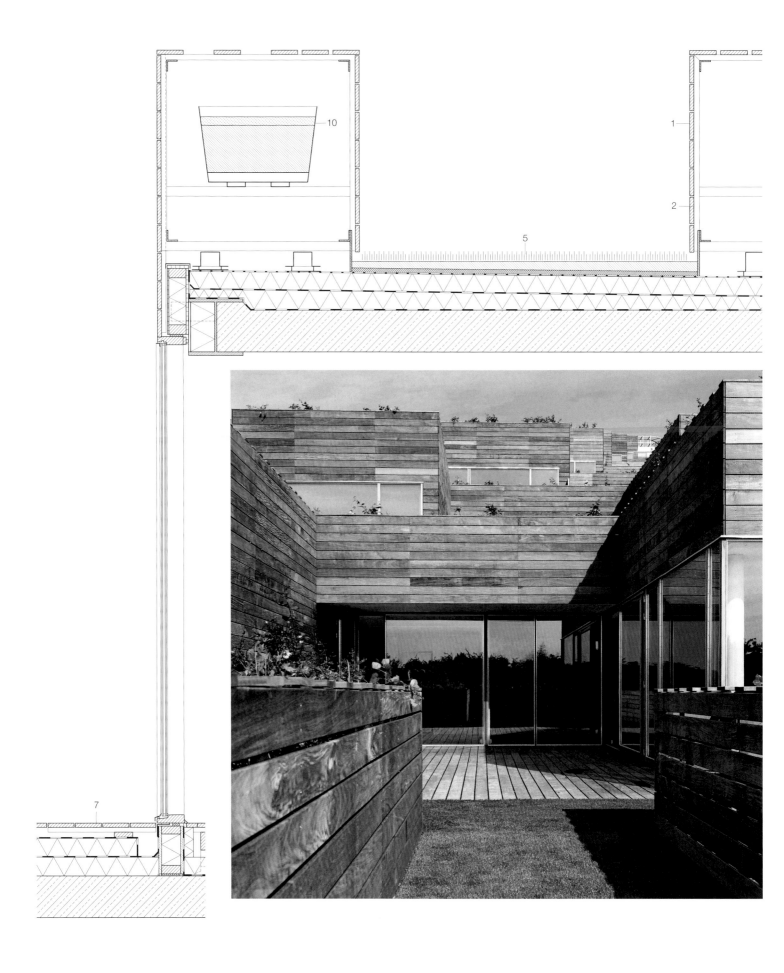

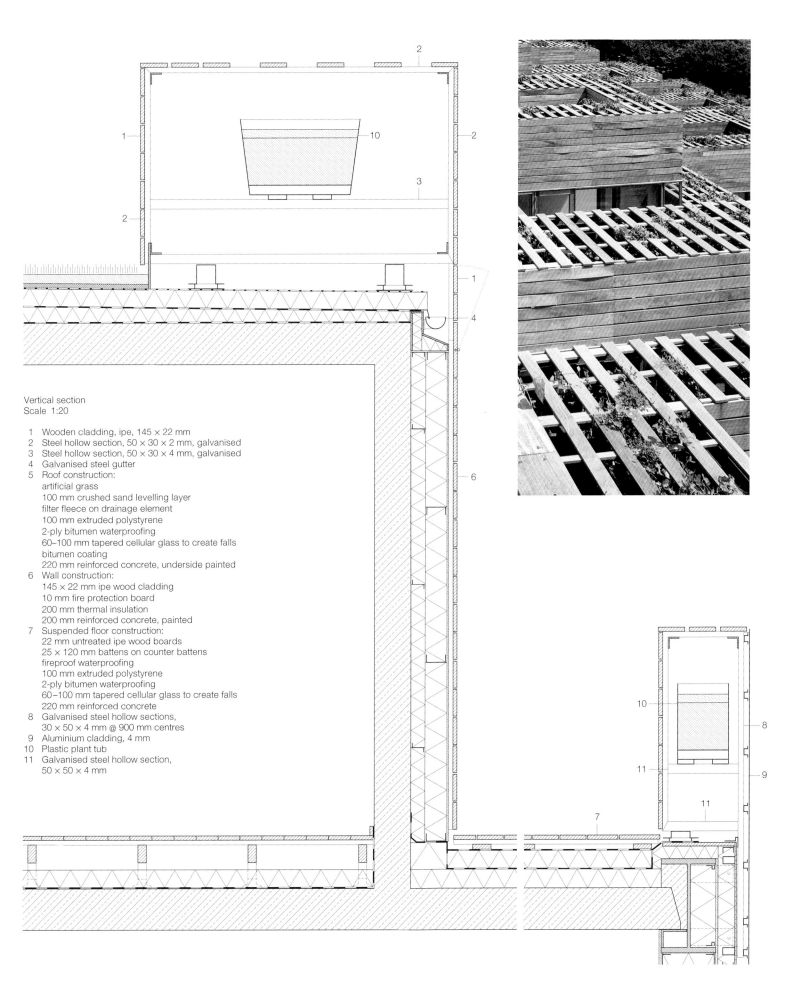

Vertical section
Scale 1:20

1 Wooden cladding, ipe, 145 × 22 mm
2 Steel hollow section, 50 × 30 × 2 mm, galvanised
3 Steel hollow section, 50 × 30 × 4 mm, galvanised
4 Galvanised steel gutter
5 Roof construction:
   artificial grass
   100 mm crushed sand levelling layer
   filter fleece on drainage element
   100 mm extruded polystyrene
   2-ply bitumen waterproofing
   60–100 mm tapered cellular glass to create falls
   bitumen coating
   220 mm reinforced concrete, underside painted
6 Wall construction:
   145 × 22 mm ipe wood cladding
   10 mm fire protection board
   200 mm thermal insulation
   200 mm reinforced concrete, painted
7 Suspended floor construction:
   22 mm untreated ipe wood boards
   25 × 120 mm battens on counter battens
   fireproof waterproofing
   100 mm extruded polystyrene
   2-ply bitumen waterproofing
   60–100 mm tapered cellular glass to create falls
   220 mm reinforced concrete
8 Galvanised steel hollow sections,
   30 × 50 × 4 mm @ 900 mm centres
9 Aluminium cladding, 4 mm
10 Plastic plant tub
11 Galvanised steel hollow section,
   50 × 50 × 4 mm

Example 12

Production and offices building

Baar, Switzerland, 2002

Architects:
Barkow Leibinger Architekten, Berlin
Frank Barkow, Regine Leibinger
Project team:
Michael Eidenbenz, Karsten Ruf, Mari Fujita,
Alexandra Ultsch, Maik Westhaus
Structural engineers:
Scepan, Baar
Building physics consultants:
Martinelli + Menti AG, Meggen

A factory for metalworking machinery has existed on this site on the edge of an industrial estate in Baar, not far from Lake Zug, since 1963. A new production and offices building has now been built, forming an L-shaped extension to the existing factory. The idea for the striking black-and-white facades of the new building was taken from the surrounding timber-frame houses of this Swiss town. The horizontal "folds" in the facades, different in each storey, lend them a degree of relief and thus create a reference to the surrounding rolling landscape. The smaller, offices wing is "telescoped" into the production bay and this in turn into the slope of the hillside. The main entrance is located at the junction between these two sections and a gallery on the upper floor here links the two parts internally. This is the location of the meeting rooms, from where customers have a view of the machinery in the production area. It is also possible to gain access to the roof of the office wing, which can be used by the staff during breaks.

A roof with extensive planting covers both the office wing and the production area. The edges of the roof, clad in black anodised sheet aluminium, employ a ventilated form of construction. The true depth of the roof construction, with its thick thermal insulation, is concealed from the observer by setting the parapet 1 m back from the edge of the roof on all sides.

The use of various metals for both facade and roof finishes, and also internally, reflects the identity of the client.

• Green roof
• Narrow roof edge with overhang

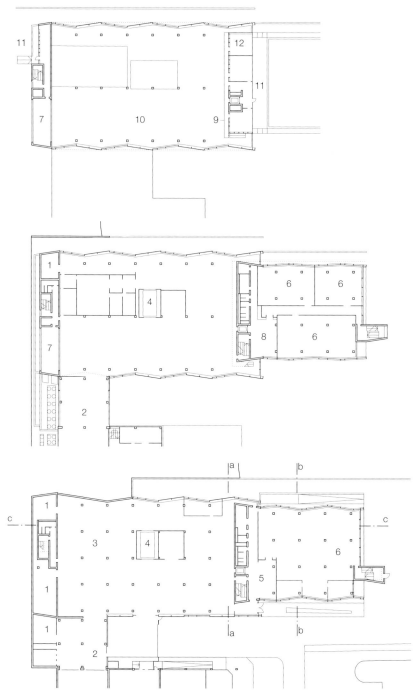

Plans · Sections
Scale 1:1000
Location plan
Scale 1:4000

1 Plant room
2 Link to existing building
3 Production area
4 Goods lift
5 Reception
6 Office
7 Works manager
8 Lobby
9 Gallery
10 Void
11 Rooftop terrace
12 Meeting room
13 Production, existing
14 Production, new

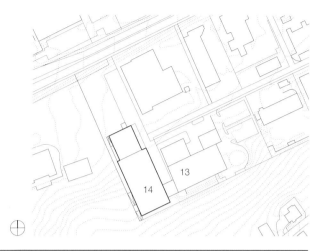

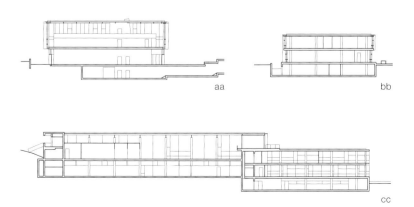

aa   bb

cc

Example 12

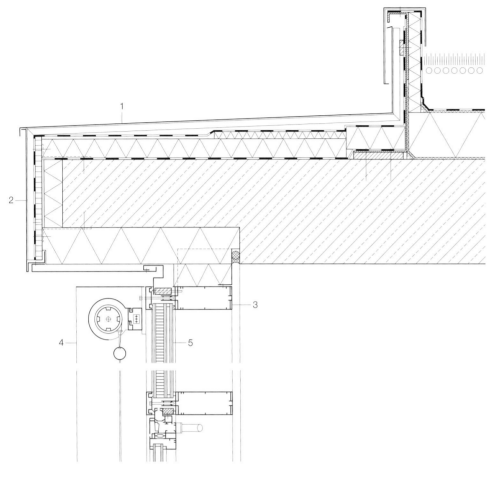

Sections
Offices wing · Production area
Scale 1:10

1   Overhang construction:
    3 mm black anodised sheet aluminium
    25 mm air space, waterproofing
    65–90 mm PU insulation for falls
    waterproofing, 180 mm reinforced concrete
    120 mm thermal insulation
    3 mm black anodised sheet aluminium
2   Sheet aluminium, 3 mm, black anodised finish, water-
    proofing, 32 mm glulam, 60 mm thermal insulation
3   Post-and-rail construction, 155 x 60 mm extruded
    aluminium sections, black anodised finish
4   Aluminium angle, 190 x 60 mm, black anodised finish
5   Low E glazing: 8 mm tough. safety glass + 22 mm
    cavity with aluminium honeycomb panel + 8 mm
    tough. safety glass with low E coating + 10 mm
    krypton-filled cavity + 8 mm lam. safety glass
6   Roof construction:
    extensive planting on 120 mm substrate
    3-ply EPDM waterproofing, root-resistant
    glass fleece separating membrane
    120 mm incompressible thermal insulation
    glass fleece separating membrane, vapour barrier
    105 mm perforated trapezoidal profile sheeting
    300 mm steel channels
7   Overhang construction:
    3 mm black anodised sheet aluminium
    25 mm air space, waterproofing
    65–90 mm PU insulation for falls, waterproofing
    105 mm perforated trapezoidal profile sheeting
8   Sheet aluminium, 3 mm, black anodised finish, 155 mm
    air space, 140 mm thermal insulation
9   Post-and-rail construction, 180 x 80 x 7 mm steel
    hollow sections, black stove-enamelled finish
10 Aluminium angle, 230 x 80 mm, black anodised finish
11 Low E glazing: 12 mm tough. safety glass with low E
    coating + 16 mm cavity + 10 mm float glass
12 Steel I-section, 600 mm

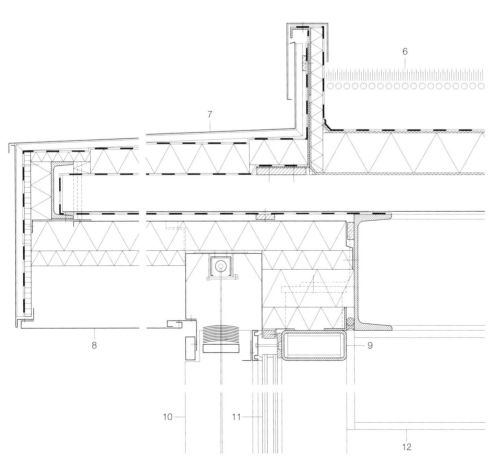

R&D centre

Maranello, Italy, 2004

Architect:
Massimiliano Fuksas, Rome
Interior architect:
Doriana O. Mandrelli
Project team:
Giorgio Martocchia, Defne Dilber Stolfi, Adele
Savino, Fabio Cibinel, Dario Binarelli, Gianluca
Brancaleone, Nicola Cabiati, Andrea Marazzi
Structural engineer:
Gilberto Sarti, Rimini

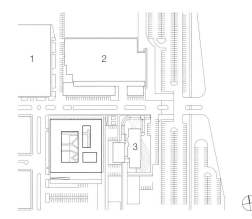

Location plan
Scale 1:5000
Sections
Scale 1:750

1 Paint shop
2 Engines shop
3 Wind tunnel

This new R&D centre is situated in the centre of
the site occupied by the Italian sports car manufacturer Ferrari. The fundamental idea behind
the design was to integrate nature into this
group of buildings dedicated to high technology
and in so doing create a pleasant working
environment. All office areas are organised as
flexible open-plan zones – visual contact
between the teams is important.
The interplay of light, water and plants has given
rise to a building with scenic qualities, a building whose precise contours make it appear
amazingly lightweight. The two lower floors are
positioned around a central courtyard with
bamboo plants. Lightwells provide illumination
for the basement where the development of new
cars takes place in several large rooms. The
topmost floor, which cantilevers 7 m beyond the
entrance facade, is raised clear of the rest of
the building on isolated columns and the staircase cores. The roof of the ground floor below
is a readily accessible, open recreation area,
the only interruption being the two glass boxes
for the conference rooms. This roof surface is in
the form of a water basin which through evaporation ensures that the floors below remain cool in
summer. A system of walkways crosses the
water to link the conference rooms, which shim-
mer in the Ferrari colours of red and yellow. The
light reflected from the water is in turn reflected
by the aluminium cladding to the soffit of the
second floor, which is arranged around three
large openings that allow light to reach the
areas below. The steel structure here has been
resolved into bridge-like, accessible Vierendeel
girders. Two steel staircases positioned diago-
nally in the central courtyard link the various
levels.

• Green roof
• Water basin on the roof

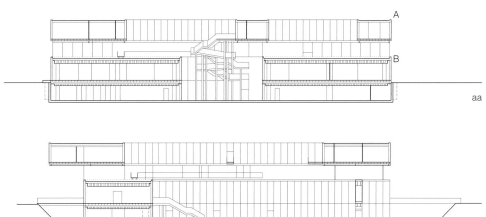

aa

bb

Example 13

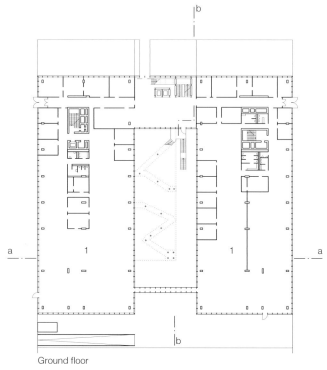

Ground floor

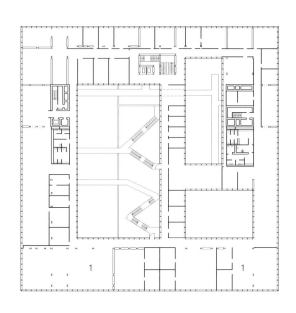

2nd floor

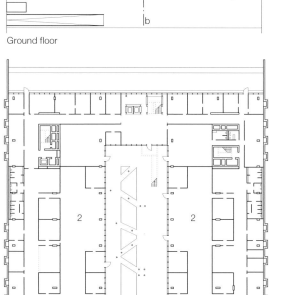

Basement

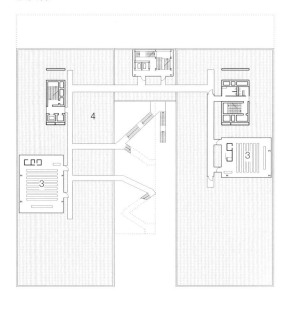

1st floor

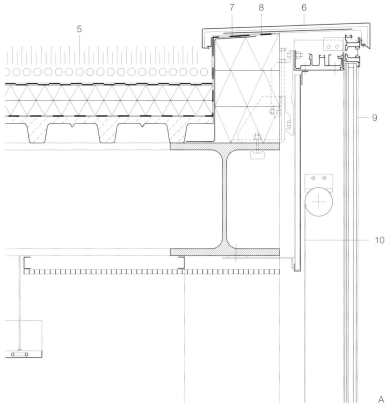

A

Plans
Scale 1:1000

1 Open-plan office
2 Car hall
3 Conference room
4 Water-filled rooftop garden

Details
Scale 1:10

5 Roof to 2nd floor:
extensive planting on substrate
PVC waterproofing
separating membrane
2 No. 40 mm bitumen-faced glass
wool board
vapour barrier
screed laid to falls, 0.5%
70 mm trapezoidal profile sheeting with concrete topping
HEB 300 main beams
400 mm suspended cooling ceiling

6 Sheet aluminium, 2 mm
7 Galvanised sheet steel
8 Mineral wool
9 Insulating glass in aluminium
frame, 10 mm lam. safety glass +
16 cavity + 12 mm lam. safety glass
10 Anti-glare roller blind
11 Roof to ground floor:
100 mm water basin
gravel
PVC waterproofing
2 No. 40 mm glass wool board
vapour barrier
400 mm reinforced concrete
400 mm suspended cooling ceiling

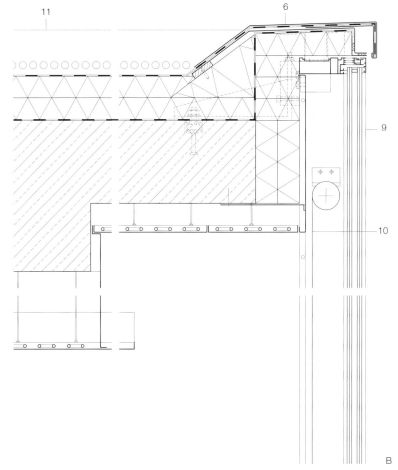

B

Example 14

Residential complex

Beijing, China, 2008

Architects:
Steven Holl Architects, New York
Project team:
Li Hu, Hideki Hirahara, Yenling Chen, Chris
McVoy, Tim Bade
Structural engineers:
Guy Nordenson & Associates, New York
China Academy of Building Research, Beijing

The gigantic "Linked Hybrid" building complex fits neatly into a site not far from Beijing's old city walls. The aim of this 220 000 m² project is to express modern urban living in a city of the 21st century. It is intended to introduce more individualism into the city, which is dominated by standardised, repetitive housing for the masses.

The 728 apartments of this complex provides homes for 2500, all designed according to feng shui principles. Diverse internal layouts ensure individual design freedoms. Conceived as a city within a city, this group of buildings includes numerous public and commercial amenities aimed at easing daily life for the residents. In the centre of the site there is a spacious park with large ponds, bridges and islands. The whole complex also invites visitors from outside and is intended to serve as a public space. All eight apartment towers are interconnected by bridges between the 12th and 18th storeys, which form almost a complete ring and offer the residents diverse views and visual relationships. The roofs of the bridges are made from steel sections, the loadbearing layer is a steel-concrete composite system.

Lifts provide access to a fitness centre, swimming pool, cafés and galleries. The integration of sustainable elements is an important part of the planning concept. Some 660 geothermal heat pumps in boreholes as deep as 100 m are intended to regulate the temperatures in the complex with no additional energy input. All the roofs are designed with areas of planting. And the roofs to the low-rise buildings – shops, hotel, school and multi-screen cinema – are all in the form of accessible roof gardens. Spacious private gardens have been allocated to the penthouse apartments in the towers.

- Roof gardens
- Green roofs

aa

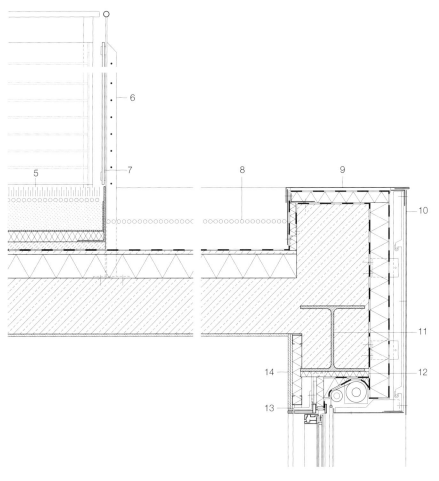

Section
Scale 1:1250
Location plan
Scale 1:2500

1 Residential tower
2 Hotel
3 Cinema
4 Recreational amenities

Vertical section through roof garden
Scale 1:20

5 Roof construction:
  planting on 150 mm substrate
  filter fleece
  55 mm drainage layer
  40 mm screed
  waterproofing
  levelling layer, 20 mm screed
  120 mm thermal insulation
  300 mm reinforced concrete
6 Balustrade uprights, 50 × 60 mm
  steel T-sections
7 Balustrade infill panel, 8 mm safety glass
8 Gravel
9 Sheet steel parapet capping
  aluminium clip
  waterproofing
  55 mm thermal insulation
  waterproofing
10 Aluminium panel, 3 mm
  100 × 165 mm steel angle
  waterproofing
  100 mm thermal insulation
  waterproofing
11 Steel I-section 350 × 350 mm
12 Insulating panel, 25 mm, fire-retardant
13 Steel angle, 100 × 100 mm
14 Plasterboard, 2 No. 12.5 mm,
  50 × 50 mm steel channel frame,
  50 mm thermal insulation

Example 14

Section through bridge
Scale 1:20

1 Roof construction:
   2-ply waterproofing
   80 mm thermal insulation
   max. 50 mm screed laid to falls
   70 mm trapezoidal profile sheeting
   with concrete topping
2 Sheet aluminium, 3 mm, painted
3 Steel channel, 60 × 250 mm
4 Steel I-section
5 Steel hollow section, 200 × 300 mm
6 Insulating glass: 6 mm tough. safety
   glass + 12 mm cavity + 2 No. 6,.5 mm
   lam. safety glass

7 Suspended ceiling,
   20 mm sound insulation panel
8 Floor construction:
   rubber floor covering
   20 mm screed
   70 mm trapezoidal profile sheeting with
   concrete topping
9 Steel hollow section, 150 × 500 mm
10 Panel: 100 mm rigid foam thermal insula-
    tion with sheet aluminium facing both sides
11 Cover plate: 15 × 900 mm steel on
    50 × 50 mm steel angles
12 Soffit panel, sheet aluminium

Grammar school

Dallgow-Döberitz, Germany, 2005

Architects:
Grüntuch Ernst Architekten, Berlin
Armand Grüntuch, Almut Ernst
Project team:
Florian Fels, Erik Behrends, Olaf Menk,
Jacob van Ommen, Volker Raatz
Structural engineers:
GTB Berlin
Roof waterproofing and planting:
Sachverständigenbüro Drefahl, Berlin

On the boundary between the open country-side and the urban sprawl around Berlin, this colourful new grammar school functions both as a link and as a contrast to the small-scale rural structures. The clear contours of the building are also straightforward and functional internally as well, and wherever possible the fair-face concrete of the box-frame construction has been left exposed. The yellowish green facade creates a friendly atmosphere both inside and outside. Perforated sheet metal, clear glass panels, printed glass and glass-faced metal panels make up the colourful patchwork. Some panels filter the view, others are transparent, yet others translucent. Bordered by L-shaped classroom wings are the key elements in the two-storey design – the large courtyard at ground level and the rooftop terrace, both of which are used by the pupils between lessons. A wide outdoor staircase, which doubles as seating, leads from the courtyard to the rooftop terrace with its wooden seating zones. The school gymnasium, partly below ground level, and the school hall rise above the terrace level and are illuminated via rooflights. The accessible roofscape joins together the rooftop terraces of all the blocks and creates diverse external relationships and visual links. These areas offer the pupils high-quality external amenities: as sun deck, communication zone or platform for viewing the surrounding landscape, the courtyard below and the covered main entrance. The rooftop structures introduce some variety into the area but also serve practical functions: the spacious wooden seating areas on the roof house the rooflights for the gymnasium below; and two periscope systems not only channel daylight into the corridor below but also create a visual link between above and below.

• Roof accessible to pupils
• Daylight via rooftop structures

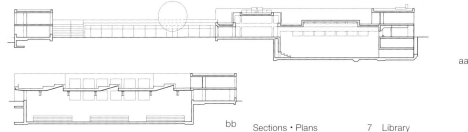

aa

bb

cc

Sections · Plans
Scale 1:1000

1 Classroom
2 Changing room
3 Gymnasium
4 School hall
5 Kitchen
6 Cafeteria
7 Library
8 Courtyard
9 Office
10 Staff room
11 Classroom
12 Seating zone on rooftop terrace
13 Rooflight to school hall
14 Rooflight/periscope

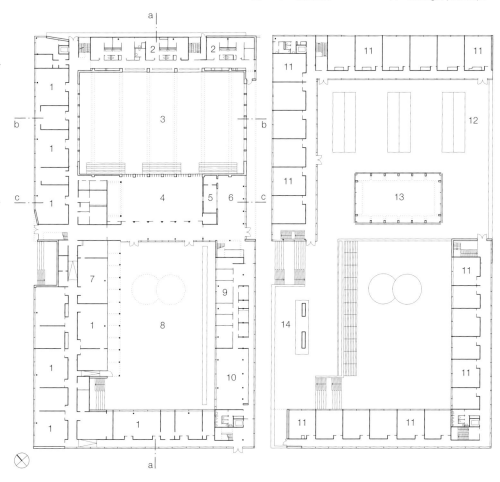

Example 15

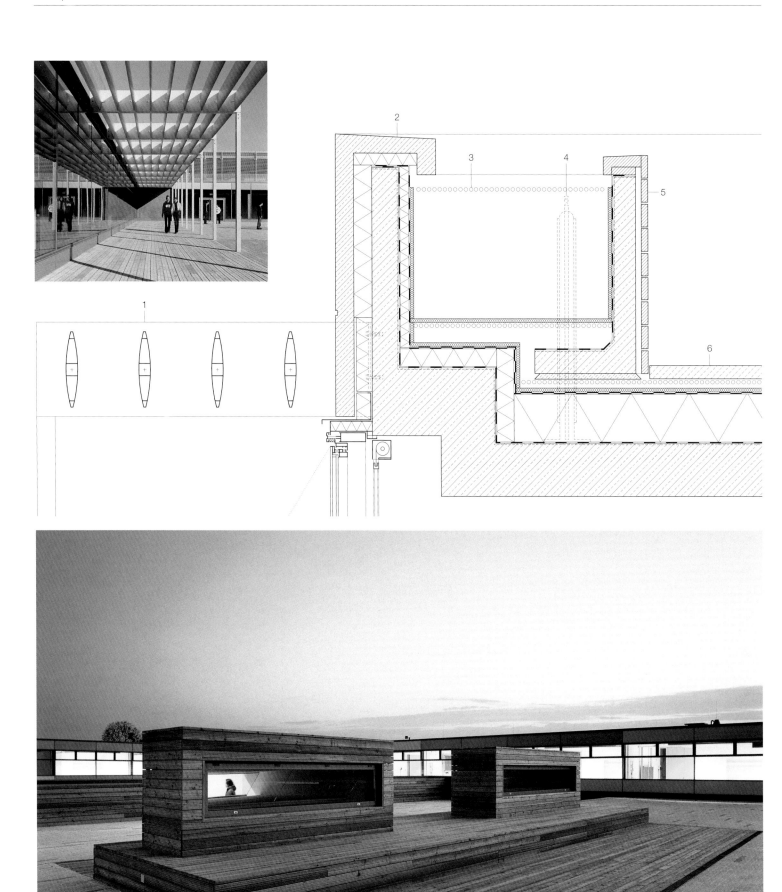

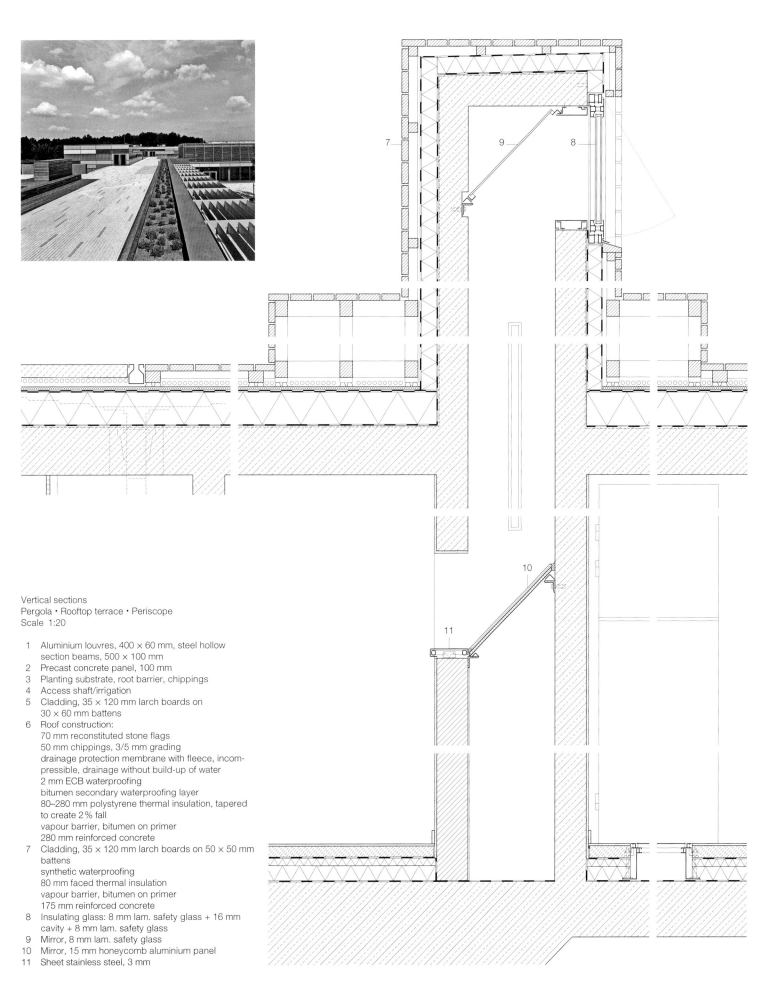

Vertical sections
Pergola · Rooftop terrace · Periscope
Scale 1:20

1 Aluminium louvres, 400 × 60 mm, steel hollow
   section beams, 500 × 100 mm
2 Precast concrete panel, 100 mm
3 Planting substrate, root barrier, chippings
4 Access shaft/irrigation
5 Cladding, 35 × 120 mm larch boards on
   30 × 60 mm battens
6 Roof construction:
   70 mm reconstituted stone flags
   50 mm chippings, 3/5 mm grading
   drainage protection membrane with fleece, incompressible, drainage without build-up of water
   2 mm ECB waterproofing
   bitumen secondary waterproofing layer
   80–280 mm polystyrene thermal insulation, tapered
   to create 2 % fall
   vapour barrier, bitumen on primer
   280 mm reinforced concrete
7 Cladding, 35 × 120 mm larch boards on 50 × 50 mm
   battens
   synthetic waterproofing
   80 mm faced thermal insulation
   vapour barrier, bitumen on primer
   175 mm reinforced concrete
8 Insulating glass: 8 mm lam. safety glass + 16 mm
   cavity + 8 mm lam. safety glass
9 Mirror, 8 mm lam. safety glass
10 Mirror, 15 mm honeycomb aluminium panel
11 Sheet stainless steel, 3 mm

Example 15

Vertical sections
Rooftop seating zones · Roof to gymnasium
Scale 1:20

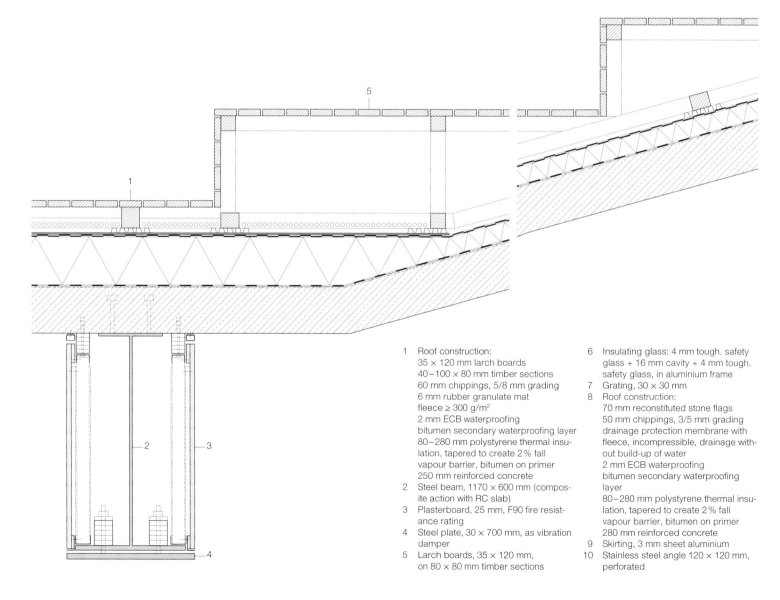

1 Roof construction:
   35 × 120 mm larch boards
   40–100 × 80 mm timber sections
   60 mm chippings, 5/8 mm grading
   6 mm rubber granulate mat
   fleece ≥ 300 g/m²
   2 mm ECB waterproofing
   bitumen secondary waterproofing layer
   80–280 mm polystyrene thermal insulation, tapered to create 2 % fall
   vapour barrier, bitumen on primer
   250 mm reinforced concrete
2 Steel beam, 1170 × 600 mm (composite action with RC slab)
3 Plasterboard, 25 mm, F90 fire resistance rating
4 Steel plate, 30 × 700 mm, as vibration
   damper
5 Larch boards, 35 × 120 mm,
   on 80 × 80 mm timber sections

6 Insulating glass: 4 mm tough. safety
   glass + 16 mm cavity + 4 mm tough.
   safety glass, in aluminium frame
7 Grating, 30 × 30 mm
8 Roof construction:
   70 mm reconstituted stone flags
   50 mm chippings, 3/5 mm grading
   drainage protection membrane with
   fleece, incompressible, drainage without build-up of water
   2 mm ECB waterproofing
   bitumen secondary waterproofing
   layer
   80–280 mm polystyrene thermal insulation, tapered to create 2 % fall
   vapour barrier, bitumen on primer
   280 mm reinforced concrete
9 Skirting, 3 mm sheet aluminium
10 Stainless steel angle 120 × 120 mm,
   perforated

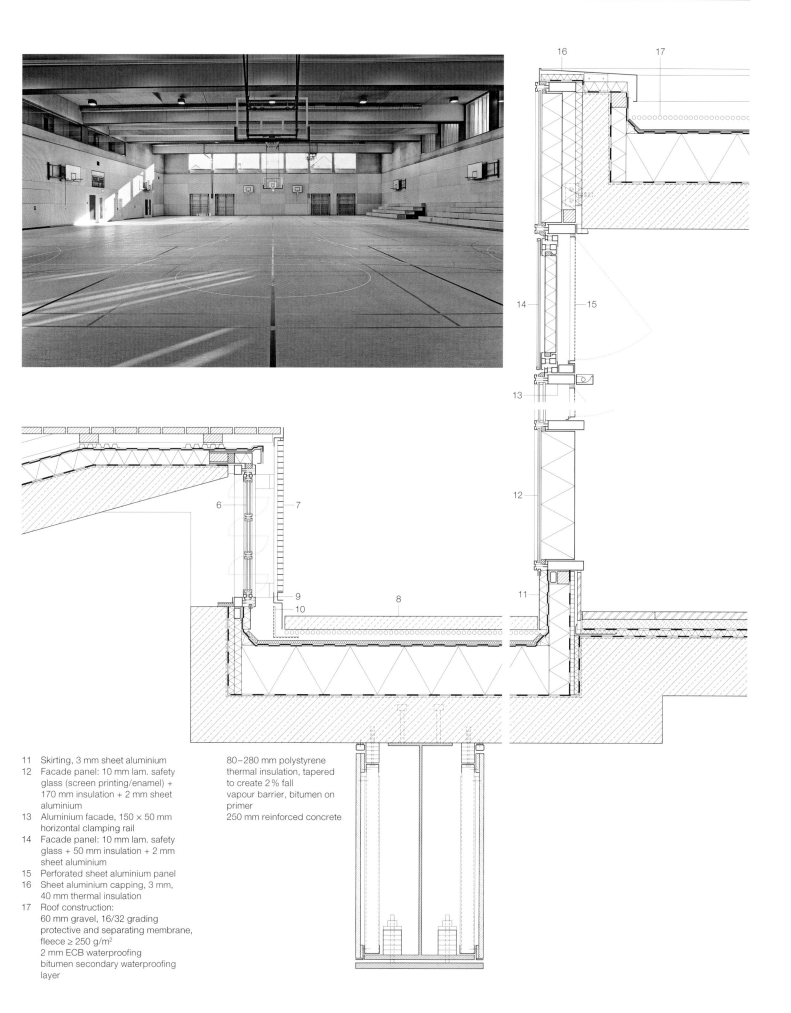

11 Skirting, 3 mm sheet aluminium
12 Facade panel: 10 mm lam. safety
   glass (screen printing/enamel) +
   170 mm insulation + 2 mm sheet
   aluminium
13 Aluminium facade, 150 × 50 mm
   horizontal clamping rail
14 Facade panel: 10 mm lam. safety
   glass + 50 mm insulation + 2 mm
   sheet aluminium
15 Perforated sheet aluminium panel
16 Sheet aluminium capping, 3 mm,
   40 mm thermal insulation
17 Roof construction:
   60 mm gravel, 16/32 grading
   protective and separating membrane,
   fleece ≥ 250 g/m²
   2 mm ECB waterproofing
   bitumen secondary waterproofing
   layer

80–280 mm polystyrene
thermal insulation, tapered
to create 2 % fall
vapour barrier, bitumen on
primer
250 mm reinforced concrete

Example 16

Kindergarten

Tokyo, Japan, 2007

Architects:
Tezuka Architects, Tokyo
Takaharu Tezuka, Yui Tezuka
Masahiro Ikeda Co. Ltd, Tokyo
Project team:
Chie Nabeshima, Asako Komparu, Kousuk
Suzuki, Naoto Murakaji, Shigefumi Araki,
Shuichi Sakuma, Ryuya Maoi
Structural engineers:
Masahiro Ikeda Co. Ltd., Tokyo

The kindergarten in Tachikawa, a suburb of
Tokyo, actually consists of little more than a
roof. On plan it looks like an arena or a course
for go-kart races. Its oval form with the central
open area at ground level grew out of the architects' idea to create a building without any
wasted corners, a building that strengthens the
community spirit, a building with spacious but
clearly structured open areas. Despite its
shape, the kindergarten is, however, anything
but introverted. In the open play areas beneath
the flat roof, it is not only the group rooms that
are interlinked without barriers; all the facade
elements are fully glazed and can be slid aside
so that the children can also play inside the
building, in the shade, as though they were
playing outside.
The width of the annular oval varies. But as the
height of the roof above the ground differs
between inside and outside, the roof forms a
platform with an almost constant slope. This
corresponds to a three-dimensional, gently
undulating hyperbolic surface. A delicate balustrade of vertical bars enables the children (and
there are 500 in this establishment!) to dangle
their legs safely over the edge of the roof. The
gutter along the eaves is very wide so that it
does not become clogged by falling leaves.
Four downspouts are enough to discharge all
the rainwater into large containers in the inner
courtyard. Openings have been left in the roof
to allow the trees to grow through. It is difficult
for a child to climb a tree from the bottom, but
much easier from the roof and so bespoke
safety nets were fabricated because the trees
are a great attraction for the children.

• Accessible roof as playground
• Roof overhang with integral gutter

Roof structure
Plan
Scale 1:800

1   Entrance gate
2   Staff room
3   Group room
4   School building
    (existing)

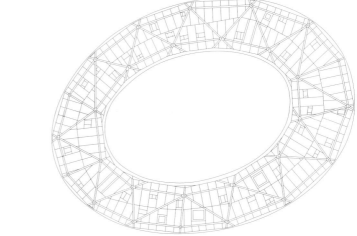

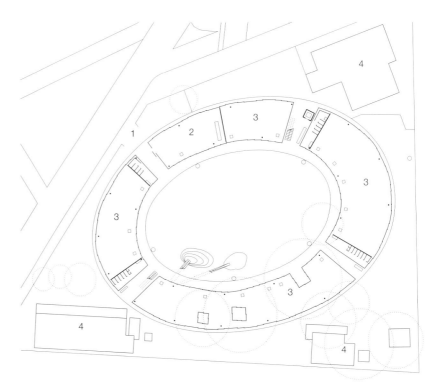

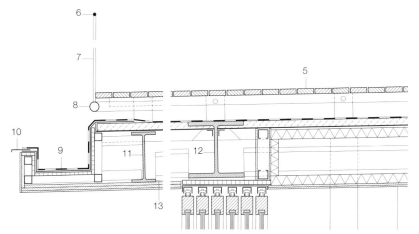

5 Roof construction:
20 mm cherry wood boards
60 × 45 mm battens @ 450 mm centres
pedestals, steel brackets welded to steel structure
EPDM waterproofing
50 mm concrete topping, non-loadbearing
trapezoidal profile sheeting
50 mm thermal insulation
200 mm space for services
50 mm thermal insulation
15 mm plasterboard
9 mm plasterboard acoustic ceiling, perforated
6 Handrail: ⌀ 16 mm steel bar

7 Uprights: ⌀ 13 mm steel bars @ 123 mm centres
8 Balustrade support: ⌀ 42 mm steel tube, 90 × 90 mm hot-dip galvanised steel hollow sections @ 900 mm centres
9 Gutter construction:
EPDM waterproofing
sheet steel bent to suit
40 × 40 mm steel hollow section
2 No. 6 mm fibre-cement board
15 mm plasterboard
10 Aluminium angle
11 Secondary beam, IPE 260
12 Primary beam, HEA 300
13 Secondary beam, IPE 140
14 Steel tube, ⌀ 19 mm, wire rope
15 Safety net around tree trunk:
6 mm vinyl net, 60 mm apertures

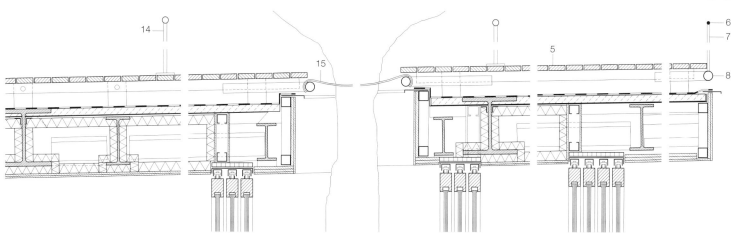

Example 17

St. Dominic Centre

Munich, Germany, 2008

Architects:
meck architekten, Munich
Andreas Meck
Project team:
Wolfgang Amann, Peter Fretschner (project
manager), Susanne Frank, Johannes Dörle,
Alexander Sälzle, Werner Schad (tender docu-
ments), Wolfgang Kusterer (site manager)
Structural engineers:
Statoplan, Munich
Building physics consultants:
Müller BBM, Planegg

Nordheide is a new district for 5000 inhabitants
in the city of Munich. Its spiritual focal point is
formed by the St. Dominic Centre with its chapel,
community club, kindergarten and Catholic
Church care services. The building is visible from
afar and is immediately conspicuous because
of its clay brickwork, which presents a stark
contrast to the neighbouring housing.
The centre with its large internal courtyard was
conceived as a peaceful place. And the successful realisation of this contemplative architecture
can be attributed to the use of high-quality peat-
fired clay bricks on all surfaces – walls, floors
and soffits. This material radiates a natural aura
and a certain sensuality. Particularly irregular
bricks were deliberately selected in order to
lend the surfaces vitality and relief. The clay
brickwork is implemented in such a way that it
underscores the idea of a single volume that
has been dissected. Accordingly, the plan form
of the building plays a crucial role. Laid flat, not
on edge, the clay bricks are used as a floor finish
in the main entrance and the internal courtyard,
also as a finish on the inaccessible parts of the
roof and as a floor covering for the three rooftop
terraces, which belong to the caretaker's apartment, the youth centre and the Catholic Church
care services. The rooftop terraces are enclosed
by high walls to provide private outdoor areas
that still permit a view of the surroundings. The
design of the parapet coping as a precast concrete unit faced with brick slips emphasizes the
uniform appearance and the distinct geometrical
forms.

• Rooftop terraces
• Clay brickwork for roofs and facades
• Inverted roof

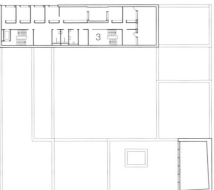

West elevation
Plans · Sections
Scale 1:1000

1   Chapel
2   Community club
3   Catholic Church services
    centre
4   Kindergarten
5   Caretaker's apartment
6   Rooftop terrace for care-
    taker
7   Youth centre
8   Rooftop terrace for youth
    centre
9   Youth centre group room

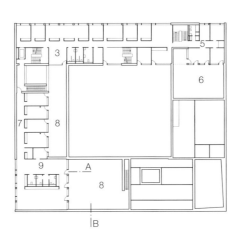

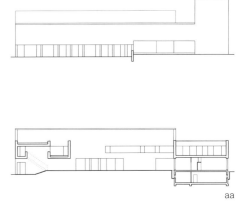

aa

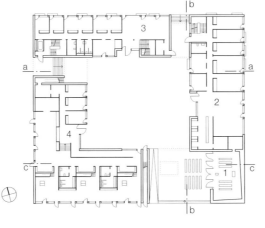

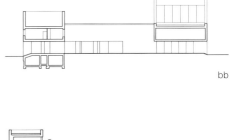

bb

cc

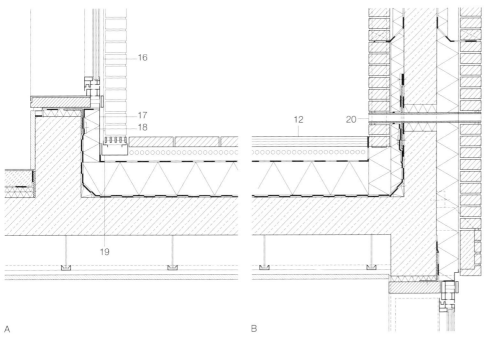

A

B

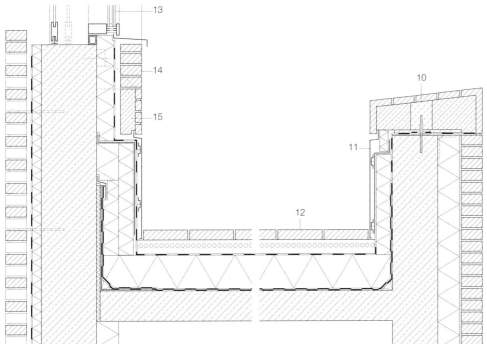

C

Vertical sections
Scale 1:20

10 Precast concrete coping with brick slip facing, 24 × 61.5 × 200 and 17.5 × 61.5 × 200 mm, on mortar bed, fixed in place with grouted threaded bars
11 Sheet copper skirting, 1 mm
   EPS thermal insulation, 80 mm
12 Roof construction:
   Wittmunder peat-fired clay bricks, laid flat, sand joints
   80 mm drainage layer, 8/16 mm rounded, washed gravel
   separating membrane, diffusion-permeable
   180 mm EPS thermal insulation, incompressible, bonded to substrate
   elastomeric bitumen waterproofing, laid in rubber-modified bitumen
   elastomeric bitumen built-up roofing
   160 or 200 mm reinforced concrete
13 Glazing: 16 mm lam. safety glass with grey pattern (screen printing/enamel) + 16 mm cavity + 8 mm tough. safety glass (with deep engraving), partially ink baked in engraving; inner pane with platinum-sputtered halftone pattern
14 Wall construction:
   facing leaf: Wittmunder peat-fired clay bricks, 115 × 61.5 × 200 mm, 10 mm joints
   diffusion-permeable sheet
   120 mm mineral fibre insulation
   300 mm reinforced concrete
   50 mm mineral fibre insulation
   vapour barrier
   30 mm cavity
   facing bricks in decorative bond, 115 × 61.5 × 200 mm
15 Precast concrete lintel, with 115 × 240 mm clay brick slips on one side
16 Post-and-rail facade
17 Sheet aluminium, coated
18 Thermal insulation, 80 mm
19 Drainage channel
20 Overflow: welded sheet copper pipe, 55 × 240 × 1.5 mm, with 300 × 500 mm waterstop flange integrated into waterproofing

Example 18

Hospital extension

Munich, Germany, 1994

Architects:
Schunck Ulrich Krausen, Munich
Project team:
Nobert Krausen, Stephan Will, Robert Kellner,
Iris Heckl, Heinz Grünberger
Structural engineers:
Höllerer, Schäfer & Partner, Munich

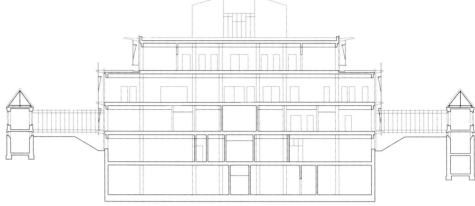

The municipal hospital, built around 1910, is
typical of the pavilion-style hospital that evolved
around the end of the 18th century. Besides
being embedded in a park-type landscape,
this hospital is characterised by its obvious
structures and the clear arrangement of its net-
work of paths. The extension does not try to
steal the limelight from the original buildings. It
achieves this not through imitation, but rather
through a contemporary adaptation of the
municipal and architectural concept that even
today is still viable. So the new structure is not
a brick structure attached firmly to the ground
but instead an additional, lightweight configuration. The facade to the modern frame structure
takes up ideas from the architecture of the
existing buildings in the form of its double-leaf
design and fine articulation. The detail the top
of the outer leaf is formed by the obligatory
safety net around the helicopter landing pad on
the roof, but at the same time it forms a transparent canopy marking the termination of the
facade.

The depth of the roof had to be kept to a minimum in order to fit in with the existing buildings;
the construction of the helicopter landing pad
helps in this respect. It consists of reinforced
concrete slabs approx. 5 × 5 m which in order
to transfer the shear forces are anchored
together and were cast in situ directly on the
waterproofing. Their thickness increases
towards the centre of the pad in order to create
an approx. 1 % fall. Hot-water heating loops
and cable ducts for lighting are integrated into
the slabs.

The roof surfaces in front of the second floor
with their extensive planting help to improve the
climate locally. These are in the form of inverted
roofs constructed without any falls, with the
waterproofing material laid directly to the structural roof slab.

• Helicopter landing pad
• Green roof

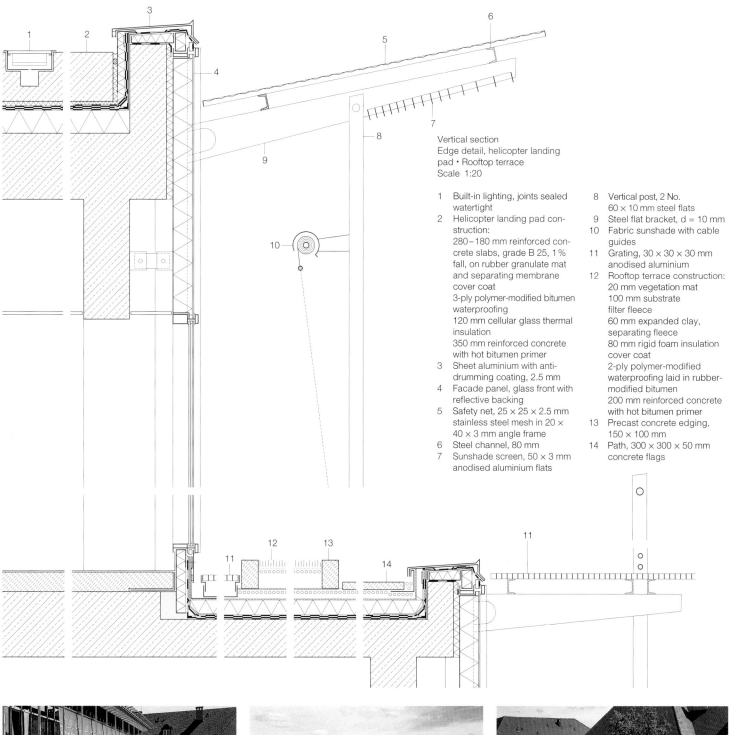

Vertical section
Edge detail, helicopter landing
pad · Rooftop terrace
Scale 1:20

1 Built-in lighting, joints sealed
   watertight
2 Helicopter landing pad construction:
   280–180 mm reinforced concrete slabs, grade B 25, 1 %
   fall, on rubber granulate mat
   and separating membrane
   cover coat
   3-ply polymer-modified bitumen
   waterproofing
   120 mm cellular glass thermal
   insulation
   350 mm reinforced concrete
   with hot bitumen primer
3 Sheet aluminium with anti-
   drumming coating, 2.5 mm
4 Facade panel, glass front with
   reflective backing
5 Safety net, 25 × 25 × 2.5 mm
   stainless steel mesh in 20 ×
   40 × 3 mm angle frame
6 Steel channel, 80 mm
7 Sunshade screen, 50 × 3 mm
   anodised aluminium flats

8 Vertical post, 2 No.
   60 × 10 mm steel flats
9 Steel flat bracket, d = 10 mm
10 Fabric sunshade with cable
   guides
11 Grating, 30 × 30 × 30 mm
   anodised aluminium
12 Rooftop terrace construction:
   20 mm vegetation mat
   100 mm substrate
   filter fleece
   60 mm expanded clay,
   separating fleece
   80 mm rigid foam insulation
   cover coat
   2-ply polymer-modified
   waterproofing laid in rubber-
   modified bitumen
   200 mm reinforced concrete
   with hot bitumen primer
13 Precast concrete edging,
   150 × 100 mm
14 Path, 300 × 300 × 50 mm
   concrete flags

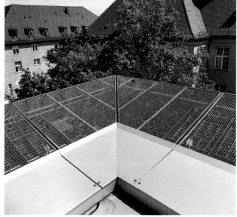

Statutory instruments, directives, standards

The EU has issued directives for a number of products, the particular aim of which is to ensure the safety and health of users. These directives must be implemented in the EU member states in the form of compulsory legislation and regulations.

The directives themselves do not contain any technical details, but instead only lay down the mandatory underlying requirements. The corresponding technical values are specified in associated sets of technical rules (e.g. codes of practice) and in the form of EN standards harmonised throughout Europe.

Generally, the technical rules provide advice and information for everyday activities. They are not statutory instruments, but rather give users decision-making aids, guidelines for implementing technical procedures correctly and/or practical information for turning legislation into practice. The use of the technical rules is not compulsory; only when they have been included in government legislation or other statutory instruments do they become mandatory, or when the parties to a contract include them in their conditions.

In Germany the technical rules include DIN standards, VDI directives and other publications such as the Technical Rules for Hazardous Substances.

The standards are divided into product, application and testing standards. They often relate to just one specific group of materials or products and are based on the corresponding testing and calculation methods for the respective materials and components. The latest edition of a standard – which should correspond with the state of the art – always applies. A new or revised standard is first published as a draft for public discussion before (probably with revisions) it is finally adopted as a valid standard.

The origin and area of influence of a standard can be gleaned from its designation:

- DIN plus number (e.g. DIN 4108) is essentially a national document (drafts are designated with "E" and preliminary standards with "V").
- DIN EN plus number (e.g. DIN EN 335) is a German edition of a European standard – drawn up by the European Standardisation Organisation CEN – that has been adopted without amendments.
- DIN EN ISO (e.g. DIN EN ISO 13786) is a standard with national, European and worldwide influence. Based on a standard from the International Standardisation Organisation ISO, a European standard was drawn up, which was then adopted as a DIN standard.
- DIN ISO (e.g. DIN ISO 2424) is a German edition of an ISO standard that has been adopted without amendments.

The following compilation represents a selection of statutory instruments, directives and standards that reflects the state of the art regarding building materials and building material applications as of June 2010.

Part B Structure

Loadbearing structure
DIN 1045-1 Concrete, reinforced and prestressed concrete structures – Part 1: Design and construction. Aug 2008

DIN 1045-2 Concrete, reinforced and prestressed concrete structures – Part 2: Concrete – Specification, properties, production and conformity – Application rules for DIN EN 206-1. Aug 2008

DIN 1045-3 Concrete, reinforced and prestressed concrete structures – Part 3: Execution of structures. Aug 2008

DIN 1045-4 Concrete, reinforced and prestressed concrete structures – Part 4: Additional rules for the production and conformity control of prefabricated elements. Jul 2001

DIN 1055-1 Actions on structures – Part 1: Densities and weights of building materials, structural elements and stored materials. Jun 2002

DIN1055-3 Actions on structures – Part 3: Self-weight and imposed load in building. Mar 2006

DIN 1055-4 Actions on structures – Part 4: Wind loads. Mar 2005

DIN 1055-5 Actions on structures – Part 5: Snow loads and ice loads. Jul 2005

DIN 4223 Prefabricated reinforced components of autoclaved aerated concrete. Dec 2003

DIN 18190-4 Waterproof sheeting for the waterproofing of buildings; waterproof sheeting with inlay of metal foil. Oct 1992

DIN 52129 Uncoated bitumen saturated sheeting. Nov 1993

DIN 52131 Bitumen waterproof sheeting for fusion welding. Nov 1995

DIN 52133 Polymer bitumen waterproof sheeting for fusion welding. Nov 1995

DIN 52143 Bitumen roofing felt with glass fleece base; terms and definitions, designation, requirements. Aug 1985

DIN 59231 Corrugated sheets and roofing sheets, surface coated. Nov 2003

DIN EN 494 Fibre-cement profiled sheets and fittings. Jun 2007

DIN EN 1991-1-1 Eurocode 1: Actions on structures – Part 1-1: General actions; densities, self-weight, imposed loads for buildings. Oct 2002

DIN EN 1991-1-3 Eurocode 1: Actions on structures – Part 1-3: General actions – snow loads. Sept 2004

Loadbearing decks
DIN 1045-1 Concrete, reinforced and prestressed concrete structures – Part 1: Design and construction. Aug 2008

DIN 1052 Design of timber structures – General rules and rules for buildings. Dec 2008

DIN 4074 Strength grading of wood. Dec 2008

DIN 4102-4/A1 Fire behaviour of building materials and building components – Part 4: Synopsis and application of classified building materials, components and special components; Amendment A1. Nov 2004

DIN 4223-2 Prefabricated reinforced components of autoclaved aerated concrete – Part 2: Design and calculation of structural components. Dec 2003

DIN 4223-4 Prefabricated reinforced components of autoclaved aerated concrete – Part 4: Design and calculation of structural components; Application of components in structures. Dec 2003

DIN 18334 German construction contract procedures (VOB) – Part C: General technical specifications in construction contracts (ATV) – Carpentry and timber construction works. Apr 2010

DIN 18800-5 Steel structures – Part 5: Composite structures of steel and concrete – Design and construction. Mar 2007

DIN 18807-1/A1 Trapezoidal sheeting in building – Trapezoidal steel sheeting – General requirements and determination of loadbearing capacity by calculation; Amendment A1. May 2001

DIN 18807-3 Trapezoidal sheeting in building; trapezoidal steel sheeting; structural analysis and design. Jun 1987

DIN EN 300 Oriented strand boards (OSB). Sept 2006

DIN EN 312 Particleboards – Specifications. Jun 2009

DIN EN 636 Plywood – Specifications. Nov 2003

DIN EN 1992-1-1 Eurocode 2: Design of concrete structures – Part 1-1: General rules and rules for buildings. Oct 2005

DIN EN 1994-1-1 Eurocode 4: Design of composite steel and concrete structures – Part 1-1: General rules and rules for buildings. Jul 2006

DIN EN 1995 Eurocode 5: Design of timber structures. Sept 2008

DIN EN 12602 Prefabricated reinforced components of autoclaved aerated concrete. Aug 2008

DIN EN 13986 Wood-based panels for use in construction. Mar 2005

DIN EN 14782 Self-supporting metal sheet for roofing, external cladding and internal lining. Mar 2006

DIN EN 14509 Self-supporting double skin metal faced insulating panels – Factory-made products. Feb 2007

Technische Regeln für die Verwendung von linienförmig gelagerten Verglasungen (TRLV). Aug 2006

Technische Regeln für die Bemessung und die Ausführung punktförmig gelagerter Verglasungen (TRPV). Aug 2006

Part C Building physics

Thermal insulation
DIN 4108-1 Thermal insulation in buildings; quantities and units. Aug 1981

DIN 4108-2 Thermal protection and energy economy in buildings – Part 2: Minimum requirements for thermal insulation. Jul 2003

DIN 4108-3 Thermal protection and energy economy in buildings – Part 3: Protection against moisture subject to climate conditions. Jul 2001

DIN V 4108-4 (pre-standard) Thermal insulation and energy economy in buildings – Part 4: Hygrothermal design values. Jun 2007

DIN V 4108-6 (pre-standard) Thermal protection and energy economy in buildings – Part 6: Calculation of annual heat and energy use. Mar 2004

DIN E EN 410 (draft standard) Glass in building – Determination of luminous and solar characteristics of glazing. Jul 2010

DIN EN ISO 6946 Building components and building elements – Thermal resistance and thermal transmittance – Calculation method (ISO 6946:2007). Apr 2008

DIN EN ISO 7345 Thermal insulation – Physical quantities and definitions. Jan 1996

DIN EN ISO 10211 Thermal bridges in building construction – Heat flows and surface temperatures – Detailed calculations. Apr 2008

DIN EN ISO 13788 Hygrothermal performance of building components and building elements – Internal surface temperature to avoid critical surface humidity and interstitial condensation – Calculation methods. Nov 2001

DIN E EN 13164 (draft standard) Thermal insulation products for buildings – Factory-made extruded polystyrene foam (XPS) products. May 2010

DIN EN 13363 Solar protection devices combined with glazing – Calculation of solar and light transmittance. Sept 2007

Energieeinsparverordnung (EnEV) Verordnung über energieeinsparenden Wärmeschutz und energieeinsparende Anlagentechnik bei Gebäuden. Mar 2009

Moisture control
DIN 4108-1 Thermal insulation in buildings; quantities and units. Aug 1981

DIN 4108-2 Thermal protection and energy economy in buildings – Part 2: Minimum requirements to thermal insulation. Jul 2003

DIN 4108-3 Thermal protection and energy economy in buildings – Part 3: Protection against moisture subject to climate conditions. Jul 2001

DIN V 4108-4 (pre-standard) Thermal insulation and energy economy in buildings – Part 4: Hygrothermal design values. Jun 2007

DIN V 4108-6 (pre-standard) Thermal protection and energy economy in buildings – Part 6: Calculation of annual heat and energy use. Mar 2004

DIN E 41087 (draft standard) Thermal insulation and energy economy in buildings – Part 7: Airtightness of buildings, requirements, recommendations and examples for planning and performance. Jan 2009

DIN 18531-1 Waterproofing of roofs – Sealings for non-utilised roofs – Part 1: Terms and definitions, requirements, design principles. May 2010

DIN 18531-2 Waterproofing of roofs – Sealings for non-utilised roofs – Part 2: Materials. May 2010

DIN 18531-3 Waterproofing of roofs – Sealings for non-utilised roofs – Part 3: Design, handling of materials, execution of sealings. May 2010

DIN 18531-4 Waterproofing of roofs – Sealings for non-utilised roofs – Part 4: Maintenance. May 2010

DIN E 68800-2 (draft standard) Wood preservation – Part 2: Preventive constructional measures in buildings. Nov 2009

DIN EN ISO 13788 Hygrothermal performance of building components and building elements – Internal surface temperature to avoid critical surface humidity and interstitial condensation – Calculation methods. Nov 2001

DIN EN 15026 Hygrothermal performance of building components and building elements – Assessment of moisture transfer by numerical simulation. Jul 2007

Fire
DIN 4102-1 Fire behaviour of building materials and building components – Part 1: Building materials; concepts, requirements and tests. May 1998

DIN 4102-2 Fire Behaviour of Building Materials and Building Components; Building Components; Definitions, Requirements and Tests. Sept 1977

DIN 4102-4/A1 Fire behaviour of building materials and building components – Part 4: Synopsis and application of classified building materials, components and special components; Amendment A1. Nov 2004

DIN 4102-7 Fire behaviour of building materials and building components – Part 7: Roofing; definitions, requirements and testing. Jul 1998

DIN 18230 Structural fire protection in industrial buildings – Part 1: Analytically required fire resistance time. Jun 2008

DIN 18234 Fire safety of large roofs for buildings – Fire exposure from below. Sept 2003

DIN EN 1363 Fire resistance tests. Oct 1999

DIN E EN 1366 (draft standard) Fire resistance tests for service installations. Sept 2008

DIN EN 13501-1 Fire classification of construction products and building elements – Part 1: Classification using data from reaction to fire tests. Jan 2010

DIN EN 13501-2 Fire classification of construction products and building elements – Part 2: Classification using data from fire resistance tests, excluding ventilation services. Feb 2010

DIN EN 13501-5 Fire classification of construction products and building elements – Part 5: Classification using data from external fire exposure to roofs tests. Feb 2010

DIN V ENV 1187 (pre-standard) Test methods for external fire exposure to roofs. Oct 2006

Fachkommission Bauaufsicht der ARGEBAU (pub.): Muster-Richtlinie über den baulichen Brandschutz im Industriebau. Mar 2000

Sound insulation

DIN 4109 Sound insulation in buildings. Oct 2006

DIN 18005 Noise abatement in town planning. Jul 2002

DIN 18041 Acoustic quality in small to medium-sized rooms. May 2004

DIN EN 12354-1 Building acoustics – Estimation of acoustic performance of buildings from the performance of products – Part 1: Airborne sound insulation between rooms. Dec 2000

DIN EN 12354-2 Building acoustics – Estimation of acoustic performance of buildings from the performance of elements – Part 2: Impact sound insulation between rooms. Sept 2000

DIN EN 12354-3 Building acoustics – Estimation of acoustic performance of buildings from the performance of elements – Part 3: Airborne sound insulation against outdoor sound. Sept 2000

DIN EN 12354-4 Building acoustics – Estimation of acoustic performance of buildings from the performance of products – Part 4: Transmission of indoor sound to the outside. Apr 2001

DIN EN ISO 140-3 Acoustics – Measurement of sound insulation in buildings and of building elements – Part 3: Laboratory measurements of airborne sound insulation of building elements. Mar 2005

DIN EN ISO 140-5 Acoustics – Measurement of sound insulation in buildings and of building elements – Part 5: Field measurement of airborne sound insulation of facade elements and facades. Dec 1998

DIN EN ISO 140-7 Acoustics – Measurement of sound insulation in buildings and of building elements – Part 7: Field measurements of impact sound insulation of floors. Dec 1998

DIN EN ISO 140-18 Acoustics – Measurement of sound insulation in buildings and of building elements – Part 18: Laboratory measurement of sound generated by rainfall on building elements. Feb 2007

DIN EN ISO 717-1 Acoustics – Rating of sound insulation in buildings and of building elements – Part 1: Airborne sound insulation. Nov 2006

DIN ISO 9613-2 Acoustics – Attenuation of sound during propagation outdoors – Part 2: General method of calculation. Oct 1999

Richtlinien für den Verkehrslärmschutz an Bundesfernstraßen. May 1997

VDI 2062 Blatt 1 (draft) Technical rule: Shock and vibration isolation conceptions and principles. Oct 2009

VDI 2062 Blatt 2 Technical rule: Vibration insulation – Insulation elements. Nov 2007

VDI 2720 Blatt 1 Technical rule: Noise control by barriers outdoors. Mar 1997

VDI 3727 Blatt 1 Technical rule: Noise control by means of damping of structure-borne noise; physical fundamentals and estimating procedures. Feb 1984

VDI 3727 Blatt 2 Technical rule: Noise control by means of damping of structure-borne noise; recommended applications. Nov 1984

VDI 3833 Blatt 1 Technical rule: Dynamic damper and dynamic vibration absorber – Dynamic damper. May 2009

VDI 3833 Blatt 2 Technical rule: Dynamic damper and dynamic vibration absorber – Dynamic vibration absorber and dynamic vibration absorption. Dec 2006

Federal Ministry of Justice (pub.): Vorläufige Berechnungsmethode für den Umgebungslärm an Schienenwegen. May 2006

Part D Design principles

Materials

DIN 4102-1 Fire behaviour of building materials and building components – Part 1: Building materials; concepts, requirements and tests. Aug 1998

DIN 4102-4/A1 Fire behaviour of building materials and building components – Part 4: Synopsis and application of classified building materials, components and special components; Amendment A1. Nov 2004

DIN 4108-10 Thermal insulation and energy economy in buildings – Part 10: Application-related requirements for thermal insulation materials – Factory-made products. Jun 2008

DIN 7864-1 Sheets of elastomers for waterproofing; terms of delivery. Apr 1984

DIN 18531-1 Waterproofing of roofs – Sealings for non-utilised roofs – Part 1: Terms and definitions, requirements, design principles. May 2010

DIN 18531-2 Waterproofing of roofs – Sealings for non-utilised roofs – Part 2: Materials. May 2010

DIN EN ISO 10456 Building materials and products – Hygrothermal properties – Tabulated design values and procedures for determining declared and design thermal values. May 2010

DIN EN 12390-8 Testing hardened concrete – Part 8: Depth of penetration of water under pressure. Jul 2009

DIN EN 13162 Thermal insulation products for buildings – Factory-made mineral wool (MW) products. Feb 2009

DIN E EN 13163 (draft standard) Thermal insulation products for buildings – Factory-made expanded polystyrene (EPS) products. May 2010

DIN E EN 13164 (draft standard) Thermal insulation products for buildings – Factory-made extruded polystyrene foam (XPS) products. May 2010

DIN E EN 13165 (draft standard) Thermal insulation products for buildings – Factory-made rigid polyurethane foam (PU) products. May 2010

DIN E EN 13167 (draft standard) Thermal insulation products for buildings – Factory-made cellular glass (CG) products. Feb 2009

DIN E EN 13171 (draft standard) Thermal insulation products for buildings – Factory-made wood fibre (WF) products. May 2010

DIN EN 13707 Flexible sheets for waterproofing – Reinforced bitumen sheets for roof waterproofing. Oct 2009

DIN EN 13948 Flexible sheets for waterproofing – Bitumen, plastic and rubber sheets for roof waterproofing – Determination of resistance to root penetration. Jan 2008

DIN EN 13956 Flexible sheets for waterproofing – Plastic and rubber sheets for roof waterproofing. Apr 2007

DIN V 20000-201 (pre-standard) Use of building products in construction works – Part 201: Adaption standard for flexible sheets for waterproofing according to European standards for the use as waterproofing of roofs. Jan 2009

Zentralverband des Deutschen Dachdeckerhandwerks e.V. (pub.): Regeln für Abdichtungen – mit Flachdachrichtlinie. Cologne, 2008

Zentralverband des Deutschen Dachdeckerhandwerks e.V. (pub.): Fachregeln für Metallabdeckungen im Dachdeckerhandwerk. Cologne, 2006

Flat roof construction

Berufsgenossenschaft der Bauwirtschaft (pub.): Unfallverhütungsvorschrift BGV C22 (bisherige VBG 37): Bauarbeiten (Prävention Hochbau). Berlin, 1997

Bundesinnungsverband des Glaserhandwerks (pub.): Technische Richtlinie des Glaserhandwerks. Nr. 1: Dichtstoffe für Verglasungen und Anschlussfugen. Düsseldorf, 2009

DIN 1045-2 Concrete, reinforced and prestressed concrete structures – Part 2: Concrete – Specification, properties, production and conformity – Application rules for DIN EN 206-1. Aug 2008

DIN1055-3 Actions on structures – Part 3: Self-weight and imposed load in building. Mar 2006

DIN 1055-4 Actions on structures – Part 4: Wind loads. Mar 2005

DIN 1072 Road and foot bridges; design loads. Dec 1985

DIN 1986-100 Drainage systems on private ground – Part 100: Specifications in relation to DIN EN 752 and DIN EN 12056. May 2008

DIN 4108-2 Thermal protection and energy economy in buildings – Part 2: Minimum requirements for thermal insulation. Jul 2003

DIN 4108-3 Thermal protection and energy economy in buildings – Part 3: Protection against moisture subject to climate conditions. Jul 2001

DIN 4108-10 Thermal insulation and energy economy in buildings – Part 10: Application-related requirements for thermal insulation materials – Factory-made products. Jun 2008

DIN 4426 Equipment for building maintenance – Safety requirements for workplaces and accesses – Design and execution. Sept 2001

DIN 18195-1 Waterproofing of buildings – Part 1: Principles, definitions, attribution of waterproofing types. Aug 2000

DIN 18531-1 Waterproofing of roofs – Sealings for non-utilised roofs – Part 1: Terms and definitions, requirements, design principles. May 2010

DIN 18531-2 Waterproofing of roofs – Sealings for non-utilised roofs – Part 2: Materials. May 2010

DIN 18545-1 Glazing with sealants; rebates; requirements. Feb 1992

DIN 18545-2 Sealing of glazing with sealants – Part 2: Sealants, designation, requirements, testing. Dec 2008

DIN 18545-3 Glazing with sealants; rebates; glazing systems. Feb 1992

DIN 18915 Vegetation technology in landscaping – Soil working. Aug 2002

DIN 18917 Vegetation technology in landscaping – Turf and seeding. Aug 2002

DIN 18919 Vegetation technology in landscaping – Care of vegetation during development and maintenance in green areas. Aug 2002

DIN E 18035-4 (draft standard) Sports grounds – Part 4: Sports turf areas. May 2007

DIN EN 752 Drain and sewer systems outside buildings. Apr 2008

DIN EN 1253-1 Gullies for buildings – Part 1: Requirements. Sept 2003

DIN EN 12056-1 Gravity drainage systems inside buildings – Part 1: General and performance requirements. Jan 2001

DIN EN 12056-2 Gravity drainage systems inside buildings – Part 2: Sanitary pipework, layout and calculation. Jan 2001

DIN EN 12056-3 Gravity drainage systems inside buildings – Part 3: Roof drainage, layout and calculation. Jan 2001

Forschungsgesellschaft Landschaftsentwicklung Landschaftsbau e.V. (FLL) (pub.): Empfehlungen zu Planung und Bau von Verkehrsflächen auf Bauwerken. Bonn, 2005

Forschungsgesellschaft Landschaftsentwicklung Landschaftsbau e.V. (FLL) (pub.): Richtlinien für die Planung, Ausführung und Pflege von Dachbegrünungen – Dachbegrünungsrichtlinie. Bonn, 2008

Technische Regeln für die Verwendung von linienförmig gelagerten Verglasungen (TRLV). Aug 2006

Technische Regeln für die Bemessung und die Ausführung punktförmig gelagerter Verglasungen (TRPV). Aug 2006

Zentralverband des Deutschen Dachdeckerhandwerks e.V. (pub.): Grundregel für Dachdeckungen, Abdichtungen und Außenwandbekleidungen. Cologne, 2008

Zentralverband des Deutschen Dachdeckerhandwerk e.V. (pub.): Regeln für Abdichtungen – mit Flachdachrichtlinie. Cologne, 2008

Zentralverband des Deutschen Dachdeckerhandwerks e.V. (pub.): Regeln für Dachdeckungen. Cologne, 2009

Zentralverband des Deutschen Dachdeckerhandwerks e.V. (pub.): Fachregeln für Metallabdeckungen im Dachdeckerhandwerk. Cologne, 2006

Zentralverband des Deutschen Dachdeckerhandwerks e.V. (pub.): Regeln für Metallarbeiten. Cologne, 2006

Bibliography

General
Griffin, C. W.; Fricklas, R. L.: Manual of Low-Slope Roof
  Systems, 3rd ed. New York, 1996

Part A   Introduction

Alberti, Leon Battista: Ten Books on Architecture. Florence,
  1485, (engl. trans: Leoni, James, 1755)
Bock, Ralf: Adolf Loos. Works and Projects. Milan, 2007
Bösinger, Willy; Girsberger, Hans: Le Corbusier – Œuvre
  complète. Basel, 1999
Bosmann, Jos (ed.): Le Corbusier und die Schweiz.
  Zurich, 1987
Curtis, William J. R.: Modern Architecture Since 1900.
  Oxford, 1982
Döcker, Richard: Terrassentyp. Stuttgart, 1929
Gollwitzer, G.; Wirsing, W.: Dachgärten + Dachterrassen.
  Munich, 1962
Hoffmann, Ot: Handbuch für begrünte und genutzte
  Dächer. Leinfelden-Echterdingen, 1987
Lustenberger, Kurt: Adolf Loos. Zurich, 1994
Piper, Jan: Die Natur der hängenden Gärten. In: Daidalos
  No. 23, 1987
Marperger, Paul Jacob: Altanen. Eine Werbeschrift für
  das flache Dach. Pub. by Friedrich Bock & Georg
  Gustav Wieszner, Nuremberg, 1930
Van Doesburg, Théo: Über Europäische Architektur.
  Basel, 1990
Zimmermann, Claire: Mies van der Rohe. Cologne, 2006

Part B   Structure

Loadbearing structure
Herzog, Thomas et al.: Timber Construction Manual.
  Munich/Basel, 2004
Schneider, Klaus-Jürgen (ed.): Bautabellen für Ingenieure.
  Neuwied, 2004
Schneider, Klaus-Jürgen; Volz, Helmut; Widjaja, Eddy:
  Entwurfshilfen für Architekten und Bauingenieure:
  Vorbemessung, Faustformeln, Tragfähigkeitstafeln,
  Beispiele. Berlin, 2010

Loadbearing decks
Arbeitsgemeinschaft Holz e.V. in cooperation with Holz-
  absatzfonds, Absatzförderungsfonds der deutschen
  Forst & Holzwirtschaft: Konstruktive Holzwerkstoffe.
  Düsseldorf, 2001
Arbeitsgemeinschaft Holz e.V. in cooperation with Holz-
  absatzfonds, Absatzförderungsfonds der deutschen
  Forst & Holzwirtschaft: Konstruktive Vollholzprodukte.
  Munich, 2000
Bathon, Leander; Bletz, Oliver: Flachdächer in Holz-Beton-
  Verbundbauweise. In: Holzbau – die Neue Quadriga,
  03/2007, p. 25ff.
Bauen mit Stahl (pub.): Stahlbau Arbeitshilfe 46. Sandwich-
  elemente. Düsseldorf, 2000
Berner, Klaus: Selbsttragende und aussteifende Sand-
  wichbauteile. Möglichkeiten für kleinere und mittlere
  Gebäude. In: Stahlbau 05/2009, pp. 298ff.
Berner, Klaus; Raabe, Oliver: Bemessung von Sandwich-
  bauteilen. Pub. by Industrieverband für Bausysteme im
  Metallleichtbau e.V. (IFBS), Düsseldorf, 2006
Bindseil, Peter: Stahlbetonfertigteile. Konstruktion –
  Berechnung – Ausführung. Cologne, 2007
Böttcher, Marc: Dach und Fassadenelemente aus Stahl.
  Erfolgreich Planen und Konstruieren. Pub. by Stahl-
  Informations-Zentrum. Düsseldorf, 2007
Deutscher Ausschuss für Stahlbeton (DAfStb) (pub.):
  Erläuterungen zur DAfStb-Richtlinie "Wasserundurch-
  lässige Bauwerke aus Beton". Berlin, 2006
Dürr, Markus; Kathage, Karsten; Saal, Helmut: Schubsteifig-
  keit zweiseitig gelagerter Stahltrapezbleche. In: Stahl-
  bau 04/2006, pp. 280ff.
European Convention for Constructional Steelwork
  (ECCS): European Recommendations for Sandwich
  Panels 2000. CIB Publication 257. Brussels, 2001

Haldimann, Matthias: Structural Use of Glass. IABSE
  Structural Engineering Documents 10. Zurich, 2008
Herzog, Thomas et al.: Timber Construction Manual.
  Munich/Basel, 2004
Hierlein, Elisabeth: Betonfertigteile im Geschoss und
  Hallenbau. Grundlagen für die Planung. Pub. by Fach-
  vereinigung deutscher Betonfertigteilbau e.V. (FDB).
  Düsseldorf, 2009
Kech, Johann; Schwarze, Knut: Bemessung von Stahl-
  trapezprofilen für Biegung und Normalkraft. Pub. by
  Industrieverband
  für Bausysteme im Metallleichtbau e.V. (IFBS), Düssel-
  dorf, 2009
Kech, Johann; Schwarze, Knut: Bemessung von Stahl-
  trapezprofilen nach DIN 18807 – Schubfeldbeanspru-
  chung – Konstruktion für einschalige Flachdächer. Pub.
  by Industrieverband für Bausysteme im Metallleichtbau
  e.V. (IFBS), Düsseldorf, 2007
Kolb, Josef: Systems in Timber Engineering. Basel, 2008
Lohmeyer, Gottfried: Schäden an Flachdächern und
  Wannen aus wasserundurchlässigem Beton. Schadenfreies Bauen, vol. 2. Stuttgart, 1996
Lohmeyer, Gottfried; Ebeling, Karsten: Schäden an wasserundurchlässigen Wannen und Flachdächern aus Beton.
  Schadenfreies Bauen, vol. 2. Stuttgart, 2007
Lohmeyer, Gottfried; Ebeling, Karsten: Weiße Wannen
  einfach und sicher. Konstruktion und Ausführung
  wasserundurchlässiger Bauwerke aus Beton. Düssel-
  dorf, 2009
Mönck, Willi; Rug, Wolfgang: Holzbau – Bemessung und
  Konstruktion. Berlin, 2008
Pöter, Hans: Metallleichtbaukonstruktionen: Früher und
  heute. In: Stahlbau 05/2009, pp. 288ff.
Pöter, Hans: Bausysteme aus Stahl für Dach und Fassade.
  Pub. by Stahl-Informations-Zentrum, Düsseldorf, 2010
Schulitz, Helmut C. et al.: Steel Construction Manual.
  Munich /Basel, 2000
Stahl-Informations-Zentrum: Bausysteme aus Stahl für
  Dach und Fassade. Dokumentation 558. Düsseldorf,
  2010
Weller, Bernhard et al.: Glass in Building. Munich, 2009
Werner, Hartmut: Brettstapelbauweise. In: Informations-
  dienst Holz: Holzbau Handbuch Reihe 1, Teil 17, Folge 1.
  Düsseldorf, 1997
Winter, Stefan; Schopbach, Holger: Hoch gestapelt –
  Brettstapeldecken in der Quasi-Balloon-Bauweise. In:
  Holzbau – die neue quadriga, 01/2004

Part C   Building physics

Thermal insulation
Dederich, Ludger: Flachdächer in Holzbau weise. Pub.
  by Holzabsatzfonds. Bonn, 2008
Gösele, Karl; Schüle, Walter: Schall – Wärme – Feuchte.
  Grundlagen, neue Erkenntnisse und Ausführungshin-
  weise für den Hochbau. Wiesbaden, 1997
Hauser, Gerd; Stiegel, Horst: Wärmebrücken-Atlas für den
  Mauerwerksbau. Wiesbaden, 2001
Hauser, Gerd; Stiegel, Horst: Pauschalierte Erfassung der
  Wirkung von Wärmebrücken. In: Bauphysik
  17/1995, pp. 65ff.
Künzel, Hartwig M.; Sedlbauer, Klaus: Reflektierende
  Flachdächer – Sommerlicher Wärmeschutz kontra
  Feuchteschutz. In: IBP-Mitteilung 482. Stuttgart, 2007
Künzel, Hartwig M.: Bieten begrünte Umkehrdächer einen
  dauerhaften Wärmeschutz? In: IBP-Mitteilung 271.
  Stuttgart, 1995
Künzel, Helmut: Wie ist der Feuchteeinfluss auf die Wärme-
  leitfähigkeit von Baustoffen unter heutigen Bedingungen
  zu bewerten? In: Bauphysik 11/1989, pp. 185ff.
Maßong, Friedhelm: EnEV 2009 kompakt. Über 100 Ant-
  worten auf die wichtigsten Fragen zum Energieausweis.
  Cologne, 2009
Maßong, Friedhelm: Wärmeschutz nach EnEV 2009 im
  Dach- und Holzbau. Sichere Konstruktionen, Projekte,
  Energieausweise. Cologne, 2010
Pfundstein, Margit, et al.: Insulating Materials. Basel/
  Munich, 2008
Richter, Ekkehard; Fischer, Heinz M.: Lehrbuch der Bau-
  physik. Schall – Wärme – Feuchte – Licht – Brand –
  Klima. Wiesbaden, 2008

Sedlbauer, K., Gottschling, H.: Sommerliche Temperaturbeanspruchung der Dachhaut bei belüfteten und nicht
  belüfteten Flachdächern. In: IBP-Mitteilung 357. Stuttgart, 1999

Moisture control
Finch, G.; Hubbs, B.; Bombino, R.: Osmosis and the Blistering of Polyurethane Waterproofing Membranes. 12th
  Canadian Conference on Building Science & Technology, Montreal, 2009
Geißler, Achim; Hauser, Gerd: Abschätzung des Risikopotentials infolge konvektiven Feuchtetransports. Final
  report, AIF research project No. 12764. Kassel, 2002
Geshwiler, M.: Air Pressures in Wood Frame Walls. Thermal Performance of the Exterior Envelopes of Buildings.
  7th Conference Clearwater Beach. Florida, 1998
Gösele, Karl; Schüle, Walter: Schall – Wärme – Feuchte.
  Grundlagen, neue Erkenntnisse und Ausführungshin-
  weise für den Hochbau. Wiesbaden, 1997
Künzel Hartwig. M.: Verfahren zur ein- und zweidimen-
  sionalen Berechnung des gekoppelten Wärme und
  Feuchtetransports in Bauteilen mit einfachen Kenn-
  werten. Dissertation. Stuttgart, 1994
Künzel, Hartwig. M.: Dampfdiffusionsberechnung nach
  Glaser – quo vadis? In: IBP-Mitteilung 355. Stuttgart,
  1999
Künzel, Hartwig M.; Sedlbauer, Klaus: Reflektierende
  Flachdächer – sommerlicher Wärmeschutz kontra
  Feuchteschutz. In: IBP-Mitteilung 482. Stuttgart, 2007
Künzel, Hartwig M.; Zirkelbach, Daniel: Trocknungsreser-
  ven schaffen – Einfluss des Feuchteeintrags aus Dampf-
  konvektion. In: Holzbau – die neue quadriga 01/2010,
  pp. 28ff.
Mohrmann, Martin: Feuchteschäden beim Flachdach. In:
  Holzbau – die neue quadriga, 03/2007, pp. 13ff.
Oswald, Rainer: Fehlgeleitet. Unbelüftete Holzdächer mit
  Dachabdichtungen. In: deutsche bauzeitung 07/2009,
  pp. 74ff.
Schmidt, Daniel; Winter, Stefan: Flachdächer in Holzbau-
  weise. Pub. by Informationsdienst Holz. Bonn, 2008
Sedlbauer, Klaus: Vorhersage von Schimmelpilzbildung
  auf und in Bauteilen. Dissertation. Stuttgart, 2001

Fire
Berghofer, Ernest; Hausladen, Gerhard; Giertlova, Zuzanna;
  Sonntag, Rainer: Konzeptioneller Brandschutz. Strate-
  gien für ganzheitliche Gebäudeplanung. Munich, 2004
Institut für Schadenverhütung & Schadenforschung der
  öffentlichen Versicherer e.V. (IFS) (pub.): IFS Report
  09/2006. Kiel, 2006
Mayr, Josef; Battran, Lutz: Handbuch Brandschutzatlas.
  Cologne, 2009
Schneider, Ulrich; Fransen, Jean Marc; Lebeda, Christian:
  Baulicher Brandschutz. Nationale und europäische
  Normung, Bauordnungsrecht, Praxisbeispiele. Berlin,
  2008

Sound insulation
Gösele, Karl; Schüle, Walter: Schall – Wärme – Feuchte.
  Grundlagen, neue Erkenntnisse und Ausführungshin-
  weise für den Hochbau. Wiesbaden, 1997
Scholl, Werner; Bietz, Heinrich: Integration des Holz und
  Skelettbaus in die neue DIN 4109. Final report. Stutt-
  gart, 2005
Baden-Württemberg Environment Ministry (pub.): Lärm-
  schutz für kleine Ohren. Leitfaden zur akustischen
  Gestaltung von Kindertagesstätten. Stuttgart, 2009
Weber, L.; Koch, S.: Anwendung von Spektrum-Anpas-
  sungswerten. Teil 1: Luftschalldämmung.
  In: Bauphysik 21/1999, pp. 167ff.
Weber, L.; Schreier, H.; Brandstetter, K.D.: Measurement
  of Sound Insulation in Laboratory – Comparison of Dif-
  ferent Methods. In: Proceedings – International Conference on Acoustics. Rotterdam 2009, pp. 701ff.
Weber, L.; Seidel, J.; Rotaru, D.; Zhou, X.: Messung von
  Regengeräuschen nach DIN EN ISO 140-18. In: Fortschritte der Akustik. Vol. 2. Berlin, 2006, pp. 465ff.

Part D   Design principles

Materials

Bobran, Hans W., Bobran-Wittfoht, Ingrid, Schlauch, Dirk: Flat Roofs with sealing layers – the search for a safe form of construction. In: Detail 07–08/2002, pp. 954ff.

Bobran-Wittfoht, Ingrid; Schlauch, Dirk: Sealing layers for flat roofs – the correct choice. In: Detail 05/2001, pp. 912ff.

Eiserloh, Hans Peter: Handbuch Dachabdichtung. Aufbau – Werkstoffe – Verarbeitung – Details. Cologne, 2009

Ernst, Wolfgang: Liquid applied roof waterproofing kits – an option not only for complex roofs. In: Detail 12/2006, pp. 1438ff.

Haack, Alfred; Emig, Karl-Friedrich: Abdichtungen im Gründungsbereich und auf genutzten Deckenflächen. Berlin, 2002

Hegger, Manfred et al.: Construction Materials Manual. Munich, 2006

Holzapfel, Walter: Werkstoffkunde für Dach-, Wand- und Abdichtungstechnik. Cologne, 2003

Kennzeichnungspflicht für Dachsubstrate. In: Dach + Grün 1/2009, p. 6

Köhler, Martin: Geotextilrobustheitsklassen: Eine praxisnahe Beschreibung der Robustheit von Vliesstoffen und Geweben gegenüber Einbaubeanspruchungen. In: tis Tiefbau Ingenieurbau Straßenbau 11/2007, p. 52

Mötzl, Hildegund; Zegler, Thomas: Ökologie der Dämmstoffe. Vienna/New York, 2000

Oswald; R., Spilker, R.; Liebert, G.; Sous, S.; Zöller, M.: Zuverlässigkeit von Flachdachabdichtungen aus Kunststoff und Elastomerbahnen. Stuttgart, 2008

Riegler, Rosina: Fachgerechte Ausführung und Sanierung von Flachdächern und Gründächern. Merching, 2009

Schittich, Christian et al.: Glass Construction Manual. 2nd ed. Munich/Basel, 2007

Schunck, Eberhard et al.: Roof Construction Manual – Pitched Roofs. Munich/Basel, 2003

Flat roof construction

Bobran, Hans W., Bobran-Wittfoht, Ingrid, Schlauch, Dirk: Flat Roofs with sealing layers – the search for a safe form of construction. In: Detail 07–08/2002, pp. 954ff.

Bobran-Wittfoht, Ingrid; Schlauch, Dirk: Sealing layers for flat roofs – the correct choice. In: Detail 05/2001, pp. 912ff.

Bundesverband der deutschen Zementindustrie e.V. (pub.): Flachdächer aus Zement. Zement-Merkblatt Hochbau. Hannover, 1999

Buttschardt, Tillmann: Extensive Dachbegrünungen und Naturschutz. Karlsruhe, 2001

Euro Inox (pub.): Dächer aus Edelstahl Rostfrei. Reihe Bauwesen, vol. 4. Luxembourg, 2004

Haack, Alfred; Emig, Karl-Friedrich: Abdichtungen im Gründungsbereich und auf genutzten Deckenflächen. Berlin, 2002

Holzapfel, Walter: Dächer. Erweitertes Fachwissen für Sachverständige und Baufachleute. Stuttgart, 2009

Lech, Jürgen: Dach und Bauwerksabdichtung in der Praxis. Schadensbilder, Sanierungsmöglichkeiten, Detaillösungen. Renningen, 2008

Liesecke, H. J.: Begrünung von Well- und Trapezprofilen mit einem Verbundschaumstoff. In: Dach + Grün, 01/2003

Lohmeyer, Gottfried: Flachdächer – einfach und sicher. Konstruktion und Ausführung von Flachdächern aus Beton ohne besondere Dichtungsschicht. Düsseldorf, 1993

Molitor, Patrick: Der Photovoltaik-Anlagen Projektleitfaden. Solaranlagen Grundwissen von A–Z. Hamburg, 2009

Riegler, Rosina: Fachgerechte Ausführung und Sanierung von Flachdächern und Gründächern. Merching, 2009

Schittich, Christian et al.: Glass Construction Manual. 2nd ed. Munich/Basel, 2007

Schubert, Reinhard: Dächer mit Dachabdichtungen. Bochum, 2002

Schunck, Eberhard et al.: Roof Construction Manual – Pitched Roofs. Munich/Basel, 2003

Spilker, Ralf; Oswald, Rainer: Flachdachsanierung über durchfeuchteter Dämmschicht. Aachen, 2003

Useful addresses

Europäische Vereinigung dauerhaft dichtes Dach e.V. (ddD)
Wolfratshauser Str. 45 b
82049 Pullach
Germany
www.durable-roof.com

European Convention for Contructional Steelwork (ECCS)
32, av. des Ombrages, bte 20
1200 Brussels
Belgium
www.steelconstruct.com

European Federation of Green Roof Associations
UK:
Livingroofs.org Ltd
7 Dartmouth Grove
London SE10 8AR
UK
www.efb-greenroof.eu
www.livingroofs.org

European Quality Assurance
Association for Panels and Profiles
Max-Planck-Straße 4
40237 Düsseldorf
Germany
www.epaq.eu

Fachvereinigung Bauwerksbegrünung e.V. (FBB)
Kanalstr. 2
66130 Saarbrücken
Germany
www.fbb.de

Federation of European Rigid Polyethane Foam Associations
(PU Europe)
Av. E. Van Nieuwenhuyse 6
1160 Brussels
Belgium
www.pu-europe.eu

Forschungsgesellschaft Landschaftsentwicklung
Landschaftsbau e.V. (FLL)
Colmantstr. 32
53115 Bonn
Germany
www.fll.de

Fraunhofer Institute for Building Physics (IBP)
Stuttgart Institute
Nobelstr. 12
70569 Stuttgart
Germany
www.ibp.fraunhofer.de

Fraunhofer Institute for Building Physics (IBP)
Holzkirchen Institute
Fraunhoferstr. 10
83626 Valley/Oberlaindern
Germany
www.ibp.fraunhofer.de

Glass and Glazing Federation (GGF)
54 Ayres Street
London SE1 1EU
UK
www.ggf.co.uk

Green Roofs for Healthy Cities North America (GRHC)
406 King Street East
Toronto, ON M5A 1L4
Canada
www.greenroofs.org

Gütegemeinschaft Substrate für Pflanzen e.V.
Heisterbergallee 12
30453 Hannover
Germany
www.substrate-ev.org

International Green Roof Association (IGRA)
PO Box 880127
13107 Berlin
Germany
www.igra-world.com

National Federation of Roofing Contractors Ltd.
Roofing House, 31 Worship Street
London, EC2A 2DY, UK
www.nfrc.co.uk

National Roofing Contractors Association
10255 W. Higgins Road
Suite 600
Rosemont, IL 60018-5607, USA
www.nrca.net

Zentralverband des deutschen Dachdeckerhandwerks e.V. (ZVDH)
Fachverband Dach-, Wand- & Abdichtungstechnik
Fritz-Reuter-Str. 1
50968 Cologne
Germany
www.dachdecker.de

Picture credits

The authors and publishers would like to express their sincere gratitude to all those who have assisted in the production of this book, be it through providing photos or artwork or granting permission to reproduce their documents or providing other information. Photographs not specifically credited were taken by the architects or are works photographs or were supplied from the archives of the magazine DETAIL. Despite intensive endeavours we were unable to establish copyright ownership in just a few cases; however, copyright is assured. Please notify us accordingly in such instances. The numbers refer to the figures.

Part A   Introduction

A   Christian Schittich, Munich

A 1–2   Jan Martin Klessing, Karlsruhe
A 3–5   Christian Schittich, Munich
A 6   Ruth Schittich, Munich
A 7   Burkhard Franke, Munich
A 8–10 Piper, J.: Die Natur der hängenden Gärten. In: Daidalos No. 23, 1987
A 11   Gollwitzer, Gerda; Wirsing, Werner: Dachgärten + Dachterrassen. Munich, 1962
A 12   Döcker, R.: Terrassentyp. Stuttgart, 1929
A 13   Opderbecke, A.: Der Dachdecker und Bauklempner. Leipzig, 1901, reprint
A 14   Deutsche Bauzeitung (pub.): Baukunde des Architekten. Berlin, 1893
A 15   Garnier, T.: Une cité industrielle. Paris, 1932
A 16   Frank Lloyd Wright in his Renderings 1887–1959. A. D. A. Edita, Tokyo, 1984
A 17   Weintraub, Alan; Hess, Alan: Frank Lloyd Wright – The Houses. Milan, 2005
A 18   Ausgeführte Bauten und Entwürfe von Frank Lloyd Wright. Tübingen, 1986
A 19   Adolf Loos, Albertina Graphics Collection. Vienna, 1989
A 20–21 Christian Schittich, Munich
A 22   Warncke, C.P.: De Stijl 1917–1931. Cologne, 1990
A 23   Cornelia Hellstern, Munich
A 24   Le Corbusier, Precisions on the Present State of Architecture and City Planning. Paris, 1930
A 25   Christian Schittich, Munich
A 26   Döcker, R.: Terrassentyp. Stuttgart, 1929
A 27   Stadt und Siedlung, Monatsheft zur Deutschen Bauzeitung, 1928
A 28   Bauwelt, 1926
A 31   Nerdinger, Winfried: Konstruktion und Raum in der Architektur des 20. Jahrhunderts. Munich, 2002
A 32   Christian Schittich, Munich
A 33   Eberhard Schunck, Munich
A 34   Nate Umstead, Grand Rapids, Michigan
A 35   Werner Huthmacher, Berlin
A 36   Shinkenchikusha, Tokyo
A 37   Georges Fessy, Paris
A 38   Shinkenchikusha, Tokyo

Part B   Structure

B   David Franck, Ostfildern

Loadbearing structure
B 1.1   Jan-Oliver Kunze, Berlin
B 1.6   according to DIN 10553
B 1.7   see B 1.6
B 1.8   according to DIN EN 199113
B 1.9–10 according to DIN 10555
B 1.12   according to DIN 10554
B 1.13   see B 1.12
B 1.14   see B 1.12
B 1.15   see B 1.12
B 1.16   see B 1.12
B 1.17   see B 1.12

B 1.23   Andreas Keller, Altdorf
B 1.24   Tim Bergmann and Roman Schmidt, Munich
B 1.25   WING, Hong Kong, http://commons.wikimedia.org
B 1.26   Dietmar Strauß, Besigheim
B 1.27   Serge Kreis/Carmenzind Gräfensteiner, Zurich
B 1.28   Hiroyuki Hirai, Tokyo

Loadbearing decks
B 2.11   Fischer Profil GmbH, Netphen-Deuz
B 2.14 a Mevaco GmbH, Schlierbach
B 2.14 b–c Stahl-Informations-Zentrum, Düsseldorf
B 2.19 a–c Hans-Joachim Heyer, Werkstatt für Fotografie, University of Stuttgart
B 2.19 d Holzabsatzfonds, Bonn
B 2.20 a–d see B 2.19d
B 2.35   Werner Huthmacher/arturimages
B 2.37   according to Technische Regeln für die Verwendung von linienförmig gelagerten Verglasungen (TRLV), Aug 2006
B 2.39   according to Haldimann, Matthias: Structural Use of Glass. IABSE Structural Engineering Documents 10. Zurich, 2008

Part C   Building physics

C   Werner Huthmacher/arturimages

Thermal insulation
C 1.1   Fraunhofer Institute for Building Physics, Holzkirchen
C 1.3   according to Karl Gertis, Stuttgart
C 1.4   see C 1.3
C 1.6   according to Künzel, Helmut: Wie ist der Feuchteeinfluss auf die Wärmeleitfähigkeit von Baustoffen unter heutigen Bedingungen zu bewerten? In: Bauphysik 11/1998
C 1.7   Fraunhofer Institute for Building Physics, Holzkirchen
C 1.8   see C 1.7
C 1.9   see C 1.3
C 1.11   see C 1.3
C 1.14   according to Gertis, Karl; Hauser, Gerd: Temperaturbeanspruchung von Stahlbetondächern. In: IBPMitteilung 10. Stuttgart, 1975
C 1.15   see C 1.7
C 1.16   see C 1.7
C 1.17   see C 1.7
C 1.18   according to Dederich, Ludger: Flachdächer in Holzbauweise. Pub. by Holzabsatzfonds. Bonn, 2008
C 1.21   Alwitra GmbH & Co., Trier
C 1.22   see C 1.7
C 1.23   see C 1.7
C 1.24   according to DIN EN ISO 6964
C 1.27   according to DIN 4108-2
C 1.28   see C 1.24
C 1.29   see C 1.7
C 1.30   see C 1.7
C 1.31   according to DIN EN ISO 6946
C 1.32   see C 1.7
C 1.34   according to Energieeinsparverordnung (EnEV) – Verordnung über energiesparenden Wärmeschutz und energiesparende Anlagentechnik bei Gebäuden. Mar 2009
C 1.35   Klaus Leidorf, Buch am Erlbach

Moisture control
C 2.1   WOLFIN Bautechnik, Wächtersbach
C 2.2   Fraunhofer Institute for Building Physics, Holzkirchen
C 2.3   see C 2.2
C 2.4   see C 2.2
C 2.5   see C 2.2
C 2.6 a–b see C 2.2
C 2.7   see C 2.2
C 2.8 a–b see C 2.2
C 2.9 a–b see C 2.2
C 2.10   according to DIN 4108-3
C 2.11 a–e see C 2.2
C 2.12   see C 2.2
C 2.13   see C 2.2

C 2.14 a–b according to DIN 4108-7
C 2.15 a–b see C 2.14
C 2.16 a–b see C 2.14
C 2.17   Finch, G.; Hubbs, B.; Bombino, R.: Osmosis and the Blistering of Polyurethane Waterproofing Membranes. 12th Canadian Conference on Building Science & Technology. Montreal, 2009
C 2.18 a–c see C 2.17
C 2.19   see C 2.2
C 2.20   see C 2.2
C 2.21   see C 2.2
C 2.22   see C 2.2
C 2.23 a–b according to DIN EN 15026
C 2.24   see C 2.2
C 2.25 a–b see C 2.2
C 2.26   see C 2.2
C 2.27   see C 2.2
C 2.28 a–b see C 2.2

Fire
C 3.1   WOLFIN Bautechnik, Wächtersbach
C 3.2   according to Institut für Schadenverhütung & Schadenforschung der öffentlichen Versicherer e.V. (IFS) (pub.): IFS Report 09/2006. Kiel, 2006
C 3.3   according to Schneider, Ulrich; Fransen, Jean Marc; Lebeda, Christian: Baulicher Brandschutz. Nationale und europäische Normung, Bauordnungsrecht, Praxisbeispiele. Berlin, 2008
C 3.5   according to DIN EN 13501 & DIN 4102
C 3.6   according to Model Building Code. Nov 2002
C 3.7   see C 3.3
C 3.8   according to Fachkommission Bauaufsicht der ARGEBAU (pub.): MusterRichtlinie über den baulichen Brandschutz im Industriebau. 2000–2003

Sound insulation
C 4.1   Roland Zihlmann/Fotolia
C 4.2   Fraunhofer Institute for Building Physics, Stuttgart
C 4.3   see C 4.2
C 4.4   see C 4.2
C 4.5   see C 4.2
C 4.6   Müller-BBM, Planegg
C 4.7   see C 4.2
C 4.8   see C 4.2
C 4.9   see C 4.2
C 4.10   see C 4.2
C 4.11   Roland Halbe, Stuttgart
C 4.12   see C 4.2
C 4.13   see C 4.2
C 4.14   see C 4.2

Part D   Design principles

D   Ivan Brodey, Oslo

Materials
D 1.1   Optigrün international, Krauchenwies-Göggingen
D 1.2   according to DIN 18531
D 1.3   according to DIN EN ISO 10456
D 1.6–7 Industrieverband BitumenDach & Dichtungsbahnen (vdd), Frankfurt am Main
D 1.8–10 according to Zentralverband des Deutschen Dachdeckerhandwerks e.V.; Hauptverband der Deutschen Bauindustrie e.V. (pub.): Fachregel für Abdichtungen – Flachdachrichtlinie. Cologne, 2008
D 1.11–12 WOLFIN Bautechnik, Wächtersbach
D 1.17   according to Fachvereinigung Bauwerksbegrünung e.V. (FBB); www.fbb.de/dachbegruenung/planungshinweise/pflanzenlisten/
D 1.18   Paul Bauder GmbH & Co. KG, Stuttgart
D 1.19   ZinCo GmbH, Unterensingen
D 1.22 a–f Frank Kaltenbach, Munich
D 1.25   Saint-Gobain Isover G+H AG, Ludwigshafen am Rhein

Flat roof construction
D 2.1   Knauf Insulation GmbH, Simbach am Inn
D 2.3   IB Bludau, Munich

Authors

Klaus Sedlbauer
Born 1965
Studied physics at the Ludwig Maximilian University,
Munich
1992–2000 Scientific assistant at the Fraunhofer Institute
for Building Physics (IBP) in Stuttgart and Holzkirchen
2001 Award of doctorate at the Faculty of Construction &
Surveying at the University of Stuttgart
2003 Professor of constructional building physics and
building services at the Rosenheim University of Applied
Sciences
2003 to date Head of the Fraunhofer Institute for Building
Physics and full professor for building physics at the University of Stuttgart
Member of national and international advisory boards and
expert committees (e.g. CIB, UBA Room Hygiene Com-
mission, BMVBS, DGNB). Numerous national and international publications and awards.

Eberhard Schunck
Born 1937
Studied architecture at Munich University of Technology
1961–1967 Assistant to Prof. Gerhard Weber at Munich
University of Technology
1967 to date Self-employed architect in Munich
1981–1984 Professor of building studies and design at
Augsburg University of Applied Sciences
1984–1992 Professor of building planning and design at
the University of Stuttgart
1992–2002 Full professor of construction engineering at
Munich University of Technology
Numerous publications, including "Roof Construction
Manual – Pitched Roofs", Munich/Basel, 2003.

Rainer Barthel
Born 1955
Studied construction engineering at the University of
Stuttgart
1980–1982 Assistant to Frei Otto at the Institute for Lightweight Plate & Shell Structures, University of Stuttgart,
and in Atelier Warmbronn
1983–1990 Assistant to Prof. Fritz Wenzel at the Institute
for Loadbearing Structures, University of Karlsruhe
1991 Award of doctorate at the Institute for Loadbearing
Structures, University of Karlsruhe
1990–1991 Ove Arup & Partners, London
1991–1993 Wenzel Frese Pörtner Haller, Karlsruhe
1993 to date Professor of structural design at Munich
University of Technology
1996 Founding of the Barthel & Maus consulting engineering practice, Munich
2009 to date Lecturer at the Swiss Federal Institute of
Technology Zurich (ETH)

Hartwig M. Künzel
Born 1959
Studied chemical engineering at the University of Erlangen/
Nuremberg
1987–1994 Researcher at the Fraunhofer Institute for
Building Physics (IBP) in Holzkirchen
1994 to date Head of the Hygrothermal Department at
IBP, Holzkirchen
1994 Award of doctorate at the Faculty of Civil Engineering
at the University of Stuttgart
Member or chairman of international standards committees
and expert bodies (e.g. WTA, CEN, ASHRAE). Author of
more than 250 articles in national and international journals
and proceedings.

Index

Index of names

The authors and publishers would like to thank the following sponsors for their assistance with this publication:

Adolf Würth GmbH & Co. KG, Künzelsau (D)
http://www.wuerth.de

Henkel AG & Co. KGaA
WOLFIN Bautechnik, Wächtersbach (D)
http://www.wolfin.de